T0134764

Lecture Notes in Computer Science 11559

Commenced Publication in 1973
Founding and Former Series Editors:
Gerhard Goos, Juris Hartmanis, and Jan van Leeuwen

More information about this series at http://www.springer.com/series/7409

Simon N. Foley (Ed.)

Data and Applications Security and Privacy XXXIII

33rd Annual IFIP WG 11.3 Conference, DBSec 2019
Charleston, SC, USA, July 15–17, 2019
Proceedings

 Springer

Editor
Simon N. Foley
Norwegian University of Science
and Technology
Gjøvik, Norway

ISSN 0302-9743   ISSN 1611-3349 (electronic)
Lecture Notes in Computer Science
ISBN 978-3-030-22478-3   ISBN 978-3-030-22479-0 (eBook)
https://doi.org/10.1007/978-3-030-22479-0

LNCS Sublibrary: SL3 – Information Systems and Applications, incl. Internet/Web, and HCI

This Springer imprint is published by the registered company Springer Nature Switzerland AG
The registered company address is: Gewerbestrasse 11, 6330 Cham, Switzerland

Preface

This book contains the papers that were selected for presentation and publication at the 33rd Annual IFIP WG 11.3 Conference on Data and Applications Security and Privacy (DBSec 2019) that was held in Charleston, South Carolina, USA, July 15–17, 2019.

The Program Committee accepted 21 papers out of a total of 51 papers that were submitted from 18 different countries. The papers in this book are drawn from a range of topics, including privacy, code security, security threats, security protocols, distributed systems, and mobile and Web security. The 43-member Program Committee, assisted by a further 43 external reviewers, reviewed and discussed the papers online over a period of over six weeks and with each paper receiving at least three reviews.

DBSec 2019 would not have been possible without the contributions of the many volunteers who freely gave their time and expertise. Our thanks go to the members of the Progam Committee and the external reviewers for their work in evaluating the papers. Grateful thanks are due to all the people who gave their assistance and ensured a smooth organization, in particular Csilla Farkas and Mark Daniels for their efforts as DBSec 2019 general chairs; Sabrina De Capitani di Vimercati (IFIP WG11.3 Chair) for her guidance and support, and Emad Alsuwat for managing the conference website. A special thanks goes to the invited speakers for their keynote presentations. Finally, we would like to express our thanks to the authors who submitted papers to DBSec. They, more than anyone else, are what makes this conference possible.

July 2019   Simon Foley

Organization

IFIP WG 11.3 Chair

Sabrina De Capitani di   Università degli Studi di Milano, Italy
Vimercati

General Chairs

Csilla Farkas   University of South Carolina, USA
Mark Daniels   Medical University of South Carolina, USA

Program Chair

Simon Foley   Norwegian University of Science and Technology,
Norway

Program Committee

Vijay Atluri   Rutgers University, USA
Frédéric Cuppens   IMT Atlantique, France
Nora Cuppens-Boulahia   IMT Atlantique, France
Sabrina De Capitani di   University of Milan, Italy
Vimercati
Giovanni Di Crescenzo   Perspecta Labs, USA
Wenliang Du   Syracuse University, USA
Barbara Fila   INSA Rennes, IRISA, France
Simon Foley   Norwegian University of Science and Technology,
Norway
Sara Foresti   University of Milan, Italy
Joaquin Garcia-Alfaro   Telecom SudParis, France
Stefanos Gritzalis   University of the Aegean, Greece
Ehud Gudes   Ben-Gurion University, Israel
Yuan Hong   Illinois Institute of Technology, USA
Sokratis Katsikas   Norwegian University of Science and Technology,
Norway
Florian Kerschbaum   University of Waterloo, Canada
Adam J. Lee   University of Pittsburgh, USA
Yingjiu Li   Singapore Management University, Singapore
Giovanni Livraga   University of Milan, Italy
Javier Lopez   UMA, Spain
Brad Malin   Vanderbilt University, USA
Fabio Martinelli   IIT-CNR, Italy

Sjouke Mauw	University of Luxembourg, Luxembourg
Catherine Meadows	NRL, USA
Charles Morisset	Newcastle University, UK
Martin Olivier	University of Pretoria, South Africa
Stefano Paraboschi	University of Bergamo, Italy
Günther Pernul	Universität Regensburg, Germany
Andreas Peter	University of Twente, The Netherlands
Silvio Ranise	FBK-Irst, Italy
Indrajit Ray	Colorado State University, USA
Kui Ren	State University of New York at Buffalo, USA
Pierangela Samarati	University of Milan, Italy
Andreas Schaad	WIBU-Systems, Germany
Scott Stoller	Stony Brook University, USA
Tamir Tassa	The Open University of Israel, Israel
Mahesh Tripunitara	University of Waterloo, Canada
Jaideep Vaidya	Rutgers University, USA
Vijay Varadharajan	The University of Newcastle, Australia
Lingyu Wang	Concordia University, Canada
Wendy Hui Wang	Stevens Institute of Technology, USA
Attila A Yavuz	University of South Florida, USA
Ting Yu	Qatar Computing Research Institute, Qatar
Nicola Zannone	Eindhoven University of Technology, The Netherlands

Additional Reviewers

Ahlawat, Amit
Akowuah, Francis
Alhebaishi, Nawaf
Anagnostopoulos, Marios
Asadi, Behzad
Behnia, Rouzbeh
Bui, Thang
Ceccato, Mariano
Cledel, Thomas
Dietz, Marietheres
Esquivel-Vargas, Herson
Fernandez, Gerardo
Gadyatskaya, Olga
Hitchens, Michael
Hoang, Thang
Kalloniatis, Christos
Liu, Bingyu
Luo, Meng
Mercaldo, Francesco
Michailidou, Christina
Mohammady, Meisam
Mueller, Johannes

Oqaily, Alaa
Oqaily, Momen
Ozmen, Muslum Ozgur
Puchtra, Alexander
Ramírez-Cruz, Yunior
Rizos, Athanasios
Sengupta, Binanda
Seyitoglu, Efe Ulas Akay
Tian, Yangguang
Tsohou, Aggeliki
Uuganbayar, Ganbayar
van de Kamp, Tim
van Deursen, Ton
Vielberth, Manfred
Voloch, Nadav
Wang, Han
Widel, Wojciech
Xie, Shangyu
Xu, Jiayun
Xu, Shengmin
Zhang, Mingwei

Contents

Malware

Contents

Attacks

Detecting Adversarial Attacks in the Context of Bayesian Networks

Emad Alsuwat[(⊠)], Hatim Alsuwat, John Rose, Marco Valtorta,
and Csilla Farkas

University of South Carolina, Columbia, SC 29208, USA
{Alsuwat,Alsuwath}@email.sc.edu, {Rose,Mgv,Farkas}@cse.sc.edu

Abstract. In this research, we study data poisoning attacks against Bayesian network structure learning algorithms. We propose to use the distance between Bayesian network models and the value of data conflict to detect data poisoning attacks. We propose a 2-layered framework that detects both one-step and long-duration data poisoning attacks. Layer 1 enforces "reject on negative impacts" detection; i.e., input that changes the Bayesian network model is labeled potentially malicious. Layer 2 aims to detect long-duration attacks; i.e., observations in the incoming data that conflict with the original Bayesian model. We show that for a typical small Bayesian network, only a few contaminated cases are needed to corrupt the learned structure. Our detection methods are effective against not only one-step attacks but also sophisticated long-duration attacks. We also present our empirical results.

Keywords: Adversarial machine learning · Bayesian networks ·
Data poisoning attacks · The PC algorithm · Long-duration attacks ·
Detection methods

1 Introduction

During the last decade, several researchers addressed the problem of cyber attacks against machine learning systems (see [24] for an overview). Machine learning techniques are widely used; however, machine learning methods were not designed to function correctly in adversarial settings [16,18]. Data poisoning attacks are considered one of the most important emerging security threats against machine learning systems [33,35]. Data poisoning attacks aim to corrupt the machine learning model by contaminating the data in the training phase [11]. Data poisoning was studied in different machine learning algorithms, such as Support Vector Machines (SVMs) [11,21,28], Principal Component Analysis (PCA) [9,10], Clustering [8,12], and Neural Networks (NNs) [36]. However, these efforts are not directly applicable to Bayesian structure learning algorithms.

There are two main methods used in defending against a poisoning attack: (1) robust learning and (2) data sanitization [14]. Robust learning aims to increase

© IFIP International Federation for Information Processing 2019
Published by Springer Nature Switzerland AG 2019
S. N. Foley (Ed.): DBSec 2019, LNCS 11559, pp. 3–22, 2019.
https://doi.org/10.1007/978-3-030-22479-0_1

learning algorithm robustness, thereby reducing the overall influence that contaminated data samples have on the algorithm. Data sanitization eliminates contaminated data samples from the training data set prior to training a classifier. While data sanitization shows promise to defend against data poisoning, it is often impossible to validate every data source [14].

In our earlier work [3,4], we studied the robustness of Bayesian network structure learning algorithms against traditional (a.k.a one-step) data poisoning attacks. We proposed two subclasses of data poisoning attacks against Bayesian network algorithms: (i) model invalidation attacks and (ii) targeted change attacks. We defined a novel link strength measure that can be used to perform a security analysis of Bayesian network models [5].

In this paper, we further investigate the robustness of Bayesian network structure learning algorithms against long-duration (a.k.a multi-step) data poisoning attacks (described in Sect. 3). We use the causative model proposed by Barreno et al. [6] to contextualize Bayesian network vulnerabilities. We propose a 2-layered framework to detect poisoning attacks from untrusted data sources. Layer 1 enforces "reject on negative impacts" detection [30]; i.e., input that changes the model is labeled malicious. Layer 2 aims to detect long-duration attacks; i.e., it looks for cases in the incoming data that conflict with the original Bayesian model.

The main contributions of this paper are the following: We define long-duration data poisoning attacks when an attacker may spread the malicious workload over several datasets. We study model invalidation attacks which aim to arbitrarily corrupt the Bayesian network structure. Our 2-layered framework detects both one-step and long-duration data poisoning attacks. We use the distance between Bayesian network models, B_1 and B_2, denoted as $\mathbf{ds}(B_1, B_2)$, to detect malicious data input (Eq. 3) for one-step attacks. For long-duration attacks, we use the value of data conflict (Eq. 4) to detect potentially poisoned data. Our framework relies on offline analysis to validate the potentially malicious datasets. We present our empirical results, showing the effectiveness of our framework to detect both one-step and long-duration attacks. Our results indicate that the distance measure $\mathbf{ds}(B_1, B_2)$ (Eq. 3) and the conflict measure $Conf(c, B_1)$ (Eq. 4) are sensitive to poisoned data.

The rest of the paper is structured as follows. In Sect. 2, we present the problem setting. In Sect. 3, we present long-duration data poisoning attacks against Bayesian network structure learning algorithms. In Sect. 4, we present our 2-layered detection framework and our algorithms. In Sect. 5 we present our empirical results. In Sect. 6, we give an overview of related work. In Sect. 7, we conclude and briefly discuss ongoing work.

2   Problem Setting

We focus on structure learning algorithms in Bayesian networks. Let $\mathcal{DS}^{v} = \{c_1, \ldots, c_N\}$ be a validated dataset with N case. Each case c is over attributes $x_1, \ldots, x_n$ and of the form $c = <x_1 = v_1, \ldots, x_n = v_n>$, where v_i is the value of attribute x_i. A Bayesian network model B_1 is learned by feeding a validated dataset $\mathcal{DS}^v$ into a Bayesian structure learning algorithm, BN_Algo, such as the PC algorithm, which is the most widely used algorithm for structure learning in Bayesian networks [34], as shown in Eq. 1.

$$B_1 = BN\_Algo(\mathcal{DS}^v) \tag{1}$$

The defender attempts to divide an incoming dataset, $\mathcal{DS}^p$, coming from an untrusted source, into clean and poisoned cases. The attacker aims to inject a contaminated dataset, $\mathcal{DS}^p$ with the same attributes as $\mathcal{DS}^v$ and N_1 cases, into the validated training dataset, $\mathcal{DS}^v$. A learning error occurs if $\mathcal{DS}^u$, obtained by the union of $\mathcal{DS}^v$ and $\mathcal{DS}^p$, results in a Bayesian network learning model B_2 (shown in Eq. 2), such that there is a missing link, a reversed link, or an additional link in B_2 that is not in B_1.

$$B_2 = BN\_Algo(\mathcal{DS}^u) \tag{2}$$

To estimate the impact of the poisoned dataset on the validated dataset, we define a distance function between two Bayesian network models B_1 and B_2, denoted as $\mathbf{ds}(B_1, B_2)$. Intuitively, B_1 is the validated model and B_2 is the potentially corrupted model.

Let $B_1 = (V, E_1)$ and $B_2 = (V, E_2)$ be two Bayesian network models where $V = \{x_1, x_2, \ldots, x_n\}$ and $E = \{(x_u, x_v) : x_u, x_v \in V\}$. Let B_1 be the validated model resulting from feeding $\mathcal{DS}^v$ to a Bayesian network structure learning algorithm, and B_2 be the newly learned model resulting from feeding $\mathcal{DS}^u$ to a Bayesian network structure learning algorithm. Let $e_1 = (x_u, x_v)$ be a directed edge from vertex x_u to vertex x_v, and $e_2 = (x_v, x_u)$ be a directed edge from vertex x_v to vertex x_u (e_2 is the reverse of e_1). The distance function, $\mathbf{ds}(B_1, B_2)$, is a non-negative function that measures the changes in the newly learned model B_2 with respect to the original model B_1. The distance function, $\mathbf{ds}(B_1, B_2)$, is defined as follows:

(**Distance measure**). Let Bayesian network models $B_1 = (V, E_1)$ and $B_2 = (V, E_2)$ be the results of feeding $\mathcal{DS}^v$ and $\mathcal{DS}^u$, respectively, to a Bayesian network structure learning algorithm. $\mathbf{ds}(B_1, B_2)$ is defined as the sum of distances over pairs of vertices $(x_u, x_v) \in V \times V$ as follows:

$$\mathbf{ds}(B_1, B_2) = \sum_{(x_u, x_v) \in V \times V} \mathbf{ds_{x_u x_v}}(B_1, B_2) \tag{3}$$

where $\mathbf{ds_{x_u x_v}}(B_1, B_2)$ is the distance between every pair of vertices $(x_u, x_v) \in V \times V$.

We define $\mathbf{ds_{x_u x_v}}(B_1, B_2)$ as the cost of making a change to B_1 that results in the newly learned model B_2. The function $\mathbf{ds_{x_u x_v}}(B_1, B_2)$ between the two Bayesian network models B_1 and B_2 is defined as follows [19]:

Status 1 (True Negative Edges): if $((e_1 \notin E_1 \ \&\& \ e_2 \notin E_1) \ \&\& \ (e_1 \notin E_2 \ \&\& \ e_2 \notin E_2))$, then there is no edge (neither e_1 nor e_2) between vertex x_u and vertex x_v in either models B_1 and B_2. Hence, $\mathbf{ds_{x_u x_v}}(B_1, B_2) = 0$.

Status 2 (True Positive Edges): if $((e_1 \in E_1 \ \&\& \ e_1 \in E_2) \ || \ (e_2 \in E_1 \ \&\& \ e_2 \in E_2))$, then the same edge (either e_1 or e_2) appears from vertex x_u to vertex x_v in both models B_1 and B_2. Hence, $\mathbf{ds_{x_u x_v}}(B_1, B_2) = 0$.

Status 3 (False Negative Edges): if $((e_1 \ || \ e_2 \in E_1) \ \&\& \ (e_1 \ \&\& \ e_2 \notin E_2))$, then there is an edge (either e_1 or e_2) from vertex x_u to vertex x_v in B_1 that does not exist in B_2. Without loss of generality, assume that the deleted edge from B_1 is e_1, then if the indegree of vertex x_v, denoted as $indegree(x_v)$, which is the number if edge incoming to vertex x_v, is greater than 1, then $\mathbf{ds_{x_u x_v}}(B_1, B_2) = 8$ (deleting e_1 breaks an existing v-structure and changes the Markov equivalence class); otherwise, $\mathbf{ds_{x_u x_v}}(B_1, B_2) = 4$ (deleting e_1 does not break an existing v-structure, but it changes the Markov equivalence class).

Status 4 (False Positive Edges): if $((e_1 \ \&\& \ e_2 \notin E_1) \ \&\& \ (e_1 \ || \ e_2 \in E_2))$, then there is an edge (either e_1 or e_2) from vertex x_u to vertex x_v in B_2 but not the in B_1. Without loss of generality, assume that the added edge to B_2 is e_1, then if the indegree of vertex x_v, is greater than 1, then $\mathbf{ds_{x_u x_v}}(B_1, B_2) = 8$ (adding e_1 introduces a new v-structure and changes the Markov equivalence class); otherwise, $\mathbf{ds_{x_u x_v}}(B_1, B_2) = 4$ (adding e_1 does not introduce a new v-structure, but it changes the Markov equivalence class).

Status 5 (False Positive and True Negative Edges): if $((e_1 \in E_1 \ \&\& \ e_2 \in E_2) \ \&\& \ (e_1 \in E_2 \ \&\& \ e_2 \in E_1))$, then the edge from vertex x_u to vertex x_v in B_1 is the reverse of the edge from vertex x_u to vertex x_v in B_2. Without loss of generality, assume that there is an edge, e_1, from x_u to x_v in B_1, then e_2 is the reverse of e_1 in B_2. If the indegree of vertex x_u, is greater than 1, then $\mathbf{ds_{x_u x_v}}(B_1, B_2) = 8$ (reversing e_1 introduces a new v-structure and changes the Markov equivalence class); otherwise, $\mathbf{ds_{x_u x_v}}(B_1, B_2) = 2$ (reversing e_1 does not introduce a new v-structure, but it changes the Markov equivalence class).

To investigate the coherence of an instance case, $c = \ <x_1 = v_1, \ldots, x_n = v_n>$ (or simply $<v_1, \ldots, v_n>$), in $\mathcal{DS^P}$ with the validated model B_1, we use *conflict measure*, denoted as $Conf(c, B_1)$. Conflict measure, $Conf(c, B_1)$, is defined as follows:

(Conflict measure). Let B_1 be a Bayesian network model and let $\mathcal{DS^P}$ be an incoming dataset, $Conf(c, B_1)$ is defined as the process of detecting how well a given case $<v_1, \ldots, v_n>$ fits the model B_1 according to the following equation:

$$Conf(c, B_1) = log_2 \frac{P(v_1) \ldots P(v_n)}{P(v)} \tag{4}$$

where $c = \ <v_1, \ldots, v_n>$, and $P(v)$ is the prior probability of the evidence v [31].

If $P(v) = 0$, then we conclude that there is inconsistency among the observations $<v_1, \ldots, v_n>$. If the value of $Conf(c, B_1)$ is positive, then we can conclude that $<v_1, \ldots, v_n>$ are negatively correlated (i.e., unlikely to be correlated as the model requires; $P(v_1, \ldots, v_n) < P(v_1) \times \cdots \times P(v_n)$) and thus are conflicting with the model B_1. The higher the value of $Conf(c, B_1)$ is, the more incompatibility we have between B_1 and $<v_1, \ldots, v_n>$.

In this paper, we adopt the causative model proposed by Barreno et al. [6]. Attacks on machine learning systems are modeled as a game between malicious attackers and defenders. In our setting, defenders aim to learn a validated Bayesian network model B_1 using the dataset $\mathcal{DS}^v$ with the fewest number of errors (minimum **ds** function). Malicious attackers aim to mislead the defender into learning a contaminated model B_2 using the dataset $\mathcal{DS}^u$, obtained by polluting $\mathcal{DS}^v$ with $\mathcal{DS}^p$. We assume that malicious attackers have full knowledge of how Bayesian network structure learning algorithms work. Also, we assume that attackers have knowledge of the dataset $\mathcal{DS}^v$. In addition, we assume that the poisoning percentage at which attackers are allowed to add new "contaminated" cases to $\mathcal{DS}^v$, β, is less than or equal to 0.05. The game between malicious attackers and defenders can be modeled as follows:

1. **The defender:** The defender uses a validated dataset $\mathcal{DS}^v$, to produce a validated Bayesian network model B_1.
2. **The malicious attacker:** The attacker injects a contaminated dataset, $\mathcal{DS}^p$, to be unioned with the original dataset, $\mathcal{DS}^v$, with the goal of changing the Markov equivalence class of the original validated model, B_1.
3. **Evaluation by the defender:**
   - The defender feeds the new dataset $\mathcal{DS}^u$ (Note that, $\mathcal{DS}^u = \mathcal{DS}^v \cup \mathcal{DS}^p$) to a Bayesian network structure learning algorithm, resulting in B_2.
   - The defender calculates the distance function $\mathbf{ds}(B_1, B_2)$.
   - If $\mathbf{ds}(B_1, B_2) = 0$, then Bayesian models B_1 and B_2 are identical. Otherwise, i.e., $\mathbf{ds}(B_1, B_2) > 0$, the newly learned Bayesian model B_2 is different from the original validated model B_1.
   - For each case c, the defender calculates the value of conflict measure $Conf(c, B_1)$.
   - If $Conf(c, B_1)$ is positive, then the case c conflict with the Bayesian model B_1. Otherwise, the newly incoming case is validated and added to $\mathcal{DS}^v$.

Note, that the goal of malicious attackers is to maximize the quantity $\mathbf{ds}(B_1, B_2)$. The notations used in this paper are summarized as follows:

Notation	Description
$\mathcal{DS}[x_1, \ldots, x_n]$	Schema for datasets with attributes $x_1, \ldots, x_n$
$\mathcal{DS}^v = \{c_1, \ldots, c_N\}$	Validated dataset instance with attributes $x_1, \ldots, x_n$
$\mathcal{DS}^p = \{\bar{c}_1, \ldots, \bar{c}_{N_1}\}\}$	Crafted dataset instance with attributes $x_1, \ldots, x_n$
$\mathcal{DS}^c_i = \{\bar{c}_1, \ldots, \bar{c}_{N_i}\}\}$	Contaminated dataset instance at time point i
β	Data poisoning percentage for $\mathcal{DS}^v$
λ_i	Data poisoning rate for $\mathcal{DS}^c_i$
B_1	The result of feeding $\mathcal{DS}^v$ to a learning algorithm
B_2	The result of feeding $\mathcal{DS}^u$ to a learning algorithm
$\mathbf{ds}(B_1, B_2)$	Distance function between models B_1 and B_2
$Conf(c, B_1)$	Conflict measure of how well the case c fits B_1

3   Long-Duration Data Poisoning Attacks

In our earlier work data poisoning attacks [3], we studied data poisoning attacks against Bayesian structure learning algorithms. For a Bayesian structure learning algorithms, given the dataset, $\mathcal{DS}^v$, and the corresponding model, B_1 (Eq. 1), a malicious attacker attempts to craft an input dataset, $\mathcal{DS}^p$, such that this contaminated dataset will have an immediate impact on $\mathcal{DS}^v$ and thereby on B_1. The defender periodically retrains the machine learning system to recover the structure of the new model, B_2, using $\mathcal{DS}^u$, the combination of the original dataset $\mathcal{DS}^v$ and the attacker supplied $\mathcal{DS}^p$. We call such an attack a "one-step" data poisoning attack as malicious attackers send all contaminated cases at once.

In this section, we introduce long-duration data poisoning attacks against structure learning algorithms. *Long-duration poisoning attacks* are adversarial multi-step attacks in which a malicious attacker attempts to send contaminated cases over a period of time, $t = \{1, 2, \ldots, w\}$. That is, at every time point i, a malicious attacker sends in a new dataset, $\mathcal{DS}^c_i$, which contains N_i cases, $\lambda_i N_i$ of which are corrupted cases for some $0 < \lambda_i < 1$ (λ_i is the data poisoning rate at which we allowed to add contaminated cases to $\mathcal{DS}^c_i$ at iteration i). Even though the defender periodically retrains the model, B'_2, at time i using the dataset $\mathcal{DS}^{l\text{-}d}_i$, which is equal to $\mathcal{DS}_v \cup \bigcup_{t=1}^i \mathcal{DS}^c_t$, it is not easy to detect the long-duration attack since such an attack is not instantaneous.

By the end of the long-duration poisoning attack, i.e., at time point w, the attacker would have injected $\bigcup_{t=1}^w \mathcal{DS}^c_t$ to $\mathcal{DS}^v$, resulting in a new dataset, $\mathcal{DS}^{l\text{-}d}_w$. We assume that attackers cannot add more than βN cases to $\mathcal{DS}^v$ (i.e., $0 < \bigcup_{t=1}^w \lambda_t N_t < \beta N$). When the defender retrains the model, B'_2, using the dataset $\mathcal{DS}^{l\text{-}d}_w$, the attack will dramatically affect the resulting model. Note that this attack is sophisticated since the attacker may not need to send contaminated cases with the last contaminated dataset (the wth dataset) in the long-duration attack, i.e., $\mathcal{DS}^c_w$ may trigger the attack with no poisoned cases, as our experiments show.

We propose causative, long-duration model invalidation attacks against Bayesian network structure learning algorithms. Such attacks are defined as malicious active attacks in which adversarial opponents attempt to arbitrarily corrupt the structure of the original Bayesian network model in any way. The goal of adversaries in these attacks is to poison the validated training dataset, $\mathcal{DS}^v$, over a period of time $t = \{1, \ldots, w\}$ using the contaminated dataset $\bigcup_{i=1}^w \mathcal{DS}^c_t$ such that $\mathcal{DS}^v$ will be no longer valid. We categorize causative long-duration model invalidation attacks against Bayesian network structure learning algorithms into two types: (1) Model invalidation attacks based on the notion of d-separation and (2) Model invalidation attacks based on marginal independence tests.

Causative, long-duration model invalidation attacks which are based on the notion of d-separation are adversarial attacks in which adversaries attempt to introduce a new link in any triple $(A - B - C)$ in the original Bayesian network model, B_1. The goal of the introduced malicious link, $(A - C)$, is to change the independence relations and the Markov equivalence class of B_1. Within such attacks, we can identify two subtypes: (i) Creating a New Converging

Connection (V-structure), and (ii) Breaking an Existing Converging Connection (V-structure). See Appendix A for more details.

Causative, long-duration model invalidation attacks which are based on marginal independence tests are adversarial attacks in which adversaries attempt to use marginal independence tests in order to change the conditional independence statements between variables in the original model, B_1. Such attacks can be divided into two main subtypes: (i) Removing the Weakest Edge, and (ii) Adding the Most Believable Edge yet incorrect Edge. See Appendix A for more details.

Due to space limitation, in this work, we only provide a brief description of long-duration data poisoning attacks that aim to achieve a certain attack by sending in contaminated cases over a period of time t. We refer the reader to our technical report [2] for the full algorithmic details.

4   Framework for Detecting Data Poisoning Attacks

In this section, we present our detective framework for data poisoning attacks. Our techniques build on the data sanitization approach that was proposed by Nelson et al. [30]. We extend Nelson et al. approach such that it is applicable to detect both one-step and long-duration causative attacks.

The main components of our framework are: (1) Structure learning Algorithms: the PC learning algorithm, (2) FLoD: first layer of detection, and (3) SLoD: second layer of detection.

First Layer of Detection: In the FLoD, our framework uses "Reject On Negative Impact" defense [30] to examine the full dataset ($\mathcal{DS}^\mathrm{v} \cup \mathcal{DS}^\mathrm{p}$) to detect the impact of $\mathcal{DS}^\mathrm{p}$ on $\mathcal{DS}^\mathrm{v}$. The attacker aims to use $\mathcal{DS}^\mathrm{p}$ to change the Markov equivalence class of the validated model, B_1. The first layer of detection detects the impact of adversarial attacks that aim to corrupt the model B_1 using one-step data poisoning attacks. FLoD is useful for efficiently filtering obvious data poisoning attacks.

Algorithm 1. First Layer of Detection

Input : $\mathcal{DS}^\mathrm{v} = \{c_1, \ldots, c_N\}$ and $\mathcal{DS}^\mathrm{p} = \{\bar{c}_1, \ldots, \bar{c}_{N_1}\}$
Output: $\mathbf{ds}(B_1, B_2)$

1 Generate B_1 from $\mathcal{DS}^\mathrm{v}$;
2 Generate B_2 from $\mathcal{DS}^\mathrm{v} \cup \mathcal{DS}^\mathrm{p}$;
3 Calculate $\mathbf{ds}(B_1, B_2)$ ▷ as described in section 2;
4 **if** $\mathbf{ds}(B_1, B_2) > 0$ **then**
5 | Return $\mathbf{ds}(B_1, B_2)$;
6 | Send $\mathcal{DS}^\mathrm{p}$ to be checked offline;
7 **else**
8 | Go to Algorithm 2;
9 **end**

In the FLoD, we use the distance function **ds** described in Sect. 2 as a method for detecting the negative impact of $\mathcal{DS}^p$ on the validated model B_1. If $\mathbf{ds}(B_1, B_2)$ is greater than zero, then the new incoming dataset, $\mathcal{DS}^p$, is potentiality malicious. In this case, we sent $\mathcal{DS}^p$ to be checked offline. Otherwise, we proceed with the second layer of detection, SLoD, looking for long-duration data poisoning attacks.

Algorithm 1 provides algorithmic details of FLoD detect one-step data poisoning attacks.

Second Layer of Detection: In the SLoD, our framework uses "Data Conflict Analysis" [31] to examine the newly incoming dataset $\mathcal{DS}^p$ to detect if $\mathcal{DS}^p$ has conflicting cases with the original model B_1. The Second layer of detection detects sophisticated adversarial attacks that aim to corrupt the model B_1, such as long-duration data poisoning attacks.

Algorithm 2. Second Layer of Detection

Input   : $\mathcal{DS}^v = \{c_1, \ldots, c_N\}$ and $\mathcal{DS}^p = \{\bar{c}_1, \ldots, \bar{c}_{N_1}\}$
Output: $\mathcal{DS}^v$, $\mathcal{DS}^{\mathrm{conf}}$.

1 Generate B_1 from $\mathcal{DS}^v$;
2 $\mathcal{DS}^{\mathrm{conf}} = \phi$;
3 **for** *every case c in* $\mathcal{DS}^p$ **do**
4 | Calculate $P(v)$ ▷ i.e., the probability of the evidence for c;
5 | **if** $P(v) = 0$ **then**
6 | | $\mathcal{DS}^{\mathrm{conf}} = \mathcal{DS}^{\mathrm{conf}} \cup \{c\}$ ▷ i.e., c is inconsistent with B_1;
7 | | $\mathcal{DS}^p = \mathcal{DS}^p \setminus \{c\}$ ▷ remove c from $\mathcal{DS}^p$;
8 | **else**
9 | | $Conf(c, B_1) = log_2 \frac{P(v_1)\ldots P(v_n)}{P(v)}$ ▷ calculate conflict measure for the case c;
10 | | **if** $Conf(c, B_1) > 0$ **then**
11 | | | $\mathcal{DS}^{\mathrm{conf}} = \mathcal{DS}^{\mathrm{conf}} \cup \{c\}$ ▷ i.e., c is incompatible with B_1;
12 | | | $\mathcal{DS}^p = \mathcal{DS}^p \setminus \{c\}$;
13 | | **end**
14 | **end**
15 | **if** $\mathcal{DS}^{conf} \neq \phi$ **then**
16 | | Send $\mathcal{DS}^{\mathrm{conf}}$ to be checked offline;
17 | **end**
18 | $\mathcal{DS}^v = \mathcal{DS}^v \cup (\mathcal{DS}^p \setminus \mathcal{DS}^{\mathrm{conf}})$;
19 | Return $\mathcal{DS}^v$, $\mathcal{DS}^{\mathrm{conf}}$;
20 **end**

In the SLoD, we use the value of the conflict measure $Conf(c, B_1)$ described in Sect. 2 as a method for detecting whether or not a case, c, in the newly incoming dataset, $\mathcal{DS}^p$, is conflicting with the original model B_1. If the $P(v)$ is equal to zero, then the case c is inconsistent with the validated model B_1.

If $Conf(c, B_1)$ is positive, then the case c is incompatible with the validated model B_1. In these two situations, we add inconsistent and incompatible cases to $\mathcal{DS}^{\text{conf}}$. $\mathcal{DS}^{\text{conf}}$ is then sent to be checked offline. Thereby, the model B_1 will be retrained according to the following equation: $B_1 = BN\_Algo(\mathcal{DS}^{\text{v}})$ where $\mathcal{DS}^{\text{v}} = \mathcal{DS}^{\text{v}} \cup (\mathcal{DS}^{\text{p}} \backslash \mathcal{DS}^{\text{conf}})$.

Algorithm 2 provides algorithmic details of the SLoD detect long-duration data poisoning attacks.

The process of applying our framework is summarized in Fig. 1. The workflow of our framework is described as follows: (1) A validated dataset, $\mathcal{DS}^{\text{v}}$, which is a clean training dataset that is used to recover a validated machine learning model B_1. (2) A new incoming dataset, $\mathcal{DS}^{\text{p}}$, which is coming from an untrusted source and a potentially malicious dataset, is used along with $\mathcal{DS}^{\text{v}}$ to learn B_2. (3) FLoD checks for one-step data poisoning attacks. If model change occurs (i.e., $\mathbf{ds}(B_1, B_2) > 0$), send $\mathcal{DS}^{\text{p}}$ for offline evaluation. Else, (4) SLoD checks for long-duration data poisoning attacks. If the value of conflict measure is positive (i.e., $Conf(c, B_1) > 0$), send conflicting data to offline evaluation. Else, update the validated dataset.

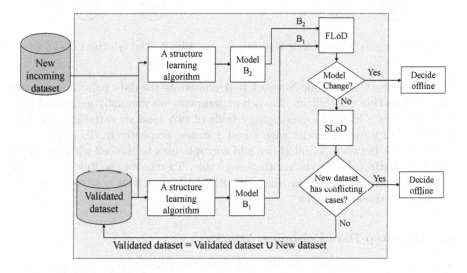

Fig. 1. Framework for detecting data poisoning attacks.

5   Empirical Results

We implemented our prototype system using the Chest Clinic Network [23]. The Chest Clinic Network was created by Lauritzen and Spielgelhalter [23] and is widely used in Bayesian network experiments. As shown in Fig. 2, Visit to Asia is a simple, fictitious network that could be used at a clinic to diagnose arriving patients. It consists of eight nodes and eight edges. The nodes are as follows:

(1) *(node A)* shows whether the patient lately visited Asia; (2) *(node S)* shows if the patient is a smoker; (3) *(node T)* shows if the patient has Tuberculosis; (4) *(node L)* shows if the patient has lung cancer; (5) *(node B)* shows if the patient has Bronchitis; (6) *(node E)* shows if the patient has either Tuberculosis or lung cancer; (7) *(node X)* shows whether the patient X-ray is abnormal; and (8) *(node D)* shows if the patient has Dyspnea. The edges indicate the causal relations between the nodes. A simple example of a causal relation is: Visiting Asia may cause Tuberculosis and so on. We refer the readers to [23] for a full description of this network.

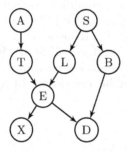
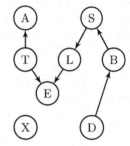

Fig. 2. The original Chest Clinic Network.   **Fig. 3.** The validated model B_1.

We used the Chest Clinic Network to demonstrate the data poisoning attacks and our detection capabilities. In each experiment, we manually generated poisoned datasets. Given the contingency table of two random variables A and B in a Bayesian network model with i and j states, respectively. To introduce a malicious link between A and B, we add corrupt cases to the cell with the highest test statistic value in the contingency table. To remove the link between A and B, we transfer cases from the cell with the highest test statistics value to the one with the lowest value.

5.1   One-Step Data Poisoning Attacks

To set up the experiment, we implemented the Chest Clinic Network using $Hugin^{TM}$ *Research 8.1*. We then used $Hugin^{TM}$ *case generator* [26,32] to generate a simulated dataset of 20,000 cases. We call this dataset $\mathcal{DS}^v$. Using the PC algorithm on dataset $\mathcal{DS}^v$ with 0.05 significance setting [26], the resulting validated structure, $B_1 = PC\_Algo(\mathcal{DS}^v)$, is given in Fig. 3. While the two networks in Figs. 2 and 3 belong to different Markov equivalence classes, we will use the validated network B_1 as the starting point of our experiment.

We evaluated the effectiveness of one-step data poisoning attacks against the validated dataset $\mathcal{DS}^v$ (i.e., against the validated model B_1). An attacker aims to use one-step data poisoning attacks to inject in a contaminated dataset $\mathcal{DS}^p$ into $\mathcal{DS}^v$, resulting in the dataset $\mathcal{DS}^u$. The defender retrains the machine learning model by feeding the new dataset $\mathcal{DS}^u$ to the PC learning algorithm ($B_2 = PC\_Algo(\mathcal{DS}^u)$), resulting in the model B_2.

We aim to study the attacker's goals, i.e., study the feasibility of one-step data poisoning attacks, which might be as follows: (i) introduce new v-structures: that is, (1) add the links $D - S$ and $S - E$ to the serial connections $D \rightarrow B \rightarrow S$ and $S \rightarrow L \rightarrow E$, respectively, and (2) add the link $A - E$ to the diverging connection $A \leftarrow T \rightarrow E$; (ii) break an existing v-structure $T \rightarrow E \leftarrow L$, i.e., shield the collider E; (iii) remove the weakest edge, i.e., remove the edge $T \rightarrow A$; and (iv) add the most believable edge, i.e., add the edge $B \rightarrow L$. (Note that, for finding the weakest link in a given causal model or the most believable link to be added to a causal model, we refer the readers to our previous works [3,5] for technical details on how to measure link strength of causal models).

In all of the scenarios, the attacker succeeded in corrupting the new model that was going to be learned by the defender, the model B_2. The attacker had to introduce a dataset $\mathcal{DS}^p$ with 67 corrupt cases (data items) to introduce the link $D - S$ in the newly learned model B_2. To introduce links $S - E$ and $A - E$ required 21 and 7 corrupt cases, respectively. To shield the collider E, the attacker only needed 4 poisoning data items. The attacker had to modify only 3 cases to break the weakest link $A - T$. To add the most believable link $B - L$ required to only 7 corrupt data items.

5.2   Long-Duration Data Poisoning Attacks

To set up the implementation of long-duration attacks, let $\mathcal{DS}^v$ be a validated training dataset with attributes $x_1, \ldots, x_n$ and N cases, and β be *data poisoning rate* at which attackers are allowed to add new "contaminated" cases to $\mathcal{DS}^v$. Let $\mathcal{DS}_i^c$ be a newly crafted dataset also with attributes $x_1, \ldots, x_n$ and N_i cases, and λ_i be data poisoning rate at which attackers allowed to add new crafted cases to $\mathcal{DS}_i^c$ (we default set $0 \leq \bigcup_{t=1}^w \lambda_i N_i \leq \beta N$).

We start by calculating τ, which is the maximum number of poisoned cases that could be added to $\mathcal{DS}^v$ over a period of time $t = \{1, \ldots, w\}$. We then learn the structure of the validated model B_1 from $\mathcal{DS}^v$ using the PC algorithm.

We then iterate w times. In each iteration t, we generate a clean dataset $\mathcal{DS}_t^{\text{clean}}$ and a poisoned dataset $\mathcal{DS}_t^p$. We let $\mathcal{DS}_t^c = \mathcal{DS}_t^{\text{clean}} \cup \mathcal{DS}_t^p$ (note that, $\mathcal{DS}_t^c$ has N_t cases, $\lambda_t N_t$ of which are poisoned). After that, we create the union of $\mathcal{DS}_t^c$ and $\mathcal{DS}^v$, resulting in $\mathcal{DS}_t^{\text{l-d}}$, which is used to learn the structure of model B_2'. Note that, in each iteration the number of cases in $\mathcal{DS}_t^p$ should be between 0 (i.e., no poisoned cases) and $\frac{\tau}{w}$, which is the maximum number of poisoned cases that could be added to $\mathcal{DS}_t^c$ in the tth iteration.

We terminate after iteration w. If $\bigcup_{t=1}^w \lambda_t N_t \leq \beta N$, we return $\mathcal{DS}_t^{\text{l-d}}$; otherwise, we print a failure message since implementing the long-duration attack on $\mathcal{DS}^v$ is not feasible.

Table 1. Results of long-duration data poisoning attacks against $\mathcal{DS}^v$.

(a) Introducing the link $A \to E$ in the diverging connection $A \leftarrow T \to E$.

Time point $t = \{1, \ldots, w\}$	$t = 1$	$t = 2$	$t = 3$	$t = 4$
Number of clean cases at time point t ($\mathcal{DS}_t^{Clean}$)	2,000	2,000	2,000	2,000
Number of crafted cases at time point t ($\mathcal{DS}_t^{Crafted}$)	3	1	3	0
$\mathcal{DS}_t^c = \mathcal{DS}_t^{Clean} \cup \mathcal{DS}_t^{Crafted}$	2,003	2,004	2,007	2,007
$\mathcal{DS}_t^{Ld} = \mathcal{DS}^v \cup \bigcup_{t=1}^w \mathcal{DS}_t^c$	14,003	16,004	18,007	20,007
Model Change	No	No	No	Yes

(b) Breaking the v-structure $T \to E \leftarrow L$.

Time point $t = \{1, \ldots, w\}$	$t = 1$	$t = 2$	$t = 3$	$t = 4$
Number of clean cases at time point t ($\mathcal{DS}_t^{Clean}$)	2,000	2,000	2,000	2,000
Number of crafted cases at time point t ($\mathcal{DS}_t^{Crafted}$)	2	2	0	0
$\mathcal{DS}_t^c = \mathcal{DS}_t^{Clean} \cup \mathcal{DS}_t^{Crafted}$	2,002	2,002	2,000	2,000
$\mathcal{DS}_t^{Ld} = \mathcal{DS}^v \cup \bigcup_{t=1}^w \mathcal{DS}_t^c$	14,002	16,004	18,004	20,004
Model Change	No	No	No	Yes

(c) Add the most believable edge, $B \to L$, to the causal model B_1.

Time point $t = \{1, \ldots, w\}$	$t = 1$	$t = 2$	$t = 3$	$t = 4$
Number of clean cases at time point t ($\mathcal{DS}_t^{Clean}$)	2,000	2,000	2,000	2,000
Number of crafted cases at time point t ($\mathcal{DS}_t^{Crafted}$)	2	2	1	2
$\mathcal{DS}_t^c = \mathcal{DS}_t^{Clean} \cup \mathcal{DS}_t^{Crafted}$	2,002	2,002	2,001	2,002
$\mathcal{DS}_t^{Ld} = \mathcal{DS}^v \cup \bigcup_{t=1}^w \mathcal{DS}_t^c$	14,002	16,004	18,005	20,007
Model Change	No	No	No	Yes

(d) Adding the link $D \to S$ to the serial connection $D \to B \to S$.

Time point $t = \{1, \ldots, w\}$	$t = 1$	$t = 2$	$t = 3$	$t = 4$
Number of clean cases at time point t ($\mathcal{DS}_t^{Clean}$)	2,000	2,000	2,000	2,000
Number of crafted cases at time point t ($\mathcal{DS}_t^{Crafted}$)	20	20	23	4
$\mathcal{DS}_t^c = \mathcal{DS}_t^{Clean} \cup \mathcal{DS}_t^{Crafted}$	2,020	2,020	2,023	2,004
$\mathcal{DS}_t^{Ld} = \mathcal{DS}^v \cup \bigcup_{t=1}^w \mathcal{DS}_t^c$	14,020	16,040	18,063	20,067
Model Change	No	No	No	Yes

(e) Adding the link $S \to E$ to the serial connection $S \to L \to E$.

Time point $t = \{1, \ldots, w\}$	$t = 1$	$t = 2$	$t = 3$	$t = 4$
Number of clean cases at time point t ($\mathcal{DS}_t^{Clean}$)	2,000	2,000	2,000	2,000
Number of crafted cases at time point t ($\mathcal{DS}_t^{Crafted}$)	7	8	5	1
$\mathcal{DS}_t^c = \mathcal{DS}_t^{Clean} \cup \mathcal{DS}_t^{Crafted}$	2,007	2,008	2,005	2,001
$\mathcal{DS}_t^{Ld} = \mathcal{DS}^v \cup \bigcup_{t=1}^w \mathcal{DS}_t^c$	14,007	16,015	18,020	20,021
Model Change	No	No	No	Yes

(f) Removing the weakest link, $T \to A$, from the causal model B_1.

Time point $t = \{1, \ldots, w\}$	$t = 1$	$t = 2$	$t = 3$	$t = 4$
Number of clean cases at time point t ($\mathcal{DS}_t^{Clean}$)	2,000	2,000	2,000	2,000
Number of crafted cases at time point t ($\mathcal{DS}_t^{Crafted}$)	1	1	1	0
$\mathcal{DS}_t^c = \mathcal{DS}_t^{Clean} \cup \mathcal{DS}_t^{Crafted}$	2,001	2,001	2,001	2,000
$\mathcal{DS}_t^{Ld} = \mathcal{DS}^v \cup \bigcup_{t=1}^w \mathcal{DS}_t^c$	14,001	16,002	18,003	20,003
Model Change	No	No	No	Yes

We assumed that $w = 4$, which means that the attacker is allowed to send in four contaminated datasets to achieve the long-duration data poisoning attack. We divided the 20,000 case dataset that was generated for one-step data poisoning attacks in Sect. 5.1 into five datasets as follows: 12,000 cases are used as $\mathcal{DS}^v$; and the rest is divided into four datasets of 2,000 cases each. We call these four datasets $\mathcal{DS}_1^{\text{Clean}}$, $\mathcal{DS}_2^{\text{Clean}}$, $\mathcal{DS}_3^{\text{Clean}}$, and $\mathcal{DS}_4^{\text{Clean}}$. Using the PC algorithm on dataset $\mathcal{DS}^v$ with 0.05 significance setting [26], the resulting validated structure, $B_1 = PC\_Algo(\mathcal{DS}^v)$, is given in Fig. 3, which is the starting point of this experiment.

We evaluated the effectiveness of long-duration data poisoning attacks against the validated dataset $\mathcal{DS}^v$ (i.e., against the validated model B_1). At every time point $t = \{1, \ldots, w\}$, the attacker injects a contaminated dataset $\mathcal{DS}_t^{\text{Crafted}}$ into $\mathcal{DS}_t^{\text{Clean}}$, resulting in the dataset $\mathcal{DS}_t^c$. This resulting dataset is then sent in as a new source of information. The defender receives $\mathcal{DS}_t^c$ and retrains the validated model, B_1, by creating the union of $\mathcal{DS}^v$ and the new incoming dataset $\mathcal{DS}_t^c$ and feeding them to the PC algorithm, resulting in the model B_2' (i.e., $B_2' = PC\_Algo(\mathcal{DS}^v \cup \mathcal{DS}_t^c)$).

The results of our experiments are presented in Table 1. In all of the scenarios, the attacker succeeded in achieving the desired modification. In our experiments, we assumed that $t = \{1, \ldots, 4\}$. For every one of the studied long-duration attacks on the dataset $\mathcal{DS}^v$ (Tables 1a, b, c, d, e, and f), the adversary had to send in the attack over 4 datasets. That is, at every time point t (for $t = 1, \ldots, 4$), the attacker had to create the union of $\mathcal{DS}_t^{\text{Clean}}$ and $\mathcal{DS}_t^{\text{Crafted}}$ resulting in $\mathcal{DS}_t^c$, which was going to be sent to the targeted machine learning system as a new source of information. The defender, on the other hand, retrained the machine learning model every time a new incoming dataset $\mathcal{DS}_t^c$ arrived.

Note that, in our experiments, long-duration attacks require the same number of contaminated cases as the one-step data poisoning attacks. An important observation is that the malicious attacker does not always have to send poisoned cases in the last dataset that will trigger the attack. For instance, in our experiments, when introducing the link $A \rightarrow E$ (Table 1a), shielding collider E (Table 1b), and removing the weakest edge (Table 1f), the last contaminated dataset, $\mathcal{DS}_4^c$, had no contaminated cases, which makes it impossible for the defender to find what caused a change in the newly learned model.

5.3 Discussion: Detecting Data Poisoning Attacks

The results of using our framework to detect one-step data poisoning attacks are presented in Table 2. Algorithm 1 succeeded to detect the negative impact (i.e., the change in the Markov equivalence class) of the new incoming dataset $\mathcal{DS}^p$ on the validated model B_1.

Table 2. Results of using FLoD to detect one-step poisoning attacks.

Attack	Attack's class	$\mathbf{ds}(B_1, B_2)$ score
Introduce the link $A \rightarrow E$	New v-structure	12
Introduce the link $D \rightarrow S$	New v-structure	24
Introduce the link $S \rightarrow E$	New v-structure	54
Introduce the link $T \rightarrow L$	Shield an existing collider	16
Remove the link $A \rightarrow T$	Delete the weakest link	4
Introduce the link $B \rightarrow L$	Add the most believable link	32

The results using our framework to detect long-duration data poisoning attacks are summarized in Table 3. Algorithm 2 succeeded to detect the long-duration impact of $\mathcal{DS}^c$ on the validated dataset $\mathcal{DS}^v$. Note, that FLoD using traditional reject on negative impact was not able to detect long-duration attacks. However, when using the SLoD, we were able to detect the conflicting cases, which are either inconsistent or incompatible with the original validated model B_1 (A detailed experiment is presented in Fig. 4). Such cases might be exploited by a malicious adversary to trigger the long-duration attack at a later time. Also, in some attacks no poisoned cases are even required to be sent with $\mathcal{DS}^c$ to trigger the long-duration attack, which is very hard to detect.

Table 3. Results of using SLoD to detect long-duration data poisoning attacks.

Attack	Attack's class	Algorithm 2 decision
Introduce $A \rightarrow E$	New v-structure	Inconsistent observations
Introduce $D \rightarrow S$	New v-structure	Incompatible observations
Introduce $S \rightarrow E$	New v-structure	Inconsistent observations
Introduce $T \rightarrow L$	Shield an existing collider	Inconsistent observations
Remove $A \rightarrow T$	Delete weakest link	Inconsistent\Incompatible observations
Introduce $B \rightarrow L$	Add most believable link	Inconsistent observations

In summary, our 2-layered approach was able to detect both one-step and long-duration attacks. Moreover, our solution did not lose all the incoming datasets; we only send conflicting cases to be checked offline. We have carried out over 200 experiments for long-duration attacks. A comprehensive description of these experiments is given in [2].

6   Related Work

In this section, we will give a brief overview of adversarial machine learning research; focusing on data poisoning. Recent surveys on adversarial machine learning can be found in [6,16,24].

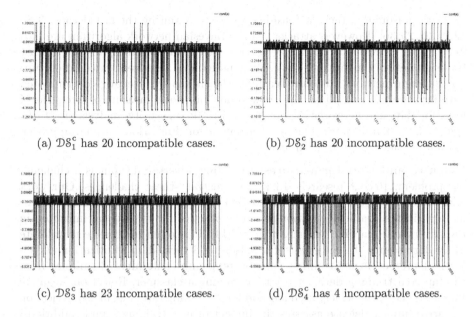

(a) $\mathcal{DS}_1^c$ has 20 incompatible cases.   (b) $\mathcal{DS}_2^c$ has 20 incompatible cases.

(c) $\mathcal{DS}_3^c$ has 23 incompatible cases.   (d) $\mathcal{DS}_4^c$ has 4 incompatible cases.

Fig. 4. The result of using **SLoD** to detect a long-duration attack that aims to introduce the link $D \rightarrow S$ in the Chest Clinic dataset, $\mathcal{DS}^v$. We present the case number in $\mathcal{DS}_t^c$ as the variable on the X-axis and the value of our conflict measure $Conf(c, B_1)$ as the variable on the Y-axis. A case is incompatible (conflicting) with the validated model B_1 if $Conf(c, B_1) > 0$.

Data Poisoning Attacks: As machine learning algorithms have been widely used in security-critical settings such as spam filtering and intrusion detection, adversarial machine learning has become an emerging field of study. Attacks against machine learning systems have been organized by [6,7,18] according to three features: Influence, Security Violation, and Specificity. Influence of the attacks on machine learning models can be either causative or exploratory. Causative attacks aim to corrupt the training data whereas exploratory attacks aim to corrupt the classifier at test time. Security violation of machine learning models can be a violation of integrity, availability, or privacy. Specificity of the attacks can be either targeted or indiscriminate. Targeted attacks aim to corrupt machine learning models to misclassify a particular class of false positives whereas indiscriminate attacks have the goal of misclassifying all false positives.

Evasion attacks and Data poisoning attacks are two of the most common attacks on machine learning systems [18]. Evasion attacks [17,20,22] are exploratory attacks at the testing phase. In an evasion attack, an adversary attempts to pollute the data for testing the machine learning classifier; thus causing the classifier to misclassify adversarial examples as legitimate ones. Data poisoning attacks [1,11,21,27,28,36] are causative attacks, in which adversaries attempt to corrupt the machine learning classifier itself by contaminating the data in the training phase.

Data poisoning attacks are studied extensively during the last decade [3, 8–12, 21, 28, 29, 36]. However, attacks against Bayesian network algorithm are limited. In our previous work, we were addressed data poisoning attacks against Bayesian network algorithms [3–5]. We studied how an adversary could corrupt the Bayesian network structure learning algorithms by inserting contaminated data into the training phase. We showed how our novel measure of strengths of links for Bayesian networks [5] can be used to do a security analysis of attacks against Bayesian network structure learning algorithms. However, our approach did not consider long-duration attacks.

Defenses and Countermeasures: Detecting adversarial input is a challenging problem. Recent research [13, 15, 25] illustrate these challenges. Our work addresses these issues in the specific context of Bayesian network structure learning algorithms. Data sanitization is a best practice for security optimization in the adversarial machine learning context [14]. It is often impossible to validate every data source. In the event of a poisoning attack, data sanitization adds a layer of protection for training data by removing contaminated samples from the targeted training data set prior to training a classifier. Reject on Negative Impact is one of the widely used method for data sanitization [6, 14, 24]. Reject on Negative Impact defense assesses the impact of new training sample additions, opting to remover or discard samples that yield significant, negative effects on the observed learning outcomes or classification accuracy [6, 14]. The base training set is used to train a classifier, after which, the new training instance is added and a second classifier is trained [6]. In this approach, classification performance is evaluated by comparing error rates (accuracy) between the original and the new, retrained classifier resulting from new sample integration [24]. As such, if new classification errors are substantially higher compared to the original or baseline classifier, it is assumed that the newly added samples are malicious or contaminated and are therefore removed in order to maximize and protect classification accuracy [6].

7   Conclusion and Future Work

Data integrity is vital for effective machine learning. In this paper, we studied data poisoning attacks against Bayesian network structure learning algorithms. We demonstrated the vulnerability of the PC algorithm against one-step and long-duration data poisoning attacks. We proposed a 2-layered framework for detecting data poisoning attacks. We implemented our prototype system using the Chest Clinic Network which is a widely used network in Bayesian networks. Our results indicate that Bayesian network structure learning algorithms are vulnerable to one-step and long-duration data poisoning attacks. Our framework is effective in detecting both one-step and long-duration data poisoning attacks, as it thoroughly validates and verifies training data before such data is being incorporated into the model.

Our ongoing work focuses on offline validation of potentially malicious datasets. Currently, our approach detects datasets that either change the

Bayesian network structure (distance measure) or in conflict with the validated model (conflict measure). We are investigating methods for (1) distinguishing actual model shift from model enrichment, i.e., our initial model was based on data that was not fully representative of the "true" distribution, and (2) determining if cases are truly conflicting or again if the initial model poorly approximates the "true" distribution. We are also investigating the applicability of Wisdom of the Crowd (WoC) [37]. Rather than human experts, we plan to use an ensemble of classifiers, i.e., take the votes of competing algorithms instead of the votes of humans. In the case of an ensemble of classifiers, one could investigate the likelihood of unexpected cases and adjust the sensitivity to anomalies by how much perturbation it causes in the model.

A   Causative, Long-duration Model Invalidation Attacks

In this Appendix, we explain the two subtypes of each of the causative long-duration attacks which are based on the notion of d-separation and marginal independence tests.

The causative long-duration attacks which are based on the notion of d-separation are divided into two main subtypes as follows:

(i) Creating a new converging connection (v-structure) attacks, in which adversaries attempt to corrupt the original Bayesian network model, B_1, by poisoning the validated dataset, $\mathcal{DS}^v$, using contaminated datasets $\bigcup_{t=1}^{w} \mathcal{DS}_t^c$. Attackers aim to introduce a new v-structure by adding the link $A \rightarrow C$ to the serial connection $A \rightarrow B \rightarrow C$, link $C \rightarrow A$ to the serial connection $A \leftarrow B \leftarrow C$, or either one of the links $A \rightarrow C$ or $C \rightarrow A$ to the diverging connection $A \leftarrow B \rightarrow C$ in B_1.

(ii) Breaking an existing converging connection (v-structure) attacks, in which malicious attackers attempt to corrupt the original model, B_1, by shielding existing colliders (v-structures). Such adversarial attacks can be performed by poisoning the dataset, $\mathcal{DS}^v$, over time using the poisoned datasets $\bigcup_{t=1}^{w} \mathcal{DS}_t^c$ such that new links are introduced to marry the parents of unshielded colliders in B_1 (i.e., add the link $A \rightarrow C$ to the converging connection $A \rightarrow B \leftarrow C$).

We divide the causative long-duration attacks which are based on marginal independence tests into two main subtypes:

(i) Removing the weakest edge attacks, in which adversarial opponents attempt to poison the validated learning dataset, $\mathcal{DS}^v$, using contaminated datasets, $\bigcup_{t=1}^{w} \mathcal{DS}_t^c$, over a period of time t with the ultimate goal of removing weak edges. Note that, a weak edge in a Bayesian model, B_1, is the easiest edge to be removed from B_1. We use our previously defined link strength measure to determine such edges [5].

(ii) Adding the most believable yet incorrect edge attacks, in which adversaries can cleverly craft their input datasets, $\bigcup_{t=1}^{w} \mathcal{DS}_t^c$, over a period of time t to poison $\mathcal{DS}^{v}$ so that adding the most believable yet incorrect edge is viable. The most believable yet incorrect edge is a newly added edge to model, B_1, with the maximum amount of belief. We use our link strength measure defined in [5] to determine such edges.

References

1. Alfeld, S., Zhu, X., Barford, P.: Data poisoning attacks against autoregressive models. In: AAAI, pp. 1452–1458 (2016)
2. Alsuwat, E., Alsuwat, H., Rose, J., Valtorta, M., Farkas, C.: Long duration data poisoning attacks on Bayesian networks. Technical report, University of South Carolina, SC, USA (2019)
3. Alsuwat, E., Alsuwat, H., Valtorta, M., Farkas, C.: Cyber attacks against the PC learning algorithm. In: Alzate, C., et al. (eds.) ECML PKDD 2018. LNCS (LNAI), vol. 11329, pp. 159–176. Springer, Cham (2019). https://doi.org/10.1007/978-3-030-13453-2_13
4. Alsuwat, E., Valtorta, M., Farkas, C.: Bayesian structure learning attacks. Technical report, University of South Carolina, SC, USA (2018)
5. Alsuwat, E., Valtorta, M., Farkas, C.: How to generate the network you want with the PC learning algorithm. In: Proceedings of the 11th Workshop on Uncertainty Processing (WUPES 2018), pp. 1–12 (2018)
6. Barreno, M., Nelson, B., Joseph, A.D., Tygar, J.D.: The security of machine learning. Mach. Learn. 81(2), 121–148 (2010)
7. Barreno, M., Nelson, B., Sears, R., Joseph, A.D., Tygar, J.D.: Can machine learning be secure? In: Proceedings of the 2006 ACM Symposium on Information, Computer and Communications Security, pp. 16–25. ACM (2006)
8. Biggio, B., et al.: Poisoning complete-linkage hierarchical clustering. In: Fränti, P., Brown, G., Loog, M., Escolano, F., Pelillo, M. (eds.) S+SSPR 2014. LNCS, vol. 8621, pp. 42–52. Springer, Heidelberg (2014). https://doi.org/10.1007/978-3-662-44415-3_5
9. Biggio, B., Didaci, L., Fumera, G., Roli, F.: Poisoning attacks to compromise face templates. In: 2013 International Conference on Biometrics (ICB), pp. 1–7. IEEE (2013)
10. Biggio, B., Fumera, G., Roli, F., Didaci, L.: Poisoning adaptive biometric systems. In: Gimel'farb, G., et al. (eds.) SSPR /SPR 2012. LNCS, vol. 7626, pp. 417–425. Springer, Heidelberg (2012). https://doi.org/10.1007/978-3-642-34166-3_46
11. Biggio, B., Nelson, B., Laskov, P.: Poisoning attacks against support vector machines. In: Proceedings of the 29th International Conference on International Conference on Machine Learning, pp. 1467–1474. Omnipress (2012)
12. Biggio, B., Pillai, I., Rota Bulò, S., Ariu, D., Pelillo, M., Roli, F.: Is data clustering in adversarial settings secure? In: Proceedings of the 2013 ACM Workshop on Artificial Intelligence and Security, pp. 87–98. ACM (2013)
13. Carlini, N., Wagner, D.: Adversarial examples are not easily detected: bypassing ten detection methods. In: Proceedings of the 10th ACM Workshop on Artificial Intelligence and Security, pp. 3–14. ACM (2017)
14. Chan, P.P., He, Z.M., Li, H., Hsu, C.C.: Data sanitization against adversarial label contamination based on data complexity. Int. J. Mach. Learn. Cybern. 9(6), 1039–1052 (2018)

15. Feinman, R., Curtin, R.R., Shintre, S., Gardner, A.B.: Detecting adversarial samples from artifacts. CoRR abs/1703.00410 (2017)
16. Gardiner, J., Nagaraja, S.: On the security of machine learning in malware C&C detection: a survey. ACM Comput. Surv. (CSUR) **49**(3), 59 (2016)
17. Goodfellow, I.J., Shlens, J., Szegedy, C.: Explaining and harnessing adversarial examples. arXiv preprint arXiv:1412.6572 (2014)
18. Huang, L., Joseph, A.D., Nelson, B., Rubinstein, B.I., Tygar, J.: Adversarial machine learning. In: Proceedings of the 4th ACM Workshop on Security and Artificial Intelligence, pp. 43–58. ACM (2011)
19. de Jongh, M., Druzdzel, M.J.: A comparison of structural distance measures for causal Bayesian network models. In: Recent Advances in Intelligent Information Systems, Challenging Problems of Science, Computer Science Series, pp. 443–456 (2009)
20. Kantchelian, A., Tygar, J., Joseph, A.: Evasion and hardening of tree ensemble classifiers. In: International Conference on Machine Learning, pp. 2387–2396 (2016)
21. Koh, P.W., Liang, P.: Understanding black-box predictions via influence functions. In: International Conference on Machine Learning, pp. 1885–1894 (2017)
22. Laskov, P., et al.: Practical evasion of a learning-based classifier: a case study. In: 2014 IEEE Symposium on Security and Privacy (SP), pp. 197–211. IEEE (2014)
23. Lauritzen, S.L., Spiegelhalter, D.J.: Local computations with probabilities on graphical structures and their application to expert systems. J. Roy. Stat. Soc. Ser. B (Methodol.) **50**, 157–224 (1988)
24. Liu, Q., Li, P., Zhao, W., Cai, W., Yu, S., Leung, V.C.: A survey on security threats and defensive techniques of machine learning: a data driven view. IEEE Access **6**, 12103–12117 (2018)
25. Lu, J., Issaranon, T., Forsyth, D.: Safetynet: detecting and rejecting adversarial examples robustly. In: 2017 IEEE International Conference on Computer Vision (ICCV), pp. 446–454, October 2017. https://doi.org/10.1109/ICCV.2017.56
26. Madsen, A.L., Jensen, F., Kjaerulff, U.B., Lang, M.: The Hugin tool for probabilistic graphical models. Int. J. Artif. Intell. Tools **14**(03), 507–543 (2005)
27. Mei, S., Zhu, X.: The security of latent Dirichlet allocation. In: Artificial Intelligence and Statistics, pp. 681–689 (2015)
28. Mei, S., Zhu, X.: Using machine teaching to identify optimal training set attacks on machine learners. In: AAAI, pp. 2871–2877 (2015)
29. Muñoz-González, L., et al.: Towards poisoning of deep learning algorithms with back-gradient optimization. In: Proceedings of the 10th ACM Workshop on Artificial Intelligence and Security, pp. 27–38. ACM (2017)
30. Nelson, B., et al.: Misleading learners: co-opting your spam filter. In: Yu, P.S., Tsai, J.J.P. (eds.) Machine Learning in Cyber Trust, pp. 17–51. Springer, Heidelberg (2009). https://doi.org/10.1007/978-0-387-88735-7_2
31. Nielsen, T.D., Jensen, F.V.: Bayesian Networks and Decision Graphs. Springer, Heidelberg (2009)
32. Olesen, K.G., Lauritzen, S.L., Jensen, F.V.: aHUGIN: a system creating adaptive causal probabilistic networks. In: Uncertainty in Artificial Intelligence, pp. 223–229. Elsevier (1992)
33. Paudice, A., Muñoz-González, L., Gyorgy, A., Lupu, E.C.: Detection of adversarial training examples in poisoning attacks through anomaly detection. arXiv preprint arXiv:1802.03041 (2018)
34. Spirtes, P., Glymour, C.N., Scheines, R.: Causation, Prediction, and Search. MIT Press, Cambridge (2000)

35. Wang, Y., Chaudhuri, K.: Data poisoning attacks against online learning. arXiv preprint arXiv:1808.08994 (2018)
36. Yang, C., Wu, Q., Li, H., Chen, Y.: Generative poisoning attack method against neural networks. arXiv preprint arXiv:1703.01340 (2017)
37. Yi, S.K.M., Steyvers, M., Lee, M.D., Dry, M.J.: The wisdom of the crowd in combinatorial problems. Cogn. Sci. **36**(3), 452–470 (2012)

AGBuilder: An AI Tool for Automated Attack Graph Building, Analysis, and Refinement

Bruhadeshwar Bezawada[1]([✉]), Indrajit Ray[2,3], and Kushagra Tiwary[2]

[1] Mahindra École Centrale, Hyderabad, India
bru@mechyd.ac.in
[2] Colorado State University, Fort Collins, CO 80523, USA
{indrajit.ray,kushagra.tiwary}@colostate.edu
[3] National Science Foundation, Alexandria, USA

Abstract. Attack graphs are widely used for modeling attack scenarios that exploit vulnerabilities in computer systems and networked infrastructures. Essentially, an attack graph illustrates a *what-if* analysis, thereby, helping the network administrator to plan for potential security threats. However, current attack graph representations not only suffer from scaling issues, but also are difficult to generate. Despite efforts from the research community there are no automated tools for generating attack graphs from textual descriptions of vulnerabilities such as those from the Common Vulnerabilities and Exposures (CVE) in the National Vulnerability Database (NVD). Additionally, there is little support for incremental updates and refinements to an attack graph model. This is needed to reflect changes to an attack graph that arise because of changes to the vulnerability state of the underlying system being modeled. In this work, we present an artificial intelligence (AI) based planning tool, *AGBuilder* – Attack Graph Builder, for automatically generating, updating and refining attack graphs. A key contribution of AGBuilder is that it uses textual descriptions of vulnerabilities to automatically generate attack graphs. Another significant contribution is that, using AGBuilder, we describe a methodology to incrementally update attack graphs when the system changes. This aspect has not been addressed in prior research and is a crucial step for achieving resiliency in the face of evolving adversarial strategies. Finally, AGBuilder has the ability to reuse smaller attack graphs, e.g., when building a network of networks, and join them together to create larger attack graphs.

Keywords: Attack graphs ·
Planning Domain Definition Language (PDDL) · AI Planning · CVE ·
NVD

1 Introduction

Cyber-attacks against safety critical and mission critical systems such as nuclear power plants are rising alarmingly. It is no longer a question of "if" but "when"

© IFIP International Federation for Information Processing 2019
Published by Springer Nature Switzerland AG 2019
S. N. Foley (Ed.): DBSec 2019, LNCS 11559, pp. 23–42, 2019.
https://doi.org/10.1007/978-3-030-22479-0_2

a system will be attacked. Thus, in order to be adequately prepared for such an eventuality, there is a need to better understand how the system can be attacked so that provisions for defense deployment can be made or, perhaps, provisions for the graceful degradation of mission services can be instantiated when all defenses have failed. Information security planning and management traditionally begins with risk assessment with the help of system mapping and dependency analysis. The outcome of this process is an identification of vulnerabilities in the system, an enumeration of the threats to critical resources arising from these vulnerabilities, and the corresponding loss expectancy. The analysis allows one to determine appropriate security controls to protect resources and minimize their susceptibility to cyber attacks.

Attack trees [2,10,17] and attack graphs [1,13,16,20,22,28] are two systematic computer security models that represent a networked system's vulnerability to malicious attacks by enumerating known vulnerabilities in the hosts or applications. They capture cause-consequence relationships between system configuration and the vulnerabilities in the form of And-Or tree (attack tree) or a directed graph (attack graph). Nodes in the tree/graph represent system states that may be of interest to the attacker. Edges connecting the nodes denote a cause-consequence relationship among the states. A key weakness of this representation is the explosion of state space. This becomes a critical drawback for analyzing large cyber-physical systems with many resources that need to be protected from a multitude of attacks.

Automated planning holds promise to reduce the number of nodes in the attack graph/tree and produce a scalable solution. Boddy et al. [3], presented Behavioral Adversary Modeling System (BAMS), a planning system that models attack scenarios and produces countermeasures to subvert the attacks in networks of large organizations. Ghosh and Ghosh [7] proposed a planner based approach for tractable representation of attack graphs and automatic generation of attack paths. However, none of these works discuss how an attack graph can be automatically constructed, refined and updated as needed from textual description of vulnerabilities such as those in the National Vulnerability Database or CVE repositories.

In this work, we model the attack graph generation and analysis problem as a *planning problem* [7] in the artificial intelligence community. When compared with logic programming approaches like [13], the planning approach lends itself to incremental updates and aggregation, which are quite useful to network administrators. We encode the attack graph in the Planner Domain Definition Language (PDDL) representation, referred to as a PDDL *domain*. However, there are some important challenges that arise in this modeling. First challenge is that, translating the attack graph/tree of a large network to the corresponding PDDL domain is an iterative process, which demands a lot of time and effort from engineers. The issue is further exacerbated by the lack of tool support to build, debug and maintain PDDL domains from textual CVE descriptions. The second challenge is that, if the underlying system is changed in any manner, for example, by installing a new application, then updating the corresponding attack graph is a computationally complex and error-prone process. The existing approaches have not addressed this situation and require generating a fresh attack graph for the changed system environment. The third challenge is that, when a PDDL domain is incrementally built more actions are added into a domain or actions already in the domain are edited. Such incremental development is found in scenarios where network administrators start by analyzing smaller parts of a network and then try to aggregate the smaller network models into a larger network model.

To address these challenges, we present a formal methodology and a corresponding tool-set, *AGBuilder* –Attack Graph Builder, designed to automatically generate PDDL based representation of attacks from textual description of vulnerabilities found in the CVE system or the NVD system. Our tool-set incorporates a natural-language processing based generator to generate a PDDL based model of attacks from vulnerabilities and support for incremental development of the PDDL model to reconcile incremental versions of PDDL domains by generating explanations for changes in plans that result from running the modified domain against a planner. Additionally, the tool-set constructs abstract syntax trees and attack graphs for the PDDL domain representation, which facilitates visualization of the domain and the planning problems.

The rest of the paper is organized as follows. In Sect. 2, we give an overview of the PDDL language and explain the modeling of attack graphs using PDDL. In Sect. 3, we give an overview of our approach. In Sect. 4, we describe the AGBuilder tool set and our methodology in detail. In Sect. 5, we describe the related work in this domain and conclude in Sect. 6.

2   Attack Graph Modeling Using PDDL

In [7], the authors provided an approach for modeling the attack graphs using AI planners. In this section, we describe this process in detail. A PDDL definition is composed of two key parts: (1) a PDDL domain definition and (2) a PDDL problem description.

PDDL Domain: A PDDL domain is a high-level description of a set of problems and the corresponding actions and constraints involved. A PDDL domain specifies the requirements it supports, the available actions, the pre-conditions and post-conditions of actions. The pre-conditions and post-conditions are expressed as first-order logic predicates. The requirements of the PDDL domain specify which features it expects the planner to support. A planner will only accept a domain if it supports all the requirements mentioned in the domain. A single PDDL domain can be used to represent multiple attacks from vulnerability databases. The PDDL domain stores pre-conditions, post-conditions and cause-effect relationships in courses of actions that represent attacks. Consider the following PDDL domain that models an attack:

```
(define (domain PAG) (:requirements :equality :disjunctive-preconditions)
(:functions (version ?software))
(:predicates (user ?User) (email-msg ?User ?Msg) ... (browser-ssl-
compromised ?Browser) (certificate-authorized ?Certificate))
(:action attacker-sends-email-with-keylogger :parameters (?User ?File ?Key-
logger) :precondition (and (user ?User) (file ?File) (has-trojan ?File) (key-
logger-trojan ?File ?Keylogger)) :effect (and (email-msg ?User bad-email)
(mail-attachment bad-email ?File))))
(:action user-visits-site :parameters (?User ?Browser ?Site) :precondition
(and (user ?User) (software ?Browser) (browser ?Browser) (site ?Site)) :effect
(and (use-software ?Browser) (user-visits-site ?User ?Site)))
(:action user-starts-email :parameters (?User ?Mailer) :precondition (and
(user ?User) (mailer ?Mailer)) :effect (and (use-software ?Mailer) (running
?Mailer)))
(:action user-reads-email :parameters (?User ?Mailer ?Msg) :precondition
(and (user ?User) (mailer ?Mailer) (use-software ?Mailer) (email-msg ?User
?Msg)) :effect (and (msg-opened ?Msg))
(:action user-presses-F1-at-vbscript-site :parameters (?User ?Browser ?Site)
:precondition (and (user ?User) (use-software ?Browser) (browser ?Browser)
(= ?Browser browser-IE) (= ?Site vbscript-link) (user-visits-site ?User
vbscript-link)) :effect (user-types F1))
(:action user-opens-attachment :parameters (?User ?Msg ?File ?Mailer) :pre-
condition (and (user ?User) (use-software ?Mailer) (mailer ?Mailer) (msg-
opened ?Msg) (mail-attachment ?Msg ?File) (file ?File)) :effect (opened
?File))
(:action key-logger-installed :parameters (?User ?File ?KeyLogger) :pre-
condition (and (user ?User) (opened ?File) (file ?File) (has-trojan ?File)
(key-logger-trojan ?File ?KeyLogger)) :effect (and (key-logger ?KeyLogger)
(installed ?KeyLogger)))
::
::
(:action user-login-with-keylogger-activated :parameters (?User ?Account
?Keylogger) :precondition (and (user ?User) (account ?Account) (key-logger
?Keylogger) (running ?Keylogger)) :effect (and (logged-in ?User ?Account)
(information-available ?User ?Account) (records ?Keylogger ?Account))))
```

The *predicates* (`user ?User`) and (`file ?file`) among others describe the
state of the world. They can either be true or false. `?User` and `?file` are formal
parameters to the predicates, user and file. They describe who the user is and
which file is being used.

Actions allow a planner to move from one state to another. An action can
have pre-conditions and post-conditions, both of which are expressed as predicates. An action can only occur in the current state if the current state supports
its preconditions, i.e., the predicates already in the current state do not negate
the predicates in the action's preconditions. The effects of an action result in the
next state.

To elaborate, the shown PDDL domain Personalized Attack Graph (PAG), contains ten actions. Each action is defined with pre-conditions and post-conditions (effects). For example, action `user-reads-email` with parameters (?User, ?Mailer and ?Msg) has preconditions (user ?User), (mailer ?Mailer), (use-software ?Mailer) and (email-msg ?User ?Msg)). These preconditions verify if the `mailer` supplied as an argument actually exists, if the user is indeed the user supplied in the argument and whether the email message has the user supplied in the argument and the message supplied in the argument. In other words, the pre-conditions test to see if those predicates are true in the current state. If the pre-conditions are true in the current state, then the post-conditions or effects are applied. The effects are (use-software ?Mailer) and (running ?Mailer). When the effects are applied, the predicates (use-software ?Mailer) and (running ?Mailer) are set to true. Each action can be a unit step towards exploiting a vulnerability.

PDDL Problem: A PDDL problem is a concrete instance of a specific PDDL domain where the general variables in a PDDL domain are replaced with concrete values and a sub-set of actions defined in the PDDL domain. A PDDL problem contains an initial state: a set of predicates that are set to true initially, and a goal state: a set of predicates that may or may not be true with the actions defined in the domain. A PDDL problem can be used to perform what-if analysis by simulating conditions of the system and testing various attack paths originating from the current simulated state of the system. Consider the following PDDL problem that corresponds to the domain "PAG".

> (*define* (problem PAG-problem1) (:*domain* PAG) (:*objects* user1 ... browser-seamonkey browser-mozilla)
> (:init (user user1) (mailer gmail) (exploit-vulnerability vulnerability-key-logger) (file file-with-trojan) (has-trojan file-with-trojan) (key-logger key-logger1) (key-logger-trojan file-with-trojan key-logger1) (site vbscript-link) (has-crafted-dialog-box vbscript-link) (vb-script-version VB-5-1) (software browser-IE) (browser browser-IE) (software browser-firefox) (= (version browser-firefox) 2) (browser browser-firefox) (information-available user1 account-bank) (account account-bank))
> (:goal (and (information-leakage account-bank))))

In the above problem, the initial state consists of predicates, (user user1) and (mailer gmail) etc. The initial state describes what is true about the system when the planner starts out. The goal state is defined as (information-leakage account-bank). This indicates that when (information-leakage account-bank) becomes true, the goal state is said to have been reached.

PDDL Planner: A PDDL planner tries to solve a PDDL problem by finding a plan that satisfies it. A successful plan is a sequence of actions from those specified in the PDDL problem for a given initial state.

PDDL Plan: A plan is a sequence of actions from the initial state in the PDDL problem to the final state in the PDDL problem. A plan can also be thought of as a sequence of transitions from the initial state to the goal state. Here is the plan produced for the domain "PAG" and problem "PAG-problem1" by using the Metric-FF planner [8]:

```
0:  ATTACKER-SENDS-EMAIL-WITH-KEYLOGGER   USER1   FILE-
WITH-TROJAN KEY-LOGGER1
1: USER-VISITS-SITE USER1 BROWSER-IE VBSCRIPT-LINK
2: USER-STARTS-EMAIL USER1 GMAIL
3: USER-READS-EMAIL USER1 GMAIL BAD-EMAIL
4:  USER-OPENS-ATTACHMENT  USER1  BAD-EMAIL  FILE-WITH-
TROJAN GMAIL
5:  KEY-LOGGER-INSTALLED  USER1  FILE-WITH-TROJAN  KEY-
LOGGER1
6:   USER-PRESSES-F1-AT-VBSCRIPT-SITE   USER1   BROWSER-IE
VBSCRIPT-LINK
7: KEY-LOGGER-ACTIVATED KEY-LOGGER1 BROWSER-IE
8:       USER-LOGIN-WITH-KEYLOGGER-ACTIVATED       USER1
ACCOUNT-BANK KEY-LOGGER1
9: ATTACKER-INTERCEPTS KEY-LOGGER1 ACCOUNT-BANK
```

The sequence of actions is "ATTACKER-SENDS-EMAIL-WITH-KEY LOGGER USER1 FILE-WITH-TROJAN KEY-LOGGER1" followed by "USER-VISITS-SITE USER1 BROWSER-IE VBSCRIPT-LINK" and so on until the last step, "ATTACKER-INTERCEPTS KEY-LOGGER1 ACCOUNT-BANK".

In essence, these steps describe that an attacker sends an email with a keylogger to a user. Once the user opens his/her browser and goes to their Gmail account, reads their email, and opens the attachment from the email sent by the attacker, the keylogger is installed on their system. When the user presses **F1** on a website with VB Script, the keylogger is activated. This allows the attacker to remotely track all of the user's keystrokes and allows the attacker to intercept the user's bank account credentials when the user visits their bank's website and tries to sign in. The PDDL domain above only has one vulnerability, but in a more realistic scenario, we can expect thousands of such vulnerabilities in a domain and at least one plan for every vulnerability, which allows us to derive attack paths.

Planning Graph: A planning graph is a layered directed graph, consisting of alternating layers of states and actions. The state layers contain predicates that are for that state. The action layers consist of actions that map pre-conditions and post-conditions. An edge from a predicate to an action indicates that the predicate is the pre-condition of that action. An edge from an action to a predicate implies that the predicate is an effect or post-condition of that action. The planning graph, therefore, represents the transitions from the initial state to the goal state by using the actions and predicates defined in the PDDL domain. A planning graph is, simply put, a more granular way to look at a PDDL plan and can help us visualize the attack path.

3   Our Approach for Automated Attack Graph Generation and Refinement

The AGBuilder tool set's workflow shown in Fig. 1 has the following key steps each of which has been developed as an independent software module:

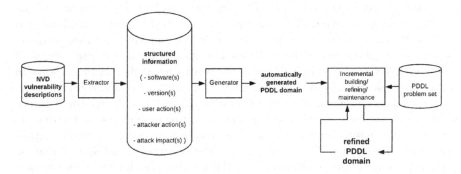

Fig. 1. Overview of the process of generating and maintaining attack graphs

- *Step 1*. The Extractor is used to extract structured information from the vulnerability descriptions. Such information is typically obtained from vulnerability databases such as NVD (https://nvd.nist.gov) or ICS-CERT (https://ics-cert.us-cert.gov).
- The structured information is used to generate a PDDL domain by the Generator module. The PDDL domain represents cause-effect relationships in the attack.
- *Step 2*. The PDDL domain is tested using different PDDL problems, which are extracted from the information in event logs of the system.[1] Each PDDL problem is meant to test at least one attack/vulnerability. These PDDL problems are stored in a database internally maintained by the AGBuilder tool.
- *Step 3*. An AI planner is used to generate one PDDL plan for each PDDL problem along with the PDDL domain. The set of actions in the PDDL planner describe an attack path in the attack graph.
- *Step 4*. Whenever the domain is updated/modified, the tool ensures that the latest version of the PDDL domain is consistent with respect to the last known stable version. The tool checks for the consistency of PDDL plans against the PDDL domain and PDDL problems, and provides feedback on plans which failed, thus helping the developer maintain the cause-effect relationships.

---

[1] An event log is record of actions taken by users and may represent the exploitation of a vulnerability.

3.1 Automatically Generating PDDL Domain from Natural Language Textual Descriptions

The automatic generation phase takes vulnerability descriptions (CVE or NVD) from a vulnerability database as input and renders an automatically generated PDDL domain as output. Our tool extends the NLP-based (natural language processing) software previously developed by us as part of a prior project [29] to extract structured information from unstructured text. The NLP algorithm implementation is based on parts-of-speech tagging model wherein lexical patterns are identified from the text and the relationships of the subjects and rules are identified from these patterns. The Stanford coreNLP POS Tagger [23] was used for tagging. A corpus of 30 descriptions were used for the manual rule generation. Based on these rules, the parts-of-speech tagging algorithm works by tagging labels of a word in the text based on its role in the sentence, like verb, noun, adjective etc, and the context of the word's usage. For instance, software names are typically tagged as proper nouns. Another rule is that software names are followed by a preposition or subordinating conjunction. A pre-stored gazette of 48709 entries, consisting of software and operating systems, is used to match a proper noun to a software name. File names can be matched using regular expressions containing period "." and so on. To identify attacker and user actions, we partition the description and based on the relative positioning of the subject, verbs and modifiers, e.g., like "through", to extract the subjects [11] and the respective actions. As a subject can be either attacker or user, sentiment analysis [6] is used to label the subject as positive or negative sentiment by considering the sentiment labels of the respective verbs and modifiers, and finally, labeling the respective subject as attacker or user. The information extracted comprises of the following: software name(s), software version(s), user action(s), attacker action(s) and attacker impact(s) Here is an example of a vulnerability description from NVD:

CVE-2010-0483: vbscript.dll in VBScript 5.1, 5.6, 5.7, and 5.8 in Microsoft Windows 2000 SP4, XP SP2 and SP3, and Server 2003 SP2, when Internet Explorer is used, allows user-assisted remote attackers to execute arbitrary code by referencing a (1) local pathname, (2) UNC share pathname, or (3) WebDAV server with a crafted .hlp file in the fourth argument (aka helpfile argument) to the MsgBox function, leading to code execution involving winhlp32.exe when the F1 key is pressed, aka "VBScript Help Keypress Vulnerability."

For the above vulnerability description, our extractor gives the following output:

Software: VBScript
Versions: [51, 56, 57, 5.8, 2000, 2003, SP2, SP2, winhlp32.exe]
Modifiers: [and 5.8]
User Action: Internet Explorer is used
User Action: the F1 key is pressed
Attack Vector: referencing a -LRB- 1 -RRB- local pathname, -LRB- 2
-RRB- UNC share pathname, or -LRB- 3 -RRB- WebDAV server with
a crafted hlp file in the fourth argument -LRB- aka helpfile argument
-RRB- to the MsgBox function, leading to code execution involving
winhlp32exe
Attack Impact: execute arbitrary code

The structured information extracted from vulnerability descriptions is then used to automatically generate the corresponding PDDL domain. Figure 2 is an example of an attack graph that is automatically generated by the Generator module of our AGBuilder toolset using the above domain and the relevant problem file, and illustrates the attack path taken to exploit the vulnerability.

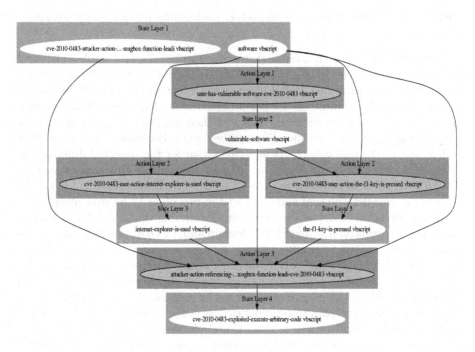

Fig. 2. Attack path generated for NVD 2010-0483

3.2   Incremental Building and Refinement of the Attack Graph

AGBuilder assumes that the PDDL domain, PDDL problem and the plan that represent the cyber threat situational awareness model are syntactically correct and valid. A system administrator focuses only on the undesired plans, i.e., plans that have changed since the last update of the domain. Figure 3 illustrates the process flow for the domain reconciliation algorithm.

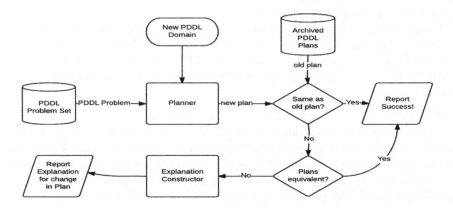

Fig. 3. Overview of incremental support for maintaining/refining attack domains

Given that the system administrator has the last known stable version of the PDDL domain, the PDDL problem set, archived plans from the last known stable version, and the new PDDL domain with bugs in it, a PDDL planner is run on each problem from the PDDL problem set and the new PDDL domain to generate one plan per problem. The tool then matches each pair of a newly generated plan and its corresponding archived plan. For each pair of plans that do not match, the tool constructs an explanation for the observed difference in plans. Sometimes plans may have a different sub-sequence of actions such as $A \rightarrow B \rightarrow C \rightarrow D$ as opposed to $A \rightarrow C \rightarrow B \rightarrow D$. If actions B and C can be executed in parallel, the tool will consider the two plans to be equivalent.

As a motivating example, for incrementally refining the model, let us assume that a planner runs on a stable PDDL domain and a PDDL problem and creates a plan $(user - opens - email - client \rightarrow user - opens - email \rightarrow user - downloads - keylogger - in - attachment \rightarrow user - enters - username - and - password - on - site1 \rightarrow user - enters - username - and - password - on - site2)$ where the actions in the plan are defined in the domain. If some modification to the domain causes the plan to be generated as $(user - opens - email \rightarrow user - opens - email - client \rightarrow user - downloads - keylogger - in - attachment \rightarrow user - enters - username - and - password - on - site1 \rightarrow user - enters - username - and - password - on - site2)$, this new plan may be considered an undesirable plan by the developer since the user now opens the email before even opening the email client. Now, to debug the domain definition and fix the plan, the developer needs to track the state of the system starting with the initial state defined in the PDDL problem and trace all the actions in the plan and their effects on the state of the system, to figure out what caused the observed difference in the plans.

Further, it is possible that a new plan (*user* − *opens* − *email* − *client* → *user* − *opens* − *email* → *user* − *downloads* − *keylogger* − *in* − *attachment* → *user* − *enters* − *username* − *and* − *password* − *on* − *site*2 → *user* − *enters* − *username* − *and* − *password* − *on* − *site*1) is observably different, but it might still be equivalent to the old plan (*user* − *opens* − *email* − *client* → *user* − *opens* − *email* → *user* − *downloads* − *keylogger* − *in* − *attachment* → *user* − *enters* − *username* − *and* − *password* − *on* − *site*1 → *user* − *enters* − *username* − *and* − *password* − *on* − *site*2. This is possible if *user* − *enters* − *username* − *and* − *password* − *on* − *site*1 and *user* − *enters* − *username* − *and* − *password*−*on* − *site*2, are independent actions, i.e., actions that do not affect each other in anyway and can be executed in parallel. If the goal of the PDDL plan is for the user's credentials from site1 and site2 to be compromised in no particular order, it should not make a difference if the user's credentials for site1 are compromised before his/her credentials for site2 and vice-versa. If the actions *user* − *enters* − *username* − *and* − *password* − *on* − *site*1 and *user* − *enters* − *username* − *and* − *password*−*on* − *site*2 do not have any preconditions or effects in common, then they can be considered independent. Therefore, even though the two plans are observably different, they are equivalent. This would imply that neither of the two is an undesired plan. The developer would still need to manually look at the code to make this deduction.

With incremental development, some inadvertent changes to the old actions in the PDDL domain can also lead to unexpected outcomes in terms of the plans produced. At some point, the interaction of the actions in the PDDL domain can get very challenging to visualize for the developer. If multiple developers collaborate on a PDDL domain, and some unintended changes are made to the domain (for instance, adding a new pre-condition to an action or deleting an action), this change in the domain can cause a different plan to be generated by the planner. It can get very hard to manually inspect this change in plan and deduce what caused the change.

Our tool assists the PDDL domain developer ensure that subsequent versions of the PDDL domain are consistent by constructing explanations for the observed changes in the old and new plan by creating two planning graphs: (1) a planning graph using the last stable version of the domain, the PDDL problem and the last stable version of the plan, and (2) planning graph using the new domain, the PDDL problem and the new plan. It then traverses the two planning graphs layer by layer and determines what caused the change in the plan. It infers whether the cause for the change is the addition/removal/replacement of a predicate in the set of pre-conditions or post-conditions, or the addition/removal/replacement of an action. This is then reported as an explanation for the observed change in plans. Next, we describe the details of the modules in AGBuilder and their functionality.

4 AGBuilder Modules

We categorize AGBuilder into its *knowledge base components* and *processing modules*.

4.1   Knowledge Base Components

Input Data: Input to AGBuilder includes the presumably faulty domain which needs to be fixed, the most recent known stable version of the domain, the set of PDDL problems pertaining to the faulty domain, and the archived PDDL plans.

Most Recent Stable Domain: This is the last known stable version of the domain that produces all the plans as expected by the user.

Faulty Domain: A domain file is considered faulty because of changes made to the most recently stable version of the domain. The most recently stable version of the domain produces all the plans as expected but the faulty domain produces unexpected/undesired plans. Unexpected/undesired plans are plans with unexpected sequences of actions in them, or sequences of actions that are different from those produced in plans from the last known stable version of the domain.

PDDL Problem Set: The PDDL domain is accompanied by a set of problems. The PDDL problems are used for generating plans. The tool uses the planner, the faulty domain, and PDDL problem set to generated plans for the faulty domain.

Archived PDDL Plans: The archived PDDL plans can be thought of as the reference output. There is one archived plan per problem in the problem set. The archived PDDL plans were generated using the most recently stable version of the PDDL domain. If the plans generated using a PDDL domain and the problem set match those in the archived set, then the plans are treated as correct and the PDDL domain used to generate those plans is considered "stable" and not "faulty".

4.2   AGBuilder Processing Modules

AGBuilder has a parsing module, a planning graph construction module, and a module for generating explanations.

Parsing Module: The parsing module has a built-in lexer and parser that reads a PDDL domain or problem into an abstract syntax tree (AST). The AST structure is useful as it allows the tool to run queries on it to find actions that contain a predicate for determining if two or more actions are independent of each other. AGBuilder uses a parser and a lexer. The parser and lexer were generated using a BNF representation of PDDL 3.1 and ANTLR [14], a popular parser generator for Java.

Planning Graph Construction Module: We use the Metric-FF the PDDL Planner to generate plans [8]. However, any other new planner can also be used, which gives the tool a distinct advantage of keeping up to date with minimal effort. The planning graph construction module takes as input, a PDDL domain, a PDDL problem and a plan generated using the PDDL domain and the PDDL problem. It converts the domain and the problem into ASTs and then uses the ASTs and the plan to create a planning graph.

1. Initialize the first layer as the set of predicates in the initial conditions from the PDDL Problem's AST.
2. For every action a_i in the sequence of actions in the plan:
   (a) Initialize the next layer as an action layer (if it is not initialized already), and set it as the current layer.
   (b) If the current action layer is not the first action layer and the current action a_i is "independent" with respect to every single action in the previous action layer:
      i. Add current action a_i to previous action layer and set previous action layer as current layer.
   (c) Initialize the next layer as state layer (if it is not initialized already) and set it as the current layer.
   (d) For every precondition of action a_i in the previous state layer, set a directed edge from the pre-condition in the previous state layer to the action a_i in the previous action layer.
   (e) Apply the postconditions of action a_i to the current state layer.
   (f) For every postcondition of the action a_i in the current state layer, set a directed edge from the action a_i in the previous action layer to the postcondition in the current state layer.

This module determines whether two actions are independent. The criteria for two actions to be independent is that the negative effects (negated predicates) of either action should not have any intersection with the preconditions or the positive effects (predicates that aren't negated) of the other action. This is formally expressed in the following relationship as: actions a1 and a2 are independent if and only if:

$$effects^- (a1) \cap (preconditions(a2) \cup effects^+(a2)) = \{\phi\}$$

and

$$effects^- (a2) \cap (preconditions(a1) \cup effects^+(a1)) = \{\phi\}$$

The relationship implies that two independent actions don't interact with each other in any way, so they can be executed simultaneously, or can be in the same layer of a planning graph. For a set of actions to be independent, every possible pair of actions in that set needs to satisfy the aforementioned relationship for independence. Thus, for a plan $A \rightarrow B \rightarrow C \rightarrow D$ where B and C are independent according to the aforementioned criteria, the planning graph would have A in the first action layer, B and C in the second action layer and D in the third action layer. Another plan $A \rightarrow C \rightarrow B \rightarrow D$ would also have A in the first action layer, B and C in the second action layer, and D in the third action layer. This is how it is determined that two plans with differing sub-sequences of actions are equivalent.

Module for Constructing Explanations: The module for constructing explanations is the main module of AGBuilder. It takes one problem at a time from the PDDL problem set, generates two planning graphs, one for the last known stable version of the domain and the archive plan, and another for the faulty domain and the new plan. It then compares these two planning graphs to deduce if two plans are actually different or if they are equivalent but have different sub-sequences of actions. If the plans are different, it uses the planning graphs to deduce an explanation for causes of the change in the new plan. The main algorithm for generating explanations in AGBuilder is as follows: For each PDDL problem x, in the problem set $\{X\}$:

1. Run planner on PDDL problem and the new (faulty) domain to generate a plan.
2. If the plan produced has a different sequence of actions than the archived plan:
   (a) Run the faulty domain and the stable version of the domain through the parsing module to get two ASTs.
   (b) Run the PDDL problem x_i through the parsing module to get an AST for the problem.
   (c) Construct one planning graph using the stable domain AST, problem x_i AST and the archived plan of the table domain on the problem x_i.
   (d) Traverse both the graphs simultaneously, one layer at a time, and find differences in actions or states at each layer. If a difference is observed in the corresponding layers of the two planning graphs, report the differences observed and halt any further traversal of planning graphs. The process of constructing explanations is discussed in more detail next.

Now, the explanation construction module takes as input the planning graph from the most recent stable domain, the planning graph from the new (presumably faulty) domain, the old domain AST and the new domain AST. The algorithm for generating explanations is as follows:

1. Skip the first state layer in the two planning graphs. This is because the first state layer has the initial preconditions from the PDDL problem (which both planning graphs share), therefore they must be the same.
2. Compare the two sets of actions after the first state layer of the two planning graphs.
3. If the sets of actions in the current action layer of the old graph and the new graph are different:
   (a) Print the actions in the current layer of the new graph which are not in the corresponding layer of the old graph. These actions are extra actions in the new plan.
   (b) Print the actions in the current layer of the old graph which are not in the corresponding layer of the new graph. These actions are missing actions in the new plan.
4. If the sets of predicates in the current state layer are different:

(a) Print the predicates in the new graph which are a negation of the predicates in the old graph.

(b) For each predicate P_i in the new graph that is $\neg P_i$ in the old graph:

    i. Print the action(s) from the previous action layer which has this predicate $\neg P_i$ in its post conditions. The action(s) that had this predicate as its post condition is what contributed to the change in the observed plan.

    ii. Print the action(s) from the previous action layer of the old graph that has the predicate P_i in its post conditions for reference.

(c) Print the predicates in the current layer of the old graph which are not in the corresponding layer of the new graph. For each of these predicates:

    i. Print the action(s) from the previous action layer in the old planning graph that has this predicate in its post condition.

5. If the sets of predicates in the current state layer of the two graphs are the same and there are more action layers:

(a) Repeat step (2) with the next action layer of the two graphs as the current layers.

Finally, we present a worked out example for the domain PAG.

4.3 Working Example of Explanation Constructor

For the exercise, we consider the PDDL problem, PAG-problem1 from Sect. 2. We edited the action `user-visits-site` and included all the parameters, pre-conditions and effects from `user-starts-email` in `user-visits-site`. In essence, the two actions were combined and put in `user-visits-site`. This was to simulate an inadvertent change to the domain.

The planner runs on the last known stable version of the domain and the problem, and then on the new version of the domain and the problem. By constructing planning graphs for both instances and crawling the planning graphs, the tool displays the change in domain that caused the new plan to be different from the previous version. The original actions in the PDDL definition of PAG-problem1 are:

(1) `user-visits-site` and (2) `user-starts-email`. A new updated PDDL Domain needs is generated with the following new actions: (1) `user-visits-site` and (2) `user-starts-email`:

(:*action* user-visits-site :*parameters* (?User ?Browser ?Site ?Mailer) :*precondition* (and (user ?User) (software ?Browser) (browser ?Browser) (site ?Site) (user ?User) (mailer ?Mailer)) :*effect* (and (use-software ?Browser) (user-visits-site ?User ?Site) (use-software ?Mailer) (running ?Mailer)))

(:*action* user-starts-email :*parameters* (?User ?Mailer) :*precondition* (and (user ?User) (mailer ?Mailer)) :*effect* (and (use-software ?Mailer) (running ?Mailer)))

Finally, the updated plan from new PDDL domain for PAG-problem1 is given by:

```
0:  ATTACKER-SENDS-EMAIL-WITH-KEYLOGGER  USER1  FILE-
WITH-TROJAN KEY-LOGGER1
1: USER-VISITS-SITE USER1 BROWSER-IE VBSCRIPT-LINK GMAIL
2: USER-READS-EMAIL USER1 GMAIL BAD-EMAIL
3:  USER-OPENS-ATTACHMENT  USER1  BAD-EMAIL  FILE-WITH-
TROJAN GMAIL
4:  KEY-LOGGER-INSTALLED  USER1  FILE-WITH-TROJAN  KEY-
LOGGER1
5:  USER-PRESSES-F1-AT-VBSCRIPT-SITE  USER1  BROWSER-IE
VBSCRIPT-LINK
6: KEY-LOGGER-ACTIVATED KEY-LOGGER1 BROWSER-IE
7:     USER-LOGIN-WITH-KEYLOGGER-ACTIVATED     USER1
ACCOUNT-BANK KEY-LOGGER1
8: ATTACKER-INTERCEPTS KEY-LOGGER1 ACCOUNT-BANK
```

Below is an example output of AGBuilder on all the problems for the last known stable version of the domain and the new version of the domain and reporting changes:

```
Problems found: (1) problems/PAG-problem1.pddl: PAG-problem1
Archived plans found: (1) archivedPlans/PAG-problem1.plan

Creating planning graph for domainPAG.pddl, PAG-problem1.pddl and
PAG-problem1.plan ...done
Creating planning graph for domainPAGV2.pddl, PAG-problem1.pddl and
new plan ...done

plans different? True
Difference observed:
State Layer 1: same
Action Layer 1: same
State Layer 2: same
Action Layer 2: same
State Layer 3: different!
extra predicates found in action: USER-VISITS-SITE
extra effects:
(1) (use-software ?Mailer)
(2) (running ?Mailer)
```

As shown above, the tool runs on the two versions of the PAG domain and PAG-problem1. It crawls the two planning graphs generated using both versions of the domain, the problem and the two versions of the plans. The plan generated from the new domain does not include "USER-STARTS-EMAIL USER1 GMAIL" because "USER-VISITS-SITE USER1 BROWSER-IE VBSCRIPT-LINK GMAIL" has all of its pre-conditions and post-conditions. AGBuilder finds a disparity at state layer 3, which is after USER-VISITS-SITE was run for both domains and complaints that the predicates in the state layers were different and leads to the action which caused this change. This is how the tool generates an explanation for the change observed in plans.

5   Related Work

There is a wealth of knowledge [2,4,10,12,13,16,19,21,22,28] on attack graphs and attack trees. We discuss a few works in brief here.

In [20], demonstrated the first technique for automatic generation of attack graphs using symbolic model checking techniques. However, this approach suffers from scalability issues. In [13], the authors model the attack graph generation as a logic programming problem in Prolog. Their tool MulVAL takes in information from vulnerability databases, configuration information from each machine and the network. Once the entire information is available and encoded in the MulVAL framework, for a given host and policy configurations, an attack simulation is performed by the MulVAL scanner for policy violations or presence of exploit paths. However, even small changes in the configuration will require the simulations to be rerun again. In AGBuilder, the configuration information can be updated incrementally and generate new attack paths. Furthermore, we can combine two smaller PDDL domains of different networks and check for attack paths. While the underlying logical framework remains same, the planning domain formulation allows for simpler way to perform incremental updates and analysis. In [16], the authors describe Bayesian Attack Graphs that not only encodes cause-consequence relationship between network states but also consider the likelihoods of exploiting these relationships. However, none of the existing works did considered building attack graphs from natural language descriptions of the attack/vulnerabilities.

Prior work in the area of applying AI planning in cyber-security applications, provide evidence for the ability of PDDL being used to model attack-graphs. Mark Roberts et al. have demonstrated that PDDL can be used for modeling an attack graph for use in personalized security agents [18].

Existing tools for PDDL allow for automatically generating PDDL code from formal models, developing PDDL code in an IDE with an integrated planner, and determining whether plans generated by a planner are correct [25]. AGBuilder seeks to compliment the existing research on knowledge engineering tools by assisting in the generation and incremental development of very large and complex domains.

ItSimple [26,27] is a knowledge engineering tool that uses simple UML models and converts them into PDDL representations. Invariants (constraints) on UML classes are represented using OCL (Object Constraint Language). According to the authors, "ItSimple was designed to give support to users during the construction of a planning domain application mainly in the initial stages of the design life cycle" ItSimple 2.0 [24,27] assists users to resolve issues during requirements specification, analysis and modeling phases. ItSimple 2.0 allows users to build use case diagrams from UML, thus allowing the requirements to be represented at a high level of abstraction. ItSimple 4.0 also features a text editor for further editing after the PDDL domain has already been created, uses modeling patterns (essentially a set of common planning models in UML), time based models that describe how properties of objects change during the execution of an action, and a wizard that allows users to quickly select the initial preconditions and goal states for actions [27]. PDDLStudio is an IDE for PDDL that features document management, syntax highlighting, code completion with hints, and planner integration [15].

ModPlan is an integrated environment that allows for knowledge acquisition and domain analysis in planning applications. ModPlan helps in knowledge engineering by examining pairwise dependencies in actions and letting the user examine these dependencies [5]. It allows for plans to be visualized using Vega and validated using VAL.

VAL is used to determine if the plans generated by a planner are correct. It first examines if the domain and the problem are syntactically correct and if the objects, preconditions and goal state(s) in the problem match the domain. It then validates the plan generated by a planner for a given domain and a problem, and reports if the plan is flawed, and how it might be fixed [9].

6   Conclusion and Future Work

In this paper, we presented a tool, AGBuilder, that generates attack graphs in PDDL from textual descriptions of CVEs. The tool assists developers with incrementally developing PDDL domains for modeling attack graphs by generating explanations for why undesired plans are produced when PDDL domains are modified, and also allows for what-if analysis of attacks with the PDDL representation. The tool also allows for generating abstract syntax trees of PDDL domains and problems, and generating attack graphs as DOT files, which can then be rendered into image files, in order to visualize attack paths. We demonstrated our tool on an NVD description of a key-logger malware propagation and showed the corresponding attack graph/paths in PDDL.

Future work will look into extending the tool to perceive the current state of the system. This will allow us to estimate courses of actions that could lead to an attack in real time. Secondly, we will obtain quantifiable quality measurements of AGBuilder's scalability. Thirdly, we will investigate the feasibility of stitching together attack paths from the various attack graphs modeled by the attack graph domain to build a single consolidated attack graph. This task will require using the domain and all problem files to generate a DOT file that can then be rendered into an image file. Since the AGBuilder already has the capability of building abstract syntax trees and attack graphs for individual graphs, we anticipate this feature enhancement to be an immediate attainable goal. Finally, we will seek to integrate AGBuilder with an online IDE that we are in the process of developing. This will allow for multiple users to collaborate and build a PDDL domain.

Acknowledgment. This work is partially supported by the U.S. Department of Energy under award number DE-NE0008571 subcontracted through the Ohio State University. This material is also based upon work performed by Indrajit Ray while serving at the National Science Foundation. Research findings presented here and opinions expressed are solely those of the authors and in no way reflect the opinions the DOE, the NSF or any other federal agencies.

References

1. Ammann, P., Wijesekera, D., Kaushik, S.: Scalable, graph-based network vulnerability analysis. In: Proceedings of the 9th ACM Conference on Computer and Communications Security, CCS 2002, pp. 217–224. ACM, New York (2002). http://doi.acm.org/10.1145/586110.586140

2. Audinot, M., Pinchinat, S., Kordy, B.: Guided design of attack trees: a system-based approach. In: 2018 IEEE 31st Computer Security Foundations Symposium (CSF), pp. 61–75. IEEE (2018)

3. Boddy, M.S., Gohde, J., Haigh, T., Harp, S.A.: Course of action generation for cyber security using classical planning. In: ICAPS, pp. 12–21 (2005)

4. Cao, C., Yuan, L.-P., Singhal, A., Liu, P., Sun, X., Zhu, S.: Assessing attack impact on business processes by interconnecting attack graphs and entity dependency graphs. In: Kerschbaum, F., Paraboschi, S. (eds.) DBSec 2018. LNCS, vol. 10980, pp. 330–348. Springer, Cham (2018). https://doi.org/10.1007/978-3-319-95729-6_21

5. Edelkamp, S., Mehler, T.: Knowledge acquisition and knowledge engineering in the modplan workbench. In: Proceedings of the First International Competition on Knowledge Engineering for AI Planning, pp. 26–33 (2005)

6. Esuli, A., Sebastiani, F.: SentiWordNet: a publicly available lexical resource for opinion mining. In: LREC, vol. 6, pp. 417–422. Citeseer (2006)

7. Ghosh, N., Ghosh, S.K.: A planner-based approach to generate and analyze minimal attack graph. Appl. Intell. **36**(2), 369–390 (2012)

8. Hoffmann, J.: The Metric-FF planning system: Translating "ignoring delete lists" to numeric state variables. J. Artif. Intell. Res. **20**, 291–341 (2003)

9. Howey, R., Long, D., Fox, M.: VAL: automatic plan validation, continuous effects and mixed initiative planning using PDDL. In: 16th IEEE International Conference on Tools with Artificial Intelligence, ICTAI 2004, pp. 294–301. IEEE (2004)

10. Jhawar, R., Kordy, B., Mauw, S., Radomirović, S., Trujillo-Rasua, R.: Attack trees with sequential conjunction. In: Federrath, H., Gollmann, D. (eds.) SEC 2015. IAICT, vol. 455, pp. 339–353. Springer, Cham (2015). https://doi.org/10.1007/978-3-319-18467-8_23

11. Klein, D., Manning, C.D.: Accurate unlexicalized parsing. In: Proceedings of the 41st Annual Meeting on Association for Computational Linguistics, vol. 1, pp. 423–430. Association for Computational Linguistics (2003)

12. Lippmann, R.P., Ingols, K.W.: An annotated review of past papers on attack graphs. Technical report, Massachusetts Inst of Tech Lexington Lincoln Lab (2005)

13. Ou, X., Boyer, W.F., McQueen, M.A.: A scalable approach to attack graph generation. In: Proceedings of the 13th ACM Conference on Computer and Communications Security, pp. 336–345. ACM (2006)

14. Parr, T.J., Quong, R.W.: ANTLR: a predicated-LL(k) parser generator. Softw. Pract. Exp. **25**(7), 789–810 (1995)

15. Plch, T., Chomut, M., Brom, C., Barták, R.: Inspect, edit and debug PDDL documents: simply and efficiently with PDDL studio. In: System Demonstrations and Exhibits at ICAPS, pp. 15–18 (2012)
16. Poolsappasit, N., Dewri, R., Ray, I.: Dynamic security risk management using Bayesian attack graphs. IEEE Trans. Depend. Secure Comput. **9**(1), 61–74 (2012)
17. Ray, I., Poolsapassit, N.: Using attack trees to identify malicious attacks from authorized insiders. In: di Vimercati, S.C., Syverson, P., Gollmann, D. (eds.) ESORICS 2005. LNCS, vol. 3679, pp. 231–246. Springer, Heidelberg (2005). https://doi.org/10.1007/11555827_14
18. Roberts, M., Howe, A.E., Ray, I., Urbanska, M.: Using planning for a personalized security agent. In: Workshop on Problem Solving Using Classical Planners at 26th AAAI Conference on Artificial Intelligence (2012)
19. Sawilla, R.E., Ou, X.: Identifying critical attack assets in dependency attack graphs. In: Jajodia, S., Lopez, J. (eds.) ESORICS 2008. LNCS, vol. 5283, pp. 18–34. Springer, Heidelberg (2008). https://doi.org/10.1007/978-3-540-88313-5_2
20. Sheyner, O., Haines, J., Jha, S., Lippmann, R., Wing, J.M.: Automated generation and analysis of attack graphs. In: Proceedings 2002 IEEE Symposium on Security and Privacy, pp. 273–284. IEEE (2002)
21. Singhal, A., Ou, X.: Security risk analysis of enterprise networks using probabilistic attack graphs. In: Wang, L., Jajodia, S., Singhal, A. (eds.) Network Security Metrics, pp. 53–73. Springer, Cham (2017). https://doi.org/10.1007/978-3-319-66505-4_3
22. Sun, X., Dai, J., Liu, P., Singhal, A., Yen, J.: Using Bayesian networks for probabilistic identification of zero-day attack paths. IEEE Trans. Inf. Forensics Secur. **13**(10), 2506–2521 (2018)
23. Toutanova, K., Manning, C.D.: Enriching the knowledge sources used in a maximum entropy part-of-speech tagger. In: Proceedings of the 2000 Joint SIGDAT Conference on Empirical Methods in Natural Language Processing and Very Large Corpora: Held in Conjunction with the 38th Annual Meeting of the Association for Computational Linguistics, vol. 13, pp. 63–70. Association for Computational Linguistics (2000)
24. Vaquero, T.S., Romero, V., Tonidandel, F., Silva, J.R.: itSIMPLE 2.0: an integrated tool for designing planning domains. In: ICAPS, pp. 336–343 (2007)
25. Vaquero, T.S., Silva, J.R., Beck, J.C.: A brief review of tools and methods for knowledge engineering for planning & scheduling. In: KEPS 2011, p. 7 (2011)
26. Vaquero, T.S., Tonidandel, F., Silva, J.R.: The itSIMPLE tool for modeling planning domains. In: Proceedings of the First International Competition on Knowledge Engineering for AI Planning, Monterey, California, USA (2005)
27. Vaquero, T.S., Tonaco, R., Costa, G., Tonidandel, F., Silva, J.R., Beck, J.C.: itSIMPLE4.0: enhancing the modeling experience of planning problems. In: System Demonstration-Proceedings of the 22nd International Conference on Automated Planning & Scheduling (ICAPS-2012), pp. 11–14 (2012)
28. Wang, L., Singhal, A., Jajodia, S.: Measuring the overall security of network configurations using attack graphs. In: Barker, S., Ahn, G.-J. (eds.) DBSec 2007. LNCS, vol. 4602, pp. 98–112. Springer, Heidelberg (2007). https://doi.org/10.1007/978-3-540-73538-0_9
29. Weerawardhana, S., Mukherjee, S., Ray, I., Howe, A.: Automated extraction of vulnerability information for home computer security. In: Cuppens, F., Garcia-Alfaro, J., Zincir Heywood, N., Fong, P.W.L. (eds.) FPS 2014. LNCS, vol. 8930, pp. 356–366. Springer, Cham (2015). https://doi.org/10.1007/978-3-319-17040-4_24

On Practical Aspects of PCFG
Password Cracking

Radek Hranický$^{(\boxtimes)}$, Filip Lištiak, Dávid Mikuš, and Ondřej Ryšavý

Faculty of Information Technology,
Brno University of Technology, Brno, Czech Republic
{ihranicky,rysavy}@fit.vutbr.cz, {xlisti00,xmikus15}@stud.fit.vutbr.cz

Abstract. When users choose passwords to secure their computers, data, or Internet service accounts, they tend to create passwords that are easy to remember. Probabilistic methods for password cracking profit from this fact, and allow the attackers and forensic investigators to guess user passwords more precisely. In this paper, we present our additions to a technique based on probabilistic context-free grammars. By modification of existing principles, we show how to guess more passwords for the same time, and how to reduce the total number of guesses without significant impact on success rate.

Keywords: Password · Cracking · Security · Grammar

1 Introduction

Confidential data and user accounts for various systems and services are protected by passwords. Though a password is usually the only piece that separates a potential attacker from accessing the privileged data, users tend to choose weak passwords which are easy to remember [1]. In reaction, system administrators and software developers introduce mandatory rules for password composition, e.g., "use at least one special character." While password-creation policies force users to create stronger passwords [11,13], recent leaks of credentials from various websites showed the reality is much more bitter. People widely craft passwords from existing words [4] and often reuse the same password between multiple sites [3]. This fact may be utilized by both malicious attackers and forensic investigators who seek for evidence in password-protected data.

Traditional ways of password cracking contain a *brute-force* attack where one tries every possible sequence of characters upon a given alphabet, and a *dictionary attack* where one uses a list of existing passwords and tries each of them. The main drawback of the brute-force attack is the size of a *keyspace* (a set of all possible password candidates) which grows exponentially with the length of the password, and one does not need to "try everything" to crack the password. The dictionary attack, on the other hand, usually checks a limited number of commonly-used or previously-leaked passwords. It is possible, however, to combine both methods to perform a "smarter" cracking. The use of probability and statistics has been shown to bring substantially better results for cracking human-created passwords [9,10,15].

One approach is the use of *Markov chains* which consider probabilities that a certain character will follow after another one. The probabilities are learned from an existing password dictionary and then reused for generating password guesses [10]. The method, however, only works with individual characters and does not consider digraphs or trigraphs. To work with larger password fragments, Weir et al. proposed the use of *probabilistic context-free grammars* (PCFG) that can describe the structure of passwords in an existing (training) dictionary. Fragments described by PCFG represent finite sequences of letters, digits, and special characters. Then, by derivation using rewriting rules of the grammar, one can not only generate all passwords from the original dictionary, but produce many new ones that still respect password-creation patterns learned from the dictionary [15].

The rewriting rules of PCFG have probability values assigned accordingly to the occurrence of fragments in the training dictionary. The probability of each possible password equals the product of probabilities of all rewriting rules used to generate it. Using PCFGs, generating password guesses is deterministic and is performed in an order defined by their probabilities. Therefore, more probable passwords are generated first.

While the creation of such grammar is fast and straightforward, the application of rewriting rules takes a significant amount of processor time, and the number of generated passwords is overwhelming in comparison with the original dictionary. For example, using Weir's tool[1] with a PCFG trained on 6.5 kB *elite-hacker*[2] dataset (895 passwords) generates a 12 MB dictionary with 1.8 million passwords. However, using 73 kB *faithwriters* dataset (8,347 passwords) generates a 28 GB dictionary with over 3 billion passwords. Even more unpleasant is the time required to generate such datasets. The first 10 and first 100 passwords of *darkweb2017*[3] dataset and *darkweb2017-top100* can be both used for training and generating within 1 min on Core(TM) i7-7700K CPU. Taking first 1000 passwords requires more than a day to generate guesses on the same processor.

1.1 Contribution

Our goal was to make the PCFG-based password cracking utilizable for practical use. We identified factors that influence the time of generating password guesses. Based on Weir's Python PCFG cracker, we created an implementation in Go[4] language, which enables to parallelize the generation of terminal structures making the password generation multiple times faster. Moreover, we

---

[1] https://github.com/lakiw/pcfg_cracker.

[2] https://wiki.skullsecurity.org/index.php?title=Passwords.

[3] https://github.com/danielmiessler/SecLists/tree/master/Passwords.

[4] https://golang.org/.

proposed methods that remove specific rewriting rules from the grammar which leads to a massive speedup of password guessing and allows the process to end in a meaningful time without having a considerable impact on success rate.

1.2   Structure of the Paper

The paper is structured as follows. Section 2 provides an introduction to PCFG and discusses related work. Section 3 describes the enhancements we made to PCFG-based techniques, while Sect. 4 shows experimental results of our work. Finally, Sect. 5 concludes the paper.

2   Background and Related Work

For a long time, probability and statistics have been applied to measure password strength [8,11,13] and generate guesses in password cracking [7,9,10,15]. Major password leaks allowed to make a clearer image of how user create their passwords [2]. Such knowledge has been utilized in multiple password cracking principles and adopted to existing tools.

Narayanan et al. proposed the use of Markov chains for password guessing. The method uses conditional probability $P(A|B)$ that character A will follow after character B. The probabilities for all characters A, B are stored in a matrix obtained by the analysis of an existing password dictionary [10]. The technique was utilized in *Hashcat* tool which uses Markov chains for brute-force attacks by default. The probability matrix can be generated automatically using *Hcstatgen*[5] utility and is stored in a *.hcstat* file. Recent versions of Hashcat use LZMA compression which is indicated by *.hcstat2* file extension.

Weir et al. introduced password cracking using *probabilistic context-free grammars* (PCFG) [15]. The mathematical model is based on classic context-free grammars [5] with the only difference that each rewriting rule is assigned a probability value. The grammar is created by training on an existing password dictionary. Each password is divided into continuous fragments of letters (L), digits (D), and special characters (S). For fragment of length n, a rewriting rule of the following form is created: $T_n \rightarrow f : p$, where T is a type of the character group (L, D, S), f is the fragment itself, and p is the probability obtained by dividing the number of occurrences of the fragment by the number of all fragments of the same type and length. In addition, we add rules that rewrite the starting symbol (S) to *base structures* which are non-terminal sentential forms describing the structure of the password [15]. For example, password "p@per73" is described by base structure $L_1S_1L_3D_2$ since it consist from a single letter followed by a single special character, three letters, and two digits. Table 1 shows rewriting rules of a PCFG generated by training on two passwords: "pass!word" and "love@love". There is only one rule that rewrites S since both passwords are described by the same base structure. By using PCFG on MySpace dataset (split to training

---

[5] https://hashcat.net/wiki/doku.php?id=hashcat_utils#hcstatgen.

and testing part), Weir et al. were able to crack 28% to 128% more passwords in comparison with the default ruleset from *John the Ripper* (JtR) tool[6] using the same number of guesses.

Table 1. An example of PCFG rewriting rules

Left	→	Right	Probability
S	→	$L_4 S_1 L_4$	1
L_4	→	pass	0.25
L_4	→	word	0.25
L_4	→	love	0.5
S_1	→	@	0.5
S_1	→	!	0.5

The proposed approach, however, does not distinguish between lowercase and uppercase letters. Weir extended the original generator by adding capitalization rules like "UULL" or "ULLL" where "U" means uppercase and "L" lowercase. The rules are applied to all letter fragments which increases the number of generated guesses [14]. After adding capitalization, the notation for letter non-terminals were changed from L_n to A_n (as alphabetical) since L now stands for lowercase.

While the previous techniques consider only the syntax of passwords, Veras et al. designed a semantics-based approach which divides password fragments into categories by semantic topics like names, numbers, love, sports, etc. With JtR in *stdin* mode feeded by a semantic-based password generator, Veras achieved better success rates than using Weir's approach or the default JtR wordlist [12].

Ma et al. showed how normalization and smoothing can increase the success rate of Markov models. By training and testing on a huge number of datasets, Ma showed that the improved Markov-based guessing could bring better results than PCFGs [9].

Weir's PCFG-based technique encountered extensions as well. Houshmand et al. introduced keyboard patterns represented by additional rewriting rules that helped improve the success rate by up to 22%, proposed the use of Laplace probability smoothing, and created guidelines for choosing appropriate attack dictionaries [7]. After that, Houshmand also introduced targeted grammars that utilize information about a user who created the password [6].

The current version of Weir's PCFG Cracker consists of two separate tools: PCFG Trainer and PCFG Manager. While *PCFG Trainer* is used to create a grammar from an existing password dictionary, *PCFG Manager* generates new password guesses from the grammar - i.e., gradually applies rewriting rules to the starting symbol and derived sentential forms.

---

[6] https://www.openwall.com/john/.

At the time of writing this paper, both tools include the support for letter capitalization rules [14], keyboard patterns [7], as well as the ability to generate new password segments using Markov chains [10]. In the training phase, a user can set a *coverage* value which defines the portion of guesses to be generated using rewriting rules only while the rest is generated using Markov-based brute-force. A *smoothing* parameter allows the user to apply probability smoothing as described in [7]. Moreover, the tools contain the support for context-sensitive character sequences like "<3" or "#1" that, if present in the training data, form a separate set of rewriting rules. Such replacements can be used to describe special strings like smileys, arrows, and others.

Despite numerous improvements made by Houshmand [7], users still have to face slow password guessing speed which is currently the bottleneck of the entire process. Besides, the generating of password guesses gets progressively slower as the time goes on and, as we detected, has high memory requirements. Creating a complete wordlist of possible password candidates using PCFGs trained on leaked datasets may take many hours or even days. Moreover, current tools do not provide information about the size of the *keyspace*, i.e., the number of possible password candidates, and thus the user has no clue about how long will the process take. This obstacle has already been reported as a GitHub issue[7]. Weir, however, does not plan to resolve the issue "anytime soon."

3   Enhancements to PCFG

We focus on making PCFG-based password cracking suitable for practical use - i.e., allow the user to create a PCFG, generate a wordlist of password guesses in a short time, and start cracking immediately. To achieve this, we decided to:

- Create a faster "password generator" that could produce more guesses at the same time using the same hardware.
- Make a tool to calculate the number of possible password guesses from a PCFG. The number can help estimate the size of an output dictionary as well as the time required to generate all password candidates.
- Analyze if modification of an existing grammar can provide any help to the password guessing process. Concretely, if it accelerates the password guessing, or makes it end in a meaningful time.

To verify the success of our efforts, we study the following metrics: (a) the number of guesses per time unit, (b) the total time of password guessing, (c) the number of generated passwords, (d) the success rate for testing datasets, i.e., how many newly-generated passwords are present in existing password dictionaries.

3.1   Key Observations

By analyzing the behavior of Weir's Python PCFG Cracker on various leaked datasets, we observed the following:

---

[7] https://github.com/lakiw/pcfg_cracker/issues/9.

- The Python implementation of PCFG Manager uses a priority queue and three processes: one that fills the queue with pre-terminal structures [15], one that creates terminal structures (password guesses), and one for storage backup. No other parallelization is supported. Thus, the processor cores are not utilized well.
- Processing long base structures like $A_1D_1A_2D_2A_3D_3A_4D_4A_5D_5$ is computationally complex and wastes a lot of time even if their probabilities are insignificant.
- Rewriting rules for alpha characters (A), digits (D), and other symbols (O) have all similar probability, while rewriting rules for base structures differ more between each other.
- For capitalization of letter fragments, a grammar usually contains few (1 to 4) rules with higher probabilities while the rest have probability below 0.1 and only little impact on success rate.

3.2   Long Base Structures

For every PCFG, possible sentential forms create a tree structure where the starting symbol represents the root node, and terminal structures are leaves. Every edge stands for the application of a rewriting rule that transforms a parent node to a child node. In terms of probabilistic password cracking, terminal structures are password candidates, and base structures (e.g., $A_4D_2O_1$) are located on the second level of the tree.

In PCFG Manager, every base structure is processed by *Deadbeat dad* algorithm [14]. The goal of this algorithm is to create new children from the current node and ensure that these child nodes are inserted into the priority queue in the correct order. Deadbeat dad replaced the original *Next* function [15] and significantly reduced the size of the priority queue at the expense of computing operations [14].

We analyzed the algorithm and observed that the most expensive task is to find every possible parent of every node which is being inserted into the priority queue. In Weir's PCFG Manager, the task is resolved by a function called dd_is_my_parent that runs in iterations whose count is potentially increased by every non-terminal present in the processed base structure. The deciding factor is the number of different probabilities assigned to the rewriting rules applicable to the non-terminal. If all usable rules have the same probability value, the number of iterations is not increased. The more different probabilities are present, the more rapidly the iteration count grows, if the non-terminal is added to the base structure.

Table 2 shows the number of dd_is_my_parent iterations under different settings. For D_3 non-terminal, all rules have the same probability, and thus D_3 has no impact on the iteration count. For A_1, rewriting rules have 26 to 29 different probability values (A_1^p). As a capitalization rule for A_1, only "L" is used. One can see, the number of iterations grows almost exponentially each time A_1 is added to the base structure.

Table 2. The number of iterations of dd_is_my_parent function

base structure	$A_1^p = 26$	$A_1^p = 27$	$A_1^p = 28$	$A_1^p = 29$
A_1	103	107	111	115
$A_1 D_3$	103	107	111	115
$A_1 D_3 A_1$	15,811	17,067	18,371	19,723
$A_1 D_3 A_1 D_3$	15,811	17,067	18,371	19,723
$A_1 D_3 A_1 D_3 A_1$	1,506,286	1,688,528	1,884,906	2,095,948
$A_1 D_3 A_1 D_3 A_1 D_3$	1,506,286	1,688,528	1,884,906	2,095,948
$A_1 D_3 A_1 D_3 A_1 D_3 A_1$	120,939,106	140,790,314	162,990,446	187,717,930

In PCFGs trained on leaked password datasets, the variedness between rule probabilities is usually high, especially for shorter character fragments. For long base structures, the dd_is_my_parent function may iterate millions of times which significantly slows the password guessing process. Such structures usually have low probability values since they are in most cases created from randomly generated strings, not created by users. We assume, removing such structures from the grammar speeds up password generation several times and does not noticeably decrease success rate at cracking sessions.

3.3   Calculating the Number of Password Candidates

The calculation of possible password guesses from a PCFG is a currently missing (see footnote 7) feature that is, however, essential for tools presented in this paper. Let $size(N)$ be the number of terminal structures that can be created by applying rewriting rules on non-terminal N. For base structure $B = N_1 N_2 \ldots N_n$, the number of possible password candidates can be calculated as:

$$cnt\_base(B) = \prod_{i=1}^{n} size(N_n). \tag{1}$$

For grammar G, the total number of possible password candidates is the sum of $cnt\_base(B)$ for all base structures $B \in G$:

$$cnt\_total(G) = \sum_{B \in G} cnt\_base(B). \tag{2}$$

The file and directory structure of Weir's PCFG considers a single rewriting rule per line. All rewriting rules have non-zero probability, and thus, all are used. Therefore, $size(N)$ for non-terminal $N = T_n$ (see Sect. 2) is, in most cases, the number of lines in n.txt file located in a directory for fragments of type T. For example, $size(D_3)$ equals the number of lines in Digits/3.txt file. Since letter capitalization rules have been introduced, it is necessary to take them into consideration. Thus, $size(A_n)$ is the number of lines in Alpha/n.txt file multiplied by the number of lines in Capitalization/n.txt file.

The calculation shown above is usable for classical PCFG-based approach only, i.e., with the `--coverage` parameter of PCFG Trainer set to 1. Otherwise, Weir's PCFG Manager would create additional character fragments using brute-force and Markov chains which is out of the scope of this paper.

3.4 The New PCFG Manager

To improve the use of resources, we created an alternative[8] to Weir's PCFG Manager. We started with a simple transcription of Python sources to Go programming language that we chose because of its speed, simplicity, and compilation to machine language. Early experiments showed that our Go-based alternative using the same algorithms was about four times faster than the original solution. However, there was still enough space for optimization.

Within all steps performed by the PCFG Manager, generating password guesses from pre-terminal structures [14,15] was the most computationally complex part. Since there is no mutual dependence between the pre-terminals, we decided to modify the program and parallelize this part of the process. Our new design uses a single *goroutine* (a lightweight thread) for filling the priority queue [15] with pre-terminal structures, and one to n goroutines for generating terminals in parallel. The n can be set by a user to reflect the processor's capabilities. Moreover, we added a parameter which allows the user to limit the number of generated password guesses. We illustrate both approaches by simplified schematics that display goroutines and data transfer operations. While Fig. 1a shows the original design of Weir's PCFG Manager, the parallel version is depicted in Fig. 1b.

For synchronization and mutual communication, goroutines use a mechanism called *channels* that act as FIFO queues. A goroutine can send values to a channel or receive values from it. By default, channels are not buffered and both send and receive operations are blocking. In our solution, we use a buffered channel of size n where the sender is blocked only if the channel contains n values in the queue. Each value represents a pre-terminal structure. The main goroutine (M) implements the Deadbeat dad algorithm [14] filling the priority queue with pre-terminals. Every time a pre-terminal is created, it is sent to the buffered channel if there is enough space. Every time the channel is full, the main goroutine is suspended automatically by the send operation. There is no need to generate more pre-terminals at the time they cannot be processed. In contrast to the original version, the proposed design allows to process multiple pre-terminals and generate passwords in parallel if $n > 1$. In that case, the only apparent drawback is the possible slight change of the password order at the output. This behavior could be resolved by adding a supplementary synchronization mechanism at the output, however, at the cost of performance loss. For practical use, we do not consider this as a large obstacle since for millions of password, the changes are insignificant because the order of larger password blocks is preserved. Moreover, if the user does not set the guess limit explicitly, or if the limit is set in the

[8] https://github.com/Dasio/pcfg-manager.

PCFG Mower (see Sect. 3.5) instead of PCFG Manager, the output dictionary contains the same passwords, and the success rate would be intact.

By profiling, we later detected that even though we accelerated generating terminal structures, the new bottleneck was at the output, where simple I/O text operations slowed down the entire process. We overcame this obstacle by adding extra output buffers to goroutines that generate terminal structures. The buffers store the terminal structures and are flushed to output after the

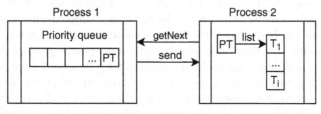

(a) Python PCFG Manager

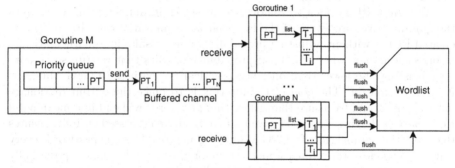

(b) Go PCFG Manager

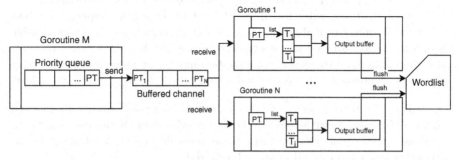

(c) Go PCFG Manager with buffered output

Fig. 1. The architecture of PCFG Manager in Python and Go (PT - pre-terminal structure, T - terminal)

entire pre-terminal is processed. The final design is illustrated in Fig. 1c and the experimental results in Sect. 4.

3.5   Grammar Filtering

To increase speed even more, we experimented with various modifications of already-trained grammars. We noticed that removing rules which rewrite the starting symbol into long base structures brings a significant speedup without higher impact on a success rate. The motivation for such filtering was discussed in Sect. 3.2. We automated the process by creating a simple script that automatically filters out all base structures longer than a user-defined maximum.

At this point, we were able to generate much more passwords per time unit. However, without a manually-defined limit for password guesses, the total amount of time required for generating was still extensive. From a practical perspective, any limit to guess count means that there is always a part of the grammar that is never used and unnecessarily wastes memory during the guess generation. Such consideration led us to speculate about reducing the size of the grammar instead of limiting guesses in PCFG Manager.

We came with an idea to remove the least significant rewriting rules from the grammar. We are aware of the fact that any removal of rules from already-created PCFG without adjusting probability values results in a mathematically incorrect grammar where the total probability of rules that rewrite some non-terminals may be lower than one. For practical use with the PCFG Manager, it does not matter. The goal of the filtering is to make the output dictionary more compact and to ensure that generating passwords will end in a meaningful time. Besides, having a reduced grammar that can be processed entirely, ensures that even the parallel run of PCFG Manager generates the same passwords every time. Nevertheless, the strongest motivation for grammar filtering is a potentially massive saving of processor time. Putting a limit before the guessing even starts prevents the Deadbeat dad algorithm from performing many useless derivation steps on trees that never form terminal passwords due to a low probability.

As denoted above, rules for alpha characters, digits, and special symbols usually have similar probabilities, thus removing them leads to a considerable loss of information which decreases the success rate. Rulesets for base structures and capitalization, on the other hand, contain many insignificant rewriting rules that can be removed safely. We created a tool called *PCFG Mower*[9] which can:

- Calculate the total number of possible password guesses from a PCFG and inform the user about achievable keyspace. Moreover, if the user knows an average speed of password guessing, it is possible to estimate the total time required for generating all password candidates.
- Filter a PCFG by performing an automatic removal of rewriting rules based on a set of options entered by the user.

---

[9] https://github.com/findo11/pcfg_mower.

Input: original grammar, *limit*, b_s, c_s
Output: reduced grammar
1: *reduce* = **true**, $i = 0$
2: **repeat**
3:     $i++$
4:     *count* = password_count()
5:     **if** *count* $\leq$ *limit* **then**
6:         *reduce* = **false**
7:     **end if**
8:     **if** *reduce* **then**
9:         Remove as many base structures as required to reduce their total probability by b_s.
10:         Remove all capitalization rules that have probability lower than $i \times c_s$.
11:     **end if**
12: **until not** *reduce*

Fig. 2. PCFG reduction algorithm

To verify our assumptions, we created a simple *PCFG reduction algorithm* that is implemented in PCFG Mower and shown in Fig. 2. The goal of the algorithm is not to provide a universal solution, but to validate or disprove that systematic PCFG filtering brings a possible benefit to password cracking. Besides the original grammar, it takes the following input parameters: *limit* defining the maximum number of password guesses to be generated, and probability values b_s, c_s. While b_s allows to set how rapidly should the algorithm remove base structures, c_s sets the same for capitalization rules. The output of the algorithm represents a PCFG which generates the maximum of *limit* password guesses.

4 Experimental Results

In this section, we demonstrate the practical benefits of our enhancements to PCFGs. For experimental purposes, we work with both original and modified datasets from real password leaks. As data sources, we used SkullSecurity (see footnote 2) pages and SecLists (see footnote 3) repository. All employed datasets are enlisted in Table 3. For shorter notation, we assign each a unique identifier (ID). The last row (def) represents the default PCFG from Weir's PCFG Cracker (see footnote 1), which is said to be trained on a random sample of million passwords from RockYou dataset.

The table shows the number of passwords in the dataset (pw count), its size, and the average password length (avg). The other columns illustrate how a PCFG trained on the dataset looks like. We show the number of rewriting rules for alpha characters (A), digits (D), other characters (O) as well as the number of rewriting rules for base structures (base) and capitalization (cap).

Table 3. Password datasets used for experiments

Dataset					PCFG				
ID	name	pw count	size	avg	A	D	O	base	cap
dw	Darkweb2017-10000.txt	10,000	82.6 kB	7	5,244	947	30	323	83
r65	rockyou-65.txt	30,290	344.5 kB	7	17,845	4,213	35	256	39
r75	rockyou-75.txt	59,187	478.9 kB	7	30,670	10,601	51	351	51
ms	myspace.txt	37,126	354.2 kB	8	22,587	4,273	133	1,574	179
tl	tuscl.txt	38,820	324.7 kB	7	26,806	6,518	71	1,290	242
pr	probab-v2-top12000.txt	12,645	100.2 kB	6	11,117	534	1	125	23
def	Random million passwords from RockYou				330,343	145,510	906	84,307	950

4.1 The Performance of PCFG Manager

At first, we measured the acceleration that can be achieved using our new PCFG
Manager in contrast with the original one from Weir et al. [15]. Table 4 shows
experimental results of generating password guesses using PCFG trained on
Darkweb2017-10000 dataset (dw), *RockYou-75* dataset (ru75), and the default
PCFG (def) used in Weir's cracker. All three experiments were performed using
a computer with Intel(R) Core(TM) i7-4700HQ CPU with 8 GB RAM. We also
decided to study the influence of disk I/O speed, so that we measured everything
using HDD and then using SSD. In all cases, we measured how many password
guesses we can generate within 3 min.

Our solution was from 8 to 40 times faster than the original one. Using
Darkweb dataset (see Fig. 3) resulted in lowest acceleration since it contains long
and complex base structures. With the default PCFG (see Fig. 4) and *Rockyou-
75* dataset (see Fig. 5), we were able to generate much more password guesses,
and the difference between HDD and SSD is more noticeable.

Table 4. No. of guesses and acceleration of PCFG manager

Training	Manager	HDD	SSD
dw	Python	3,022,923	2,948,532
	Go	24,592,908	24,609,579
	acceleration	8.14 x	8.35 x
def	Python	29,613,726	32,402,490
	Go	405,819,926	485,244,534
	acceleration	13.70 x	14.98 x
r75	Python	18,418,684	20,843,491
	Go	490,635,443	842,695,475
	acceleration	26.64 x	40.43 x

4.2 The Impact of PCFG Filtering

The second set of experiments aim to examine the effects of our attempts to reduce the grammar. Table 5 shows the results of training, modification, generating password guesses, and checking success rate using multiple datasets. The experiments were performed using Intel(R) Core(TM) i7-7700K CPU with 32 GB RAM and an SSD. Since generating password guesses using non-modified PCFGs would take hours and days, we set a time limit of 10 min to all measurements - every time the PCFG manager exceeded the 10-min interval, it was stopped.

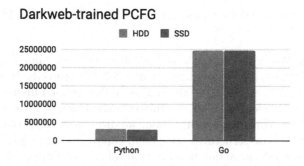

Fig. 3. No. of guesses within 3 min using Darkweb-trained PCFG

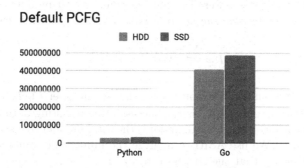

Fig. 4. No. of guesses within 3 min using Default PCFG

The first column (tr) shows which dataset we used for training to create the PCFG. For all training datasets, the first line represents generating password guesses using the original grammar - i.e., without any modification. The *longbase* modification stands for the grammar where we removed base structures longer than 10 characters (5 non-terminals). In other measurements, we used a grammar with already-removed long base structures and reduced it using *PCFG Mower* described in Sect. 3. The *mow-n* modification means that we performed *longbase* first and then we set the *limit* of the PCFG reduction algorithm

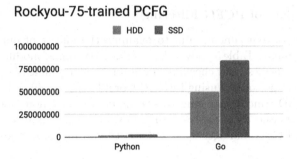

Fig. 5. No. of guesses within 3 min using Rockyou-75-trained PCFG

to n passwords. We experimented with the following *limit* values: 1,000,000,000 (1000M), 500,000,000 (500M), and 20,000,000 (20M) passwords. In all cases, the b_s and c_s constants were set to 0.001 to achieve fine-grained filtering. Since the algorithm removes selected rules, we illustrate the changes done to the grammars in each step. For every modification, we display the preserved number of rewriting rules for base structures (base) and capitalization (cap).

Next columns inform about password guessing. We display the amount of time required to generate the output dictionary (time), (or 10 m* if we reached the time 10-min limit), the size of the output dictionary (out size) and the number of its passwords in millions (mop). The rest displays the success rate of password guessing on testing datasets - i.e., the percentage telling how many generated password guesses were included in different testing datasets. The last column displays the *average success rate impact* (ASRI) which is calculated as:

$$ASRI = \frac{\sum_{i=1}^{n}(SR_i^{mod} - SR_i^{orig})}{n}$$

where SR_i^{orig} is the success rate on testing dataset number i before the modification of the PCFG, and SR_i^{mod} is the success rate on testing dataset number i after the modification of the PCFG, and n is the total number of testing datasets. In our case, $n = 4$. We use ASRI to analyze the influence of our modifications. Positive ASRI means that the success rate was improved while negative stands for decrease.

As we can see from results, removing long base structures resulted in a massive increase of password guessing speed which enabled to generate much more passwords within 10 min. We achieved the highest acceleration on *dw* and *r65* since they contain very complex passwords that create enormously long base structures. After the modification, we were able to generate over 14 times more password guesses. In contrast, training on *ms* and *tl* creates more simple grammars, and thus the speedup was not as rapid. Removing long base structures showed almost no impact on the success rate which confirms our assumption that their importance is negligible. From 16 testings, only 8 led to decrease by a maximum of 0.06%. To our surprise, the ASRI was mostly positive since in 6

Table 5. Success rates of original and modified PCFGs (* - reached the time limit)

tr	Grammar modification	base	cap	Password guesses time	out size	mop	Success rate pr	ms	dw	r65	ASRI
dw	original	323	83	10 m*	731 MB	78	45.03%	26.83%	98.27%	41.39%	
	longbase	288	83	10 m*	12 GB	1,110	45.01%	26.91%	98.35%	41.40%	+0.04%
	mow-1000M	106	40	25 s	3.3 GB	373	44.54%	24.47%	96.42%	38.36%	−1.93%
	mow-500M	106	40	25 s	3.3 GB	373	44.54%	24.47%	96.42%	38.36%	−1.93%
	mow-20M	86	32	2 s	77 MB	9	44.18%	24.12%	95.65%	38.00%	−2.39%
r65	original	256	39	10 m*	1.5 GB	151	72.34%	37.63%	88.25%	99.84%	
	longbase	223	39	10 m*	25 GB	2,210	72.30%	37.63%	88.14%	99.81%	−0.05%
	mow-1000M	161	36	3 m 31 s	11 GB	980	72.17%	37.17%	87.73%	99.61%	−0.35%
	mow-500M	123	31	1 m 31 s	4.5 GB	409	72.01%	36.62%	87.23%	99.35%	−0.71%
	mow-20M	79	20	3.5 s	130 MB	13.8	70.98%	34.26%	85.80%	97.16%	−2.47%
ms	original	1574	179	10 m*	5.7 GB	616	47.47%	93.68%	69.14%	46.42%	
	longbase	1430	179	10 m*	9.5 GB	1,030	47.45%	94.38%	69.07%	46.42%	+0.15%
	mow-1000M	110	25	3 m	9.2 GB	941	46.37%	82.40%	66.74%	43.04%	−4.54%
	mow-500M	78	20	1 m	3.1 GB	334	45.13%	79.67%	64.71%	42.62%	−6.15%
	mow-20M	21	20	2 s	126 MB	15	33.25%	61.17%	54.28%	35.58%	−18.11%
tl	original	1290	242	10 m*	4.5 GB	520	55.27%	36.87%	69.85%	43.86%	
	longbase	1158	242	10 m*	7.6 GB	870	55.23%	37.15%	69.79%	43.87%	+0.05%
	mow-1000M	91	20	2m 43s	7.5 GB	884	54.06%	30.94%	66.08%	40.37%	−3.60%
	mow-500M	48	19	1m 8s	1.8 GB	200	53.77%	29.05%	64.19%	39.39%	−4.86%
	mow-20M	24	18	2s	133 MB	17	52.08%	22.27%	55.61%	35.64%	−10.07%

cases, removing long base structures improved the success rate by up to 0.7% thanks to more passwords generated within the same time.

Next measurements analyzed grammars filtered by PCFG Mower to verify if the removal of low-probability rewriting rules brings any benefit. In all cases, the *mow* modification allowed the PCFG Manager to process the entire grammar in less than 4 min, showing that it can provide a suitable alternative to a "hard" limit for password guessing. More compact PCFGs produced smaller dictionaries. With more compact PCFGs, the generated dictionaries were smaller as well. Again, we achieved the best results with *dw* and *r65* datasets, where we were able to reduce the size from 12 GB (longbase) to 112 MB dictionary, and from 25 GB to 130 MB with a loss of success rate below 4% in all cases. For *ms* and *tl*, filtering the grammar spared time and space as well, however, the *mow-20M* limit was too strict to provide satisfactory results. For *dw*, we received the same results with *mow-1000M* and *mow-500M*. The *dw*-trained grammar contains a high number of base structures with similar probabilities. Thus, a lot of them was removed by *mow-1000M* modification, and no further filtering was necessary.

4.3   Evaluation

By modification of both PCFG Manager and existing grammars, we were able to make password guessing many times faster. What most helped the speedup was the use of a compiled programming language, the parallelization of generating terminal structures and removing rewriting rules for long base structures. For datasets we analyzed, such rules caused "more harm than good." The rewriting rules for long base structures mostly had insignificant probabilities but complicated the calculation of the computationally-complex Deadbeat dad algorithm. In all cases, the removal accelerated the password guessing dramatically.

Filtering grammars with PCFG Mower reduced the time required for password guessing rapidly. The settings, however, have to be selected wisely. With our experimental setup, we achieved the best results with PCFG Mower limit set to 500 millions of passwords. Stricter limitation produced decent results for only some cases. We assume that the success rate highly depends on the nature of selected datasets, and thus there is no universal solution.

5   Conclusion

Probabilistic methods certainly have their place in the area of password cracking. While Markov chains were adopted to existing tools a long time ago, probabilistic context-free grammars are currently more a subject of academic research than a ready-to-use technique. However, as the development of cracking methods continues by researchers, communities, and commercial subjects, the situation may change. Even the authors of Hashcat consider[10] adding support for generating "slow candidates."

From our standpoint, one of the main factors that currently complicate the use of PCFG-based techniques is the extensive amount of time required to generate password guesses. By using both analytic and experimental approach, we identified the critical spots that slowed down the entire process. We proposed methods that optimize the password guessing and allow better use of hardware resources. We experimentally proved that our new PCFG Manager is capable of generating passwords 8 to 40 times faster than the original tool from Weir et al.

Moreover, we proposed a way of PCFG filtering which provides a resource-saving alternative to a "hard" password guess limit. We showed that the systematic removal of selected rewriting rules might reduce the total amount of time required to generate password candidates without having a significant impact on the success rate. If one decides to use the filtering techniques, we recommend starting with the removal of long base structures that produce the least-probable passwords and perceptibly increase the number of necessary processor operations.

In our future research, we want to perform a more detailed analysis of the relation between PCFG filtering and the success ratio which may discover new factors that have not been revealed yet. Since the practical use of password cracking often involves a distributed environment, we currently work on distributed PCFG-based password guessing techniques which may provide a smarter alternative for a classic dictionary attack.

---

[10] https://hashcat.net/forum/thread-7903.html.

Acknowledgements. The research presented in this paper is supported by "Integrated platform for analysis of digital data from security incidents" project, no. VI20172020062 granted by Ministry of the Interior of the Czech Republic and "ICT tools, methods and technologies for smart cities" project, no. FIT-S-17-3964 granted by Brno University of Technology. The work is also supported by Ministry of Education, Youth and Sports of the Czech Republic from the National Programme of Sustainability (NPU II) project "IT4Innovations excellence in science" LQ1602.

References

1. Bishop, M., Klein, D.V.: Improving system security via proactive password checking. Comput. Secur. **14**(3), 233–249 (1995)
2. Bonneau, J.: The science of guessing: analyzing an anonymized corpus of 70 million passwords. In: 2012 IEEE Symposium on Security and Privacy, pp. 538–552, May 2012. https://doi.org/10.1109/SP.2012.49
3. Das, A., Bonneau, J., Caesar, M., Borisov, N., Wang, X.: The tangled web of password reuse. In: NDSS 2014, pp. 23–26 (2014)
4. Florencio, D., Herley, C.: A large-scale study of web password habits. In: Proceedings of the 16th International Conference on World Wide Web, WWW 2007, pp. 657–666. ACM, New York (2007). https://doi.org/10.1145/1242572.1242661
5. Ginsburg, S.: The Mathematical Theory of Context Free Languages. McGraw-Hill Book Company, New York (1966)
6. Houshmand, S., Aggarwal, S.: Using personal information in targeted grammar-based probabilistic password attacks. Advances in Digital Forensics XIII. IAICT, vol. 511, pp. 285–303. Springer, Cham (2017). https://doi.org/10.1007/978-3-319-67208-3_16
7. Houshmand, S., Aggarwal, S., Flood, R.: Next gen PCFG password cracking. IEEE Trans. Inf. Forensics Secur. **10**(8), 1776–1791 (2015)
8. Kelley, P.G., et al.: Guess again (and again and again): measuring password strength by simulating password-cracking algorithms. In: 2012 IEEE Symposium on Security and Privacy (SP), pp. 523–537. IEEE (2012)
9. Ma, J., Yang, W., Luo, M., Li, N.: A study of probabilistic password models. In: 2014 IEEE Symposium on Security and Privacy, pp. 689–704 (2014). https://doi.org/10.1109/SP.2014.50
10. Narayanan, A., Shmatikov, V.: Fast dictionary attacks on passwords using time-space tradeoff. In: Proceedings of the 12th ACM Conference on Computer and Communications Security, CCS 2005, pp. 364–372. ACM, New York (2005). https://doi.org/10.1145/1102120.1102168
11. Proctor, R.W., Lien, M.C., Vu, K.P.L., Schultz, E.E., Salvendy, G.: Improving computer security for authentication of users: influence of proactive password restrictions. Behav. Res. Methods Instrum. Comput. **34**(2), 163–169 (2002)
12. Veras, R., Collins, C., Thorpe, J.: On semantic patterns of passwords and their security impact. In: NDSS (2014)
13. Vu, K.P.L., Proctor, R.W., Bhargav-Spantzel, A., Tai, B.L.B., Cook, J., Schultz, E.E.: Improving password security and memorability to protect personal and organizational information. Int. J. Hum.-Comput. Stud. **65**(8), 744–757 (2007). https://doi.org/10.1016/j.ijhcs.2007.03.007

14. Weir, C.M.: Using probabilistic techniques to aid in password cracking attacks. Ph.D. thesis, Florida State University (2010)
15. Weir, M., Aggarwal, S., de Medeiros, B., Glodek, B.: Password cracking using probabilistic context-free grammars. In: 2009 30th IEEE Symposium on Security and Privacy, pp. 391–405 (2009). https://doi.org/10.1109/SP.2009.8

That's My DNA: Detecting Malicious Tampering of Synthesized DNA

Diptendu Mohan Kar[(✉)] and Indrajit Ray[(✉)]

Colorado State University, Fort Collins, CO 80523, USA
{diptendu.kar,indrajit.ray}@colostate.edu

Abstract. The area of synthetic genomics has seen rapid progress in recent years. DNA molecules are increasingly being synthesized in the laboratory. New biological organisms that do not exist in the natural world are being created using synthesized DNA. A major concern in this domain is that a malicious actor can potentially tweak with a benevolent synthesized DNA molecule and create a harmful organism [1] or create a DNA molecule with malicious properties. To detect if a synthesized DNA molecule has been modified from the original version created in the laboratory, the authors in [13] had proposed a digital signature protocol for creating a signed DNA molecule. It uses an identity-based signatures and error correction codes to sign a DNA molecule and then physically embed the digital signature in the molecule itself. However there are several challenges that arise in more complex molecules because of various forms of DNA mutations as well as size restrictions of the molecule itself that determine its properties, the earlier work is limited in scope. In this work, we extend the work in several directions to address these problems.

Keywords: Cyber-bio security · DNA · Identity-based signatures ·
Reed-Solomon codes · Approximate string matching ·
Pairing-based cryptography

1 Introduction

Synthesizing DNA molecules in the laboratory is quite common these days. Such a synthetic DNA molecule is often a licensed intellectual property. DNA samples are shared between academic laboratories, ordered from DNA synthesis companies and manipulated for a variety of purposes, for example, to create new biochemicals, reduce the burden of diseases, improve agricultural yields or simply to study the DNA's properties and improve upon them. There have also been instances of new biological organisms that do not exist in the natural world being created using synthesized DNA [1]. While the vast majority of such activities are pursued for beneficial purposes, there are concerns that malicious users can use the technology malevolently, for example, to make harmful biochemicals, or making existing bacteria more dangerous [1]. Recently, a DNA-based security

© IFIP International Federation for Information Processing 2019
Published by Springer Nature Switzerland AG 2019
S. N. Foley (Ed.): DBSec 2019, LNCS 11559, pp. 61–80, 2019.
https://doi.org/10.1007/978-3-030-22479-0_4

exploit was demonstrated as a proof of concept, where a synthesized DNA was used to attack a DNA sequencer that has been deliberately modified with a vulnerability [16]. Preventing such malicious use of synthesized DNA is beyond the scope of this current work. However, attribution of a physical DNA sample and establishing proof of origin can contribute significantly to deter such malicious activities.

Following the anthrax attack of 2001, there is an increased urgency to employ microbial forensic techniques to trace and track agent inventories. For instance, it has been proposed that unique watermarks be inserted in the genome of infectious agents to increase their traceability [12]. The synthetic genomics community has demonstrated the feasibility of this approach by inserting short watermarks into DNA without introducing significant perturbation to genome function [6,8,15,20]. The use of watermarks has also been proposed in order to identify genetically modified organisms (GMOs) or proprietary strains. Heider et al. [7] describe DNA-based watermarks using DNA-Crypt algorithm. This technique is applicable to provide proof of origin to a DNA molecule. However, there is a major shortcoming with all watermark based approaches. The watermark in all these works is generated from an arbitrary binary data and added to the original sequence, and so is independent of the original sequence and provides no integrity of the actual DNA sequence.

To enable effective trace back and eliminate the limitation of watermark-based approaches, Kar et al. [13] had proposed a scheme to create digital signatures of DNA molecules in living cells. The main idea is as follows: Take a DNA molecule and sequence it. The result is a string over the alphabet A, C, G, and T, representing the four nucleotide building blocks of DNA. The output of the sequencer is stored in what is called a FASTA file. For interpretability reasons, the FASTA file is annotated by the researcher to create another file called the GenBank file. The authors then use Shamir's identity-based signature scheme [23], Reed-Solomon error-correction codes [18,19] and the 16 digits Open Researcher and Contributor ID (ORCID – https://orcid.org) of the researcher to create a digital signature of the string in the FASTA file. The resulting signature is in the form of a DNA sequence which is now synthesized as a physical molecule. Finally, the signature molecule is inserted into the original DNA molecule using DNA editing tools to obtain a signed DNA molecule. When this signed molecule is shared, a receiver can sequence the signed molecule to verify that it was shared by an authentic sender and that the sequence of the original molecule has not been altered or tampered with.

However, there are significant challenges related to the placement of the signature within the molecule and various types of mutations in more complex molecules (discussed in more details in Sect. 2) that Kar et al. do not address. The current work improves the previous scheme to address these problems (Sects. 3 and 4). Moreover, we would like to shorten the size of the signature sequence as much as possible without impacting security. While biologists believe that the size of the DNA has a correlation with its properties within certain bounds, they still do not know by how much a DNA molecule can be expanded without changing the properties of interest. The current work explores other crypto-graphic algorithms towards this end (Sect. 5).

2   Limitations of Earlier Work and Current Contributions

2.1   Cyclic Shifts and Reverse Complement

In [13], the signer is *required* to send the GenBank file along with the physical DNA sample to the receiver. This is because the GenBank file is needed to align the FASTA file (which is the output of a DNA sequencer) in the same order as during the signature generation. Plasmid DNA is *cyclic* and *double-stranded*. Following DNA sequencing, any cyclic permutation of the DNA structure is possible. A sequence represented in a FASTA file, and consequently the GenBank file, is thus one of several possible linear representations of a circular structure. For example, in a FASTA file if the sequence was "ACGGTAA", and the same sample is sequenced again, the FASTA file might read as "TAAACGG".

Moreover, since DNA is composed of two complementary, anti-parallel strands, a DNA sequencer can read a sample in both the "sense" or "antisense" direction. The sequence may be represented in a FASTA file in either direction. When the sample is sequenced again, the output might be in the other direction, or what is known as the reverse complement. The reverse complement of "A" is "T" and vice-versa, and the reverse complement of "C" is "G" and vice-versa. The DNA molecule has a polarity with one end represented as 5' and the other represented as 3'. One strand adheres to its reverse complement in an anti-parallel fashion. So if the sequence is - "5'-ACGGTAA-3'", the reverse complement is "3'-TGCCATT-5'". The FASTA file will represent one strand of the DNA sequence in the 5' to 3' direction; so the FASTA file could read as "ACGGTAA" or "TTACCGT". Thus, by combining these two properties, for a DNA that contains N number of bases, the possible number of correct representations of the same sample is $2N$: N cyclic permutations plus each reverse complement.

Let us now consider the implications of this characteristic of DNA on the signature generation and verification. The sender has a sequence say "ACCGTT". The sender synthesizes the sequence and sends it to the receiver. The receiver after sequencing with an automated DNA sequencer may not have exactly "ACCGTT". It can be "TTACCG" which is a cyclic permutations. The receiver can also get something like "AACGGT" which is the reverse complement of "ACCGTT". Owing to such domain challenges, the signature verification procedure is not as simple as in digital messages.

Let us assume the signature sequence is "TTAA". (The actual signature length is 512 base pairs). In [13], the authors had defined a start and an end tag which served as delimiters for the signature. Let "ACGC" and "GTAT" be the start and end tags. For this discussion, we will use the term message to denote some linear representation of the sequence generated by a DNA sequencer. There can be three cases for including the signature sequence in the DNA sequence:

1. **Append the signature after the message:** In this case, the sender's message with the signature embedded looks like - "ACCGTT   ACGC   TTAA GTAT". The receiver, after sequencing the signed DNA sample may get something like – "GTT   ACGCTTAA   GTAT   ACC" or something else depending on which base position the sequencer considers as the beginning of the sequence. In the permutation, the DNA sequencer assumed the 4th base from the left as the start of the sequence. The message is split but the delimiters and signature are intact. The simplest way to extract the message and signature is to append the extracted sequence to itself. With the permutation, this becomes "GTT   ACGC   TTAA   GTAT   ACC || GTT   ACGC   TTAA   GTAT ACC". Now we can extract the message which will be contained between two "ACGC TTAA   GTAT" when the string is wrapped around. The receiver reconstructs the message which is "ACCGTT". The receiver can then invoke the verification. Note that this scheme works no matter which position the sequence considers as the start of the sequence.

2. **Append the signature before the message:** In this case, the sender's message with signature looks like - "ACGC   TTAA   GTAT   ACCGTT". The receiver after sequencing the DNA might get something like - "AA   GTAT ACCGTT   ACGC   TT". We observe that this is the same as the previous case. We can append the extracted sequence to itself – "AA   GTAT   ACCGTT ACGC   TT || AA   GTAT   ACCGTT   ACGC   TT. Thus we can extract the message using the same procedure as above and then invoke the verification.

3. **Append the signature between the message:** In this case, the sender's message with signature might look like - "ACC ACGC   TTAA   GTAT   GTT". The receiver after sequencing the DNA might get something like "ACGC TTAA   GTAT   GTT   ACC". The problem occurs in this scenario. Even if we append the extracted sequence, we will not be able to recover the message. After appending the sequence we get "ACGC   TTAA   GTAT   GTT   ACC || ACGC   TTAA   GTAT   GTT   ACC". We can observe that the sequence contained by the two "ACGC   TTAA   GTAT" is "GTTACC". This is not the message the sender signed. The sender signed the message on "ACCGTT". But the receiver has no way of knowing this and hence the verification will fail since the message is not the same even though there is no modification to either the message or the signature.

The problem of recovering the message only occurs when the signature is placed within the message. The other two cases when the signature is placed before or after the message works perfectly fine. However, when working with DNA molecules, it may not always be possible to place the signature at the end or the beginning of the message. This is because there can be a feature present at that location. The possible places to place the signature are most likely to be within the original sequence. For this reason the GenBank file needed to be shared. Only this way would the receiver be able to align the sequence in the same order that the sender had when he signed.

There are several reasons why we may not want to share the GenBank file. The GenBank file is created by the originator of the DNA molecule using a gene editor. Its only purpose is to annotate the DNA sequence. If the DNA is an intellectual property, then the creator of the DNA will be annotating the DNA's GenBank file with different features of different subsequences of the DNA. While the creator may be willing to divulge the property of the synthesized DNA as a whole, s/he may not be willing to divulge properties of various subsequences. Sending the GenBank file jeopardizes the latter. Moreover, gene editors maintain databases of DNA molecule properties. However, these databases may not be consistent across different editors in the sense that receivers gene editor may not have all the information about the same set of molecules that the sender's gene editor has. Finally, the GenBank file format is not the only format used by gene editors, unlike the FASTA file format. In order to not share the GenBank file with the receiver, we have changed the signature generation procedure in this work, such that the verification is not dependent on where the signer placed the signature. The details of the new signature generation procedure are explained in Sect. 3.

2.2   Mutations in Identifying Tags

In our previous work, we defined two identifying tags to demarcate the signature. The start tag was chosen as "ACGCTTCGCA" and the end tag as "GTATCCTATG". These two delimiters were chosen not just randomly but for very specific reasons. First biologists typically have some idea about what DNA sequence will not occur in their specific project. Thus they can choose delimiters from these non-occurring sequence. Second, from these possible delimiters, they will choose the ones that are simple to synthesize and assemble since DNA synthesis is expensive. Finally, they will choose a sequence that are easy to identify visually, are unlikely to develop secondary structures and have a balanced number of "A, C, G and T"s. Our domain experts selected these delimiters for this project. We also used error correction code to tolerate mutations within the DNA. However, we assumed that the start and end tag do not mutate. If they do, our previous work will fail to locate the signature and consequently, it will not be possible to verify the signature.

To overcome this limitation, in this work we propose using partial matching techniques such that the start and end tag can be located approximately. This is used in conjunction with error correction codes. Note that since the start and end tags are fixed, we know what we are searching for in the DNA molecule. For example, we may want to look for strings similar to "ACGCTTCGCA" such as "GCGCTTCGCG". The different techniques we use for achieving this are discussed in Sect. 4.

2.3   Signature Length

The length of the signature plays a very important role in this biology domain. Shorter signatures imply less cost of synthesizing the signature into a physical DNA molecule. Shorter signatures will also be less likely to impact the existing functionality and stability of the plasmid during signature embedding. Previously, we used 1024 bit keys and that resulted in 512 base-pair signature. However, 1024 bit keys are no longer considered very strong and not recommended in practice for digital signatures. Generally, 2048 bit keys are used. In our domain, this would result in a 1024 base pair signatures. This length has a higher probability of affecting the characteristics and stability of the plasmid. Furthermore, when synthesizing the signature, presently with a 512 base pair signature the cost is $46.08 - 512 base pairs at $0.09 per base pair. With a 1024 base pair signature, even if the plasmid remains stable and functional, the cost of synthesizing the signature would be $92.16. The new signature scheme with a shorter signature is described in Sect. 5.

3   DNA Signature Generation and Verification Procedure

In our DNA sign-share-validate workflow, there are three players: (i) The DNA signer will create the DNA signature and sign a DNA sequence. (ii) The verifier will use the signature to verify whether the received DNA sequence was sent by the appropriate sender and was unchanged after signing. (iii) A central authority, which is trusted, provide the signer with an encrypted token that is associated with the signer's identity. The token contains the signer's private key.

Trust Model: For this work, we assume a polynomial-time adversary, Mallory, who is trying to forge the signature of a reputed synthesized DNA molecule creator, Alice. Alice is trying to protect her IP rights/reputation as she distributes DNA molecules synthesized by her to researcher Bob. If the attacker, Mallory, is able to forge the signature of Alice then: (a) Mallory can replace the actual DNA created by Alice with her own but keep the signature intact. (b) Mallory can create her own DNA molecule and masquerade as Alice to sign it. (c) Mallory can modify parts of the signed DNA molecule created by Alice.

Use of Error Correction in DNA Signature: In the digital domain, the digital signature on a message can be used to detect integrity violations. If a violation is detected, the sender can always re-transmit the signed message without incurring much extra cost. However, in the DNA world, we are primarily shipping physical DNA samples. This implies that if a DNA signature identifies that there is an error in the signature validation, then the sample needs to be physically transported and/or synthesized again. This incurs significant cost. DNA mutation is a very natural and common phenomenon. Thus, there is a good likelihood that signature validation will fail. Moreover, associated with the problem of mutation lies the problem of sequencing. When the DNA is processed by an automated DNA sequencer, the output is not always one hundred percent correct. It is dependent on the depth of sequencing, and increased sequencing depth means higher costs. Sequencing a small plasmid to sufficient depth is relatively inexpensive, but for larger sequences, sequencing errors can be an issue. In order to overcome these limitations, we use block-based error correction codes, such as a Reed-Solomon code [19], together with signatures. The presence of error correction codes helps the receiver to locate a limited number of errors (as set by the signer) in the sequenced DNA as well as correct them. The position of the errors and the corrected values are conveyed to the verifier. The verifier can then decide if the errors are in any valuable feature of the DNA or not. If a valuable feature has been corrupted, the verifier can ask for a new shipment, else if the error was in a non-valuable area in the DNA, the verifier can disregard the error and continue to work with it.

We now describe our new DNA signature scheme. The steps are shown in Algorithm 1 (for signing) and Algorithm 2 (for verification). To avoid confusion we use the following conventions. The term *sample* is used to indicate the physical DNA molecule. The term *sequence* is used to signify the digital counterpart of a DNA molecule. This is generated by sequencing a sample in a DNA sequencer. The raw sequence (output of sequencing) is stored in a FASTA file. The annotated sequence is stored in a GenBank file. The signer creates a physical DNA sample from the signed sequence and sends the sample (only) to the verifier. The verifier sequences this sample to get another sequence that is then verified.

For ease of understanding, we denote the sequence to be signed by the string SEQUENCE, the signature by SIN, the begin and end tags as BESN and EDSN and the error correction code as ECC. Each of these strings is really a sequence of bases that can be synthesized into a physical DNA molecule and embedded in the sample. Any location reference in SEQUENCE for subsequence discussion is specific to the location within the sequence. For instance, location 3 in the string contains character Q. However, in the real sequence, the subsequence denoted by Q may occur in position 350 (for example) depending on how many bases constitute S and E.

Signature Generation: The signature generation procedure begins by scanning the GenBank file for the keyword ORIGIN and locating the actual DNA sequence. Let there exists a feature from location 1 to 3 in the sequence, which corresponds to SEQ. Next, the location of the signature placement specified by the signer is checked. If the location collides with a feature, the user is alerted to change the location. In our example, if the user had provided 2, the algorithm will alert the user that there is already a feature SEQ there and ask for a new location. If the user chooses 4 which is after Q, it will be allowed. Next, the ORCID and Plasmid ID (which are integers) are converted to the corresponding A C G T sequence by the following conversion method – [0 - AC, 1 - AG, 2 - AT, 3 - CA, 4 - CG, 5 - CT, 6 - GA, 7 - GC, 8 - GT, 9 - TA]. The reason for choosing this conversion type is that if any ORCID or Plasmid ID has repetitions e.g. if ORCID is 0000-0001-4578-9987, the converted sequence will not have a long run of a single base. Long runs of a single nucleotide can result in errors during sequencing. Let the converted ORCID and Plasmid ID sequences be ORCID and PID respectively.

To account for the problem of placing the signature within the sequence mentioned earlier in Sect. 2, the signature is generated on the hash of a tweaked version of the sequence. We left rotate a copy of the sequence by $n-1$ where n is the location within the sequence where the signature needs to be placed. For this

---

Algorithm 1. DNA Signature Algorithm Accommodating Cyclic Shifts, Reverse Complement and Mutating Tags

---

Input: The GenBank (.gb) file: file, ORCID: a 16 digit number in xxxx-xxxx-xxxx-xxxx format, Plasmid ID: a 6 digit number, Location of signature placement: number, Error tolerance limit: number (can be 0 meaning no error tolerance)

Output: Signed GenBank (.gb) and FASTA (.fa) file: file

1 Input checks e.g. correct file extension, ORCID format, integers etc.
2 Parse GenBank file. Split content and sequence based on keyword ORIGIN. Parse content to get the list of feature locations.
3 **if** *Location of signature placement NOT within a feature* **then**
4   Make the position as start of the sequence and wrap everything before the location to the end. If position is 0 or length of sequence - no wrap is needed.
5   Generate hash (SHA-256) of this sequence.
6   Generate signature on the hash.
7   Convert the signature bytes,ORCID and Plasmid_ID to ACGT sequence. Create the following string by concatenating parts:
8   BESN+ORCID+Plasmid_ID+SIN+EDSN
9   **if** *error tolerance NOT 0* **then**
10   Append MSG (shifted sequence) before BESN+ORCID+PLASMID_ID+SIN+ESN.
11   Pass SEQUENCE+BESN+ORCID+PLASMID_ID+SIN+ESN to Reed-Solomon Encoder.
12   Convert the parity bytes to ACGT. (call this ECC)
13   Signature_Sequence = BESN+ORCID+PLASMID_ID+SIN+ECC+EDSN.
14   **else**
15   Signature_Sequence = BESN+ORCID+PLASMID_ID+SIN+EDSN.
16   **if** *signature placement location is start of the original sequence* **then**
17   Final_Sequence = SEQUENCE+Signature_Sequence
18   **else if** *signature placement location is end of the original sequence* **then**
19   Final_Sequence = Signature_Sequence+SEQUENCE
20   **else**
21   part1 = prefix of SEQUENCE of length $n-1$ (where signature is to be placed at location n)
22   part2 = suffix of SEQUENCE of length $len(SEQUENCE) - n + 1$
23   Final_Sequence = part1+Signature_Sequence+part2
24   Write the Final_Sequence to a new GenBank file and FASTA file.
25 **else**
26   Alert user about collision. Allow user to input new location. Go to step 3 with new location.

---

example, the sender wants to place the signature after Q. The sequence will be shifted as – UENCESEQ. The signature is generated on the hash of the left rotated sequence UENCESEQ. The signature bits are then converted to A C G T sequence. Let this signature sequence be SIN. Let the start tag be BESN and end tag be EDSN. The signature sequence is concatenated with ORCID and PID and then placed between the start and end tags as BESN ORCID PID SIN EDSN. This entire string is then placed at the position specified by the user. We chose 4 in our example. Hence, the signed sequence looks like - SEQ BESN ORCID PID SIN EDSN UENCE.

Next, this sequence is passed into the error correction encoder. According to the number of tolerable errors specified by the user, the error correcting parity bits are generated. These parity bits are then converted to some A C G T sequence. Let this sequence be ECC. When the encoder output is generated, the sequence would look like – SEQ BESN ORCID PID SIN EDSN UENCE ECC. Next, the ECC is separated and is placed before the signature and end tag. So the final output sequence is - SEQ BESN ORCID PID SIN ECC EDSN UENCE. Note that the error correction code is generated after generating the signature sequence and combining with original sequence. Hence any error in that string can be corrected provided it is within the tolerable limit. For instance, if we put 2 as our error tolerance limit, then any 2 errors within the string SEQ BESN ORCID PID SIN ECC EDSN UENCE can be tolerated. If there is 1 error in SEQ and 1 error in SIN, or 2 errors in SIN, or 1 error in SIN and 1 error in ECC, these can be corrected. But if there are more than two errors it cannot be corrected. The final output sequence - SEQ BESN ORCID PID SIN ECC EDSN UENCE is written into another GenBank file. The descriptions are updated i.e. the locations of the signature, start, end, ecc are added and if there were features after location 4 in the original DNA, the locations of these features are also updated. This GenBank file is for reference of the sender. It is not required for signature verification and there is no need to share it with the receiver unless there are other reasons. The output sequence is now synthesized into the signed DNA sample.

Signature Verification: The signature verification procedure is described below in Algorithm 2.

The receiver sequences the shared DNA using an automated DNA sequencer. The sequence in the FASTA file might not be the in the same order when the sender signed it. That is, after sequencing the shared DNA, the FASTA file may look like - ORCID PID SIN ECC EDSN UENCE SEQ BESN which is a cyclic permutation of the sender's sequence.

The first step in the verification procedure is to extract the BESN and EDSN tags. If they are not mutated they are retrieved directly. If the tags cannot be located directly, we use Algorithm 3 to retrieve their closest matches and use them as BESN and EDSN tags. We defer the discussion on Algorithm 3 to Sect. 2.2. The verification step now will concatenate the FASTA sequence - ORCID PID SIN ECC EDSN UENCE SEQ BESN + ORCID PID SIN ECC EDSN UENCE SEQ BESN.

Now, it looks for 2 BESN tags and extracts the content between them. After obtaining the start tag, 32 bases are counted, this is the ORCID sequence, next

---

Algorithm 2. New signature verification procedure

Input: A FASTA file generated from sequencing the DNA sample received
Output: Prompt - Signature Valid or Invalid.

1 Input checks: file extension and only ACGT content.
2 Parse FASTA file and create reverse complement of the file
3 Use Algorithm 3 to get the BESN and EDSN tags.
4 **if** *(file contains BESN or EDSN) OR (reverse contains BESN or EDSN)* **then**
5   **if** *file contains BESN or EDSN* **then**
6   Create content string by appending FASTA file content thrice.
7   Get the sequence between two BESN tags. Create the following parts by counting: ORCID = first 32 chars; PLASMID_ID = next 12 chars; SIN = next 512 chars; ECC = chars between SIN and END (may be empty); MSG = chars from END to end of string.
8   **else**
        `/* When input FASTA file is in reverse complement form.    */`
9   Create content string by appending reverse complement of FASTA file content thrice.
10   Same as Step 6. i.e. get the parts from reverse complement.
11   Generate hash (SHA-256) of MSG
12   Invoke signature verification
13   **if** *signature is valid* **then**
14   Alert user about success.
15   **else**
16   Alert user about failure and start error correction procedure.
17   **if** *ECC length is 0* **then**
18   Alert user there is no ECC and correction not possible.
19   **else**
20   Create the following string from the parts: SEQUENCE+BESN+ORCID+PID+EDSN+ECC and send to Reed-Solomon decoder.
21   **if** *decoder outputs null or same as input* **then**
22   Alert user errors are more than tolerable limit.
23   **else**
24   Get the corrected parts and re-invoke verification.
25   **if** *re-verification is success* **then**
26   Alert user that verification succeeded after error correction. Compare the parts before and after error correction and display the errors.
27   **else**
28   Alert user that verification failed even after successful correction.
29 **else**
30   Alert user that BESN and EDSN tags are not present.

---

12 bases are counted, this is the plasmid ID sequence, then 512 bases are counted, this is the signature sequence. Next the substring after this signature sequence to the EDSN tag is retrieved, this is the error correction sequence. Finally, the substring between EDSN and BESN is the message for signature verification.

Until this point, we have retrieved UENCESEQ, ORCID, PID, SIN, and ECC. The UENCESEQ, ORCID and SIN is used for signature verification. With our previous signature generation method, since the message signed by the sender was SEQUENCE and the message retrieved by the verifier is ENCESEQ the hashes will be different and the validation would fail. With the new procedure, we can see that the although the sender's file contained the sequence SEQUENCE, the signature was actually generated on the shifted UENCESEQ. Due to this shift, the retrieved sequence and the sender's sequence will always be the same under any rotations. We have shifted the message of the sender to make the signature placement at the start of the message. We call this new generation scheme as force shift 0.

If the FASTA file contains the reverse complement of the sender's DNA sequence, the entire FASTA file is reverse complemented and then we look for the BESN and EDSN tags. If there is a match, we arrive at the conclusion that the FASTA file contains the reverse complement. Then we start the same verification steps on the reverse complemented FASTA sequence.

4   Allowing Mutations in Start and End Tags

The approximate matching technique, shown in Algorithm 3, breaks the entire string in which we are looking for the result into substrings of the length of the input string. Each of the broken substring in the larger string is assigned a score based on how similar it is to the input string. A match is inferred using the highest score. Now in the real DNA, we are looking for sequences of A, C, G, and T. So there might be a case that there are multiple close matches which means that there are multiple starts (or end) tags. In those cases, we use the end tags (or start tags respectively) to narrow our results. The following steps describe how the approximate matching technique works. There can be a total of four scenarios:

1. **Case 1: No mutation in either start or end tags.** - In this case, we can find the exact locations of the tags and hence approximate matching techniques are not needed. There can be mutations in any other place which will be handled by the error correction code.

2. **Case 2: Mutation in BESN tag only.** - In this case, the EDSN tag is found directly. The algorithm looks for the closest match to BESN. If there is a single match with the highest score, then we can be quite certain that the BESN tag has been located correctly. However, there can be multiple matches with close scores, i.e., there is no single stand out high score. In that case, we use the EDSN tag for further elimination of choices. We already know that the content within the start tag and the end tag is more than 556 base pairs. Hence we choose only those potential BESN tags which are at distance of 556 base

pairs/characters or more away from the EDSN tag. The logic is set to 556 or more because the length of the error correction can be 0 if the user chooses no error correction.

3. **Case 3: Mutation in EDSN tag only.** - In this case, the BESN tag is found directly. The tool looks for the closest match to EDSN. As in case 2, if there is a single match with the highest score then we can be quite certain that the EDSN tag has been located correctly. For multiple matches with close scores, we use the same logic as described in case 2 above, using the distance between the BESN and EDSN tags to be more than or equal to 556 base pairs.

4. **Case 4: Mutation in both BESN and EDSN tags.** - In this case, we try to locate the closest matches for both tags. If there is a single match with the highest score for both of them then we can be pretty certain that we have located them both correctly. Also, we invoke the criteria of length more than or equal to 556 between them for more certainty. In case of multiple potential BESN and EDSN tags, we employ the length counting criteria for each BESN and EDSN tag pair possible from the obtained results and narrow down the results.

We used the Optimal String Alignment variant of the Damerau-Levenshtein algorithm [2] as our preferred method for string matching. For a discussion on the experiments we performed to arrive at this decision please refer to Appendix 1.

5   New Identity-Based Signature Scheme with Shorter Signature Size

There are several identity-based digital signature schemes using pairings. Some of the notable schemes are: *Sakai-Kasahara* [22], *Sakai-Ohgishi-Kasahara* [21], *Paterson* [17], *Cheon* [3], and *Yi* [24]. The *Sakai-Kasahara* scheme described two types of identity-based signatures. One is El-Gamal type and the other is Schnorr type. To identify the most appropriate scheme we first implemented all the above schemes using the Java Pairing Based Cryptography library (jPBC) [5]. We then investigated the signature lengths based on different types of curves that can be used. The time to generate and validate a signature depends on the type of the curve used. We evaluated both aspects: time to sign and verify, and the size of the signature using this algorithm for all the different types of curves present in the jPBC library.

Based on the signature size and the computation cost of signature generation and verification, we identified the best scheme to be the Sakai-Kasahara Schnorr type. We now describe the Sakai-Kasahara Schnorr type identity-based signature scheme. It has four steps: setup, extract, sign and verify.

Setup: The setup generates the curve parameters. The different curves provided in the jPBC library can be used to load the parameters. Let g_1 be the generator of G_1, g_2 be the generator of G_2. A random $x \in Z_n^*$ is chosen to be the master secret. Two public keys P_1 and P_2 are calculated as - $P_1 = x \cdot g_1$ and $P_2 = x \cdot g_2$. An embedding function H is chosen such that $H(0,1)^* \rightarrow G_1$.

Algorithm 3. Approximate matching of tags

Input: Content of FASTA file: String
Output: BESN and EDSN tags: 2 Strings

1 begin = ACGCTTCGCA; end = GTATCCTATG /* hardcoded   */
2 revcomp = reverse complement of input string
3 if *input contains (*begin *and* end*)* then
4 | ⌊ BESN = begin; EDSN = end

5 else if *input contains* end *and NOT* begin then
6 | EDSN = end; Split input into substrings of length 10
7 | foreach *substring* do
8 | | ⌊ Calculate score with begin; Store each substring and score. Sort by score.
9 | if *single highest score* then
10 | | ⌊ BESN = highest score substring
11 | else if *multiple high scores* then
12 | | Calculate distance between each substring to end.
13 | | BESN = substring where distance > 556
14 | | if *multiple pairs with distance > 556.* then
15 | | | ⌊ Alert user about failure to extract tags. Exit

16 else if *input contains* begin *and NOT* end then
17 | BESN = begin; Split input into substrings of length 10
18 | Same as step 7 and 8. Replace begin with end
19 | Same as step 9. Set EDSN = highest score substring as in step 10.
20 | Same as step 11. Replace end with begin in step 12. Set EDSN as in step 13.
21 | Same as step 14 and 15.

22 else if *input does NOT contain* begin *and* end then
23 | Split input into substrings of length 10
24 | foreach *substring* do
25 | | Calculate score with both begin and end;
26 | | ⌊ Store each substring and score for both. Sort by score.
27 | if *single highest score in both* then
28 | | ⌊ BESN = highest score substring;EDSN = highest score substring;
29 | else if *multiple high scores in both* then
30 | | Calculate distance between each pair of substrings. Set BESN and EDSN where distance > 556.
31 | | if *multiple pairs with distance > 556.* then
32 | | | ⌊ Alert user about failure to extract tags. Exit

33 Repeat the same four conditions as in step 3, 5, 16 and 22 with revcomp instead of input. e.g. revcomp contains (begin and end)
34 return BESN and EDSN

Extract: Takes as input the curve parameters, the master secret key x, and a user's identity and returns the users identity-based secret key. This step is performed by the central authority for each user A with identity ID_A.

1. For an identity ID_A, calculate $C_A = H(ID_A)$. That is map the identity string to an element of G_1.
2. Calculate $V_A = x \cdot C_A$.

User A's secret key is (C_A, V_A) and is sent to the user via a secure channel.

Sign: To sign a message m, a user A with the curve parameters and the secret key (C_A, V_A) does the following:

1. Choose a random $r \in Z_n^*$. Compute $Z_A = r \cdot g_2$.
2. Compute $e = e_n(C_A, Z_A)$, where e_n is the pairing operation.
3. Compute $h = H_1(m \parallel e)$, where H_1 is a secure cryptographic hash function such as SHA-256 and $\parallel$ is the concatenation operation.
4. Compute $S = hV_A + rC_A$.

A's signature for the message m is - (h, S)

Verify: The verification procedure is as follows:

1. Compute $w = e_n(S, g_2) * e_n(C_A, -hP_2)$
2. Check $H_1(m \parallel w) \overset{?}{=} h$

The above equation works because:

$$e = e_n(C_A, Z_A) = e_n(C_A, r \cdot g_2) = e_n(C_A, g_2)^r$$
$$w = e_n(S, g_2) * e_n(C_A, -hP_2)$$
$$= e_n(hV_A + rC_A, g_2) * e_n(C_A, -hx \cdot g_2)$$
$$= e_n(hx \cdot C_A + rC_A, g_2) * e_n(C_A, g_2)^{-hx}$$
$$= e_n((hx + r) \cdot C_A, g_2) * e_n(C_A, g_2)^{-hx}$$
$$= e_n(C_A, g_2)^{hx+r} * e_n(C_A, g_2)^{-hx}$$
$$= e_n(C_A, g_2)^r$$

Hence, $h = H_1(m \parallel e) = H_1(m \parallel w)$.

The signature is a tuple (h, S) where h is the result of a hash function and is dependent on the choice of the hash function. If h is SHA-1, then length is 20 bytes, if h is SHA-256, the length is 32 bytes. The value S is an element of the group G_1. Hence its length will be dependent on the curve type and the length of the prime. There are six types of curves in the jPBC library namely – a, a1, d, e, f, and g. The different types of curves and their parameters are provided in the library as "properties" files. Table 1 summarizes the comparison of the signature length using the different curves.

Based on the signature size, the best performance is provided by the d159, f, and g149 curves. However, the length of the primes are a bit different and also

Table 1. Signature size using different curves for the Sakai-Kasahara scheme.

Curve Name	Signature Size using SHA-1 (Bytes)	Signature Size using SHA-256(Bytes)
a.properties	(20, 128) = 148	(32, 128) = 160
a1.properties	(20, 260) = 280	(32, 260) = 292
d159.properties	(20, 40) = 60	(32, 40) = 72
d201.properties	(20, 52) = 72	(32, 52) = 84
d224.properties	(20, 56) = 76	(32, 56) = 88
e.properties	(20, 256) = 276	(32, 256) = 288
f.properties	(20, 40) = 60	(32, 40) = 72
g149.properties	(20,38) = 58	(32, 38) = 70

the embedding degree is different. In the d159 curve, the prime is 159 bits and the embedding degree is 6. In the f curve, the prime is 158 bits and the embedding degree is 12. In the g149 curve, the prime is 149 bits and the embedding degree is 10. Keeping in view the small difference in signature sizes and the security related to each type, the better choice is the f curve.

The time to generate the signature and verify also depends on the type of the curve because of their properties. Table 2 summarizes the time to sign and verify using the different types of curves.

Table 2. Average time taken to sign and verify for different types of curves for the Sakai-Kasahara scheme.

Curve Name	Average time to sign (ms)	Average time to verify (ms)
a.properties	56	60
a1.properties	594	448
d159.properties	102	98
d201.properties	121	138
d224.properties	120	131
e.properties	262	214
f.properties	133	251
g149.properties	170	219

From the speed perspective, the a type curve is the fastest for generating and verifying the signature. But the size of the signature is way larger. The short signature size generating curves i.e. d159, f and g149 take a bit more time. It is, therefore, a matter of priority - signature size over speed. If we need to sign and verify a lot of messages and not care about the signature size then type A curve is a good choice. However, if the size of the signature is more important than speed like in our application, the f type curve is a better option. Also, the f type curve offers the best security among the three as its embedding degree is higher. Using this Sakai-Kasahara scheme we have reduced the signature size from 512 base pairs to 288 base pairs. The only thing it affects in our earlier algorithms is determination of BESN and EDSN in Sect. 4 when these tags mutate and we need to rely on counting base pairs to locate those tags.

Security of Scheme: Since we use well-known signature schemes that assume that no polynomial-time adversary can forge a genuine signature without knowing the secret used to sign, it trivially follows that our scheme is also secure.

6   Conclusion and Future Work

In this work, we improve the previous DNA signing scheme [13] in several directions. First, we remove the need to share the genbank file by eliminating the requirement of alignment at the sample receiver's end. The new signature generation procedure is independent of where the signer wants to place the signature. Notwithstanding any cyclic shifts or reverse complements that the receiver may get during sequencing, the signature can still be verified. To account for DNA mutations, we use error correction codes in the signature protocol to correct errors within pre-specified tolerable limits. Our second improvement is a way to locate mutated tags using approximate string matching techniques. This allows us to overcome mutation in the identifying tags and hence we can correctly recover the error correction code. This was a major problem in previous scheme.

Our third improvement is the reduction of signature size. We used pairing based cryptography to improve the previous signature scheme which generated 512 base pair signature to the Sakai-Kasahara scheme which generates 288 base pair signature. That is almost 43% gain in signature length.

One of the future directions in this work would involve signing and verifying the same DNA molecule multiple times by different users. Alice signs and sends a DNA sample to Bob and Bob validate Alice's DNA. Then Bob continues to modify it, signs it and sends it to Mallory. Can Mallory only verify Bob's signature, or is there a way for Mallory to track the entire pathway starting from Alice? It would be interesting to see if the concept of aggregate signatures can be applied in these scenarios. Also, it would be interesting to see if we put a signature on top of an existing signature whether the characteristic of the DNA changes or not. If it does not, how many signatures can be inserted before the characteristics of the original DNA molecule begin to change? Also, if we cannot put multiple signatures within the same DNA molecule, how do we remove the signature that was present before signing it again. Finally, does removing the signature also alters the property of the DNA?

Acknowledgment. This work is partly based on research supported by the Office of the Vice President of Research, Colorado State University. This material is also based upon work performed by Indrajit Ray while serving at the National Science Foundation. Research findings presented here and opinions expressed are solely those of the authors and in no way reflect the opinions of Colorado State University, the U.S. NSF or any other federal agencies. The authors would like to thank Jenna Gallegos and Jean Peccoud for their comments and suggestions.

Appendix 1 - Analysis of Distance Measures for String Matching

Various techniques exist to handle matching of similar strings. These methods measure the distance between strings using a distance equation. One of the most important works in this field is the Levenshtein distance [14]. Other notable algorithms are Damerau-Levenshtein [2,4,14], Optimal String Alignment variant of Damerau-Levenshtein (sometimes called the restricted edit distance) [2], Jaro-Winkler edit distance [11], and Jaccard index [9,10].

We used all these five algorithms for the approximate start and end tag matching. One of the reasons for using all of the above was we wanted to find out which would be most suited to the DNA domain. For testing, the FASTA file is taken as input and the start and end tag within the FASTA file are manually changed. Then we search for the location of the defined start and end tags within the mutated FASTA file. The results for each algorithm are summarized on a case by case basis in Fig. 1. As can be seen from the Figures the Jaro algorithm was fairly inaccurate with an average accuracy of only 35.12%. The Jaccard algorithm fared much better but was still imperfect with an average accuracy of only 95.18%. All of the three Levenshtein variants were perfectly accurate in their assessment. These results indicate that if accuracy was the chief concern, either of the three Levenshtein variants would be ideal choices.

Another important consideration in algorithm selection was speed. While an algorithm may be perfectly accurate in its selection of the closest match to a string this means little in practice if the algorithm has an untenable long run time. To this end, the speed of the algorithms was compared. To accomplish this each method was used to compare a series of one million random strings of a set length. A graph of the time in milliseconds (ms) for each algorithm is given in Fig. 2.

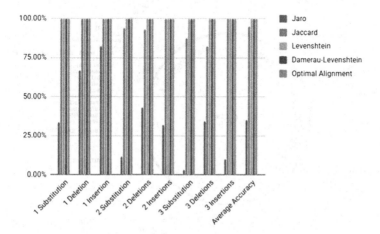

Fig. 1. Accuracy of algorithms per case as a percentage.

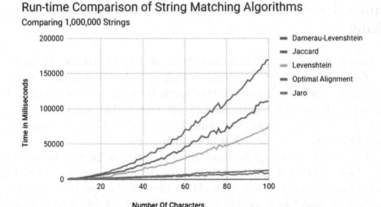

Fig. 2. Runtime analysis of various algorithms in milliseconds.

As can be seen from Fig. 2, the Jaro-Winkler and Optimal String Alignment algorithms were the quickest, each growing at very slow rates with Jaro-Winkler being slightly faster overall. Taking both of these factors into consideration Optimal String Alignment was chosen as the preferred method.

Appendix 2 - pUC19 DNA Before and After Signing

See Figs. 3 and 4.

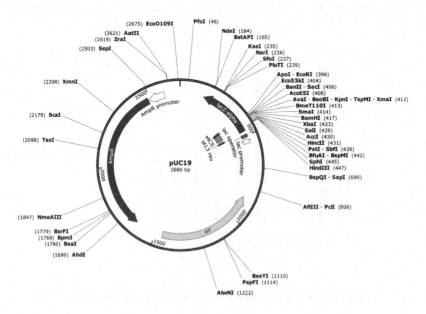

Fig. 3. View of sequenced but unsigned pUC19 in SnapGene editor

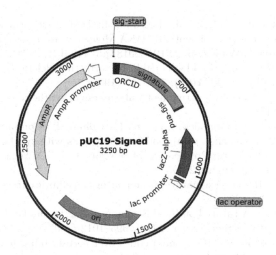

Fig. 4. View of sequenced signed pUC19 in SnapGene showing embedded signature. Note increased size of DNA

References

1. Biodefense in the Age of Synthetic Biology. National Academies of Sciences, Engineering and Medicine, Washington, D.C., June 2018
2. Damerau - Levenshtein Distance. Wikipedia, February 2019
3. Choon, J.C., Hee Cheon, J.: An identity-based signature from Gap Diffie-Hellman groups. In: Desmedt, Y.G. (ed.) PKC 2003. LNCS, vol. 2567, pp. 18–30. Springer, Heidelberg (2003). https://doi.org/10.1007/3-540-36288-6_2
4. Damerau, F.J.: A technique for computer detection and correction of spelling errors. Commun. ACM **7**(3), 171–176 (1964)
5. De Caro, A., Iovino, V.: jPBC: Java pairing based cryptography. In: Proceedings of the 16th IEEE Symposium on Computers and Communications, ISCC 2011, pp. 850–855. IEEE, Kerkyra, Corfu, Greece, 28 June–1 July 2011
6. Gibson, D.G., et al.: Creation of a bacterial cell controlled by a chemically synthesized genome. Science **329**(5987), 52–56 (2010)
7. Heider, D., Barnekow, A.: DNA-based watermarks using the DNA-Crypt algorithm. BMC Bioinf. **8**(1), 176 (2007)
8. Hutchison, C.A., et al.: Design and synthesis of a minimal bacterial genome. Science **351**(6280), aad6253 (2016)
9. Jaccard, P.: Distribution de la Flore Alpine dans le Bassin des Dranses et dans quelques régions voisines. Bulletin de la Societe Vaudoise des Sciences Naturelles **37**(140), 241–72 (1901)
10. Jaccard, P.: Etude De La Distribution Florale Dans Une Portion Des Alpes Et Du Jura. Bulletin de la Societe Vaudoise des Sciences Naturelles **37**(142), 547–579 (1901)
11. Jaro, M.A.: Advances in record-linkage methodology as applied to matching the 1985 census of Tampa, Florida. J. Am. Stat. Assoc. **84**(406), 414–420 (1989)
12. Jupiter, D.C., Ficht, T.A., Samuel, J., Qin, Q.M., de Figueiredo, P.: DNA watermarking of infectious agents: progress and prospects. PLOS Pathog. **6**(6), 1–3 (2010)

13. Kar, D.M., Ray, I., Gallegos, J., Peccoud, J.: Digital signatures to ensure the authenticity and integrity of synthetic DNA Molecules. In: Proceedings of the New Security Paradigms Workshop, NSPW 2018, pp. 110–122. ACM, Windsor (2018)

14. Levenshtein, V.I.: Binary codes capable of correcting deletions, insertions, and reversals. Sov. Phys. Dokl. **10**(8), 707–710 (1966)

15. Liss, M., et al.: Embedding permanent watermarks in synthetic genes. PLOS ONE **7**(8), 1–10 (2012)

16. Ney, P., Koscher, K., Organick, L., Ceze, L., Kohno, T.: Computer security, privacy, and DNA sequencing: compromising computers with synthesized DNA, privacy leaks, and more. In: Proceedings of the 26th USENIX Security Symposium, Vancouver, Canada, August 2017

17. Paterson, K.G.: ID-based signatures from pairings on elliptic curves. Electron. Lett. **38**(18), 1025–1026 (2002)

18. Plank, J.S., et al.: A tutorial on Reed-Solomon coding for fault-tolerance in RAID-like systems. Softw. Pract. Exper. **27**(9), 995–1012 (1997)

19. Reed, I.S., Solomon, G.: Polynomial codes over certain finite fields. J. Soc. Ind. Appl. Math. **8**(2), 300–304 (1960)

20. Richardson, S.M., et al.: Design of a synthetic yeast genome. Science **355**(6329), 1040–1044 (2017)

21. Sakai, R., Ohgishi, K., Kasahara, M.: Cryptosystems based on pairing. In: Proceedings of the 2000 Symposium on Cryptography and Information Security. Okinawa, Japan, January 2000

22. Sakai, R., Kasahara, M.: ID based cryptosystems with pairing on elliptic curve. IACR Cryptology ePrint Archive (2003)

23. Shamir, A.: Identity-based cryptosystems and signature schemes. In: Blakley, G.R., Chaum, D. (eds.) CRYPTO 1984. LNCS, vol. 196, pp. 47–53. Springer, Heidelberg (1985). https://doi.org/10.1007/3-540-39568-7_5

24. Yi, X.: An identity-based signature scheme from the Weil pairing. IEEE Commun. Lett. **7**(2), 76–78 (2003)

Mobile and Web Security

Adversarial Sampling Attacks Against Phishing Detection

Hossein Shirazi[1]([✉]), Bruhadeshwar Bezawada[2], Indrakshi Ray[1], and Charles Anderson[1]

[1] Colorado State University, Fort Collins, CO 80523, USA
{shirazi,Indrakshi.Ray}@colostate.edu, anderson@cs.colostate.edu
[2] Mahindra École Centrale, Hyderabad, Telangana, India
bru@mechyd.ac.in

Abstract. Phishing websites trick users into believing that they are interacting with a legitimate website, and thereby, capture sensitive information, such as user names, passwords, credit card numbers and other personal information. Machine learning appears to be a promising technique for distinguishing between phishing websites and legitimate ones. However, machine learning approaches are susceptible to *adversarial learning* techniques, which attempt to degrade the accuracy of a trained classifier model. In this work, we investigate the robustness of machine learning based phishing detection in the face of adversarial learning techniques. We propose a simple but effective approach to simulate attacks by generating adversarial samples through direct feature manipulation. We assume that the attacker has limited knowledge of the features, the learning models, and the datasets used for training. We conducted experiments on four publicly available datasets on the Internet. Our experiments reveal that the phishing detection mechanisms are vulnerable to adversarial learning techniques. Specifically, the identification rate for phishing websites dropped to 70% by manipulating a single feature. When four features were manipulated, the identification rate dropped to zero percent. This result means that, any phishing sample, which would have been detected correctly by a classifier model, can bypass the classifier by changing at most four feature values; a simple effort for an attacker for such a big reward. We define the concept of *vulnerability level* for each dataset that measures the number of features that can be manipulated and the cost for each manipulation. Such a metric will allow us to compare between multiple defense models.

Keywords: Phishing · Machine learning · Adversarial sampling · Classifiers

© IFIP International Federation for Information Processing 2019
Published by Springer Nature Switzerland AG 2019
S. N. Foley (Ed.): DBSec 2019, LNCS 11559, pp. 83–101, 2019.
https://doi.org/10.1007/978-3-030-22479-0_5

1 Introduction

1.1 Motivation

Phishing, as defined in [1], is an attempt to obtain sensitive information such as user-names, passwords, and credit card details by masquerading as a trustworthy entity in an electronic communication. The first recorded mention of the term is found in the hacking tool against American Online (AOL) users in the 1995 named *AOHell* while the technique was elaborated earlier in a presentation to the International HP Users group, Interix, by Felix and Hauck in 1987 [2]. Phishing attacks have shown remarkable resilience against a multitude of defensive efforts, and attackers continue to generate sophisticated phishing websites that closely mimic legitimate websites. While there were 328,000 unique attacks reported in 2007, this number almost quadrupled by 2017 [3].

Phishing, when viewed as a social-engineering attack, cannot be solved solely by educating the end users, and hence, automatic detection techniques are essential. Several defenses were proposed against phishing attacks, such as URL black-listing, keyword-based filtering, IP address filtering, and machine learning based techniques. Solutions like URL-blacklisting are no longer effective as attackers can bypass such techniques through simple URL manipulation or by hosting websites on popular free hosting services on the Internet. Machine learning based techniques appear to be a promising direction.

1.2 Problem Statement

The studies in the existing literature emphasize on feature definition or enhancing the statistical learning models to discriminate between phishing and legitimate websites. The state-of-the-art solutions for phishing detection [4–8] use engineered features based on observations made by the research experts in this domain on publicly available datasets. One crucial assumption, in existing machine learning based phishing detection approaches, is that the training data collection process is independent of the attackers' actions [9]. However, in adversarial contexts, *e.g.* phishing or spam filtering, this is far from the reality as attackers either generate noisy data samples or generate new attack samples by manipulating features of existing ones. The noisy data samples result in a classification model with low accuracy and requires a higher effort for an attacker. The manipulation of features results is a more dangerous scenario wherein an attacker can bypass an existing classifier without much effort. In this work, we explore and study the effect of adversarial sampling on phishing detection algorithms in depth, starting with some simple feature manipulation strategies, and show some surprising results that demonstrate impact on the classification accuracy with trivial feature manipulation.

1.3 Proposed Approach and Key Contributions

We gathered four separate publicly available datasets developed by other researchers and applied adversarial sampling techniques to evaluate the robustness of the trained model against artificially generated adversarial samples.

Although we do not show any solution to address this current threat, we show the vulnerability of the current approaches, and explored the robustness of the datasets against the engineered features, and the learning models. Our key contributions are as follows:

- We modeled the threat against the current defense and detection mechanism and explained the attackers' access and knowledge, which the attackers utilize to attack any given trained classifier model.
- We define the vulnerability level of phishing instances, which quantifies the attackers' efforts, and describe an approach to manipulate phishing instances and create new samples.
- We surveyed a full range of phishing detection techniques focusing on the machine learning based approaches. We showed the weakness of some well-known machine learning approaches and emphasized on how a phisher can generate new phishing website instances to evade a trained classifier in each of these approaches.
- We built an experimental setup and conducted a wide range of experiments and analyzed how vulnerable the datasets and learning model are by testing against the adversarial samples.

The rest of this paper is organized as follows. In Sect. 2, we describe a wide range of defense mechanisms against phishing attacks in the literature. Also, we describe the various adversarial attacks against the machine learning classifiers in non-phishing domains. In Sect. 3, we model the threat from three points of view: attackers' goal, knowledge, and influence. In Sect. 4, we simulated adversarial sampling attack followed by the assessing vulnerability level and quantifying the cost of the attack. In Sect. 5, we explain the results of our experiments to prove the robustness of the classifiers and datasets against these attacks. In Sect. 6, we conclude the paper and discuss some future work.

2   Related Work

2.1   Machine Learning for Phishing Detection

For phishing website detection, machine learning algorithms are well suited as they can assimilate common attack patterns such as hidden fields, keywords, and page layouts, across multiple phishing data instances and create learning models that are resilient to small variations in future unknown phishing data instances. In the prior machine learning approaches, researchers engineered novel sets of features from diverse perspectives based on public datasets of phishing and legitimate websites. While these approaches have demonstrated excellent results for detecting phishing websites, they also suffer from serious disadvantages due to adversarial sampling as we show in the following discussion.

Niakanlahiji *et al.* [4] introduced PhishMon, a scalable feature-rich framework with a series of new and existing features derived from HTTP responses, SSL certificates, HTML documents, and JavaScript files. The authors reported an accuracy of 95% on their datasets.

According to a Symantec report [10], the number of URL obfuscation based phishing attacks was up by 182.6% in 2017. Some URL obfuscation techniques used by attackers are: misspelling of the targeted domain name, using the targeted domain name in other parts of the URL like the sub-domain, adding sensitive keywords like 'login', 'secure' or 'https' etc. Sahinguz *et al.* [11] proposed a real-time detection mechanism based on Natural Language Processing (NLP) of URLs. The technique used a large dataset without requiring third-party services, and focused on features derived from URL obfuscation, and achieved an accuracy of more than 95%.

Verma *et al.* [12] defined lexical, distance, and length related features for the detection pf phishing URLs. They employed the two-sample Kolmogorov-Smirnov statistical test along with other features to detect phishing websites. They conducted a series of experiments on four large proprietary datasets and reported an accuracy of 99.3% with a false positive rate of less than 0.4%.

Jiang *et al.* [5] merged information from DNS and the URL to develop a Neural Network (DNN) with the help of NLP to detect phishing attacks. While other approaches need to specify features explicitly, this method extracts hidden features automatically. The approach relies on the information from DNS and thus, requires third-party services.

Attackers use Domain Generation Algorithms (DGA) to dynamically generate a large number of random domain names for adversarial purposes including phishing attacks. Pereira *et al.* [6] introduced an approach for detecting such domains. These domains are considered as legitimate for detection mechanisms and human analysis. The authors used a graph-based algorithm to extract the dictionaries that have been used by attackers to detect malicious domains.

While these proposed approaches are promising, they often do not considering the page content. Attackers have full control over the URL except for the Second Domain Level (SLD), and thus, they can create any URL to bypass the classifier. Also, the content of the website is the most critical factor to lure the end-users rather than the URL or domain name themselves. Therefore, any solution not considering the website content would not be useful in the real world.

Tian *et al.* [13] studied five types of domain squatting over a large DNS dataset of over 224 million registered domains. They identified 657 thousand domains that potentially targeted 702 popular websites. Using visual and Optical Character Recognition (OCR) analysis, they created a highly accurate classifier and found more than one thousand new phishing instances of which 90% of them successfully evaded well-known blacklists even after one month. The authors combined two powerful techniques: domain squatting and OCR analyses on a large dataset. The advantage of this approach is in finding new instances that evaded the current classifiers. However, there is significant cost in keeping this information current.

Shirazi *et al.* [7] observed two concerns with existing machine learning approaches: a large number of training features and bias in the type of datasets used. The study focused on the features derived from the domain name usage in phishing and legitimate websites and reported an accuracy of 97–98% on the chosen datasets. To prove the performance of the whole model, they evaluated it with unseen phishing samples from a completely different source and achieved a detection rate of 99%.

Recently, Li *et al.* [8] proposed an approach to extract the features from both URL and web page content and ran multiple machine learning techniques including GBDT, XGBoost, and LightGBM, in multiple layers, referred to as stacking approaches. The URL-based feature set includes eight features in total *e.g. using IP address, suspicious symbols, sensitive vocabulary.* The HTML based category includes features like *Alarm Window, Login Form, Length of HTML Content.* With The dataset has 20 features in total. The experiment has been conducted on three datasets, of which two are large ones with 50 K instances and the accuracy is more than 97% in all cases. Although this approach is similar to recent machine learning approaches and does not use third-party services, it is similar to other previous work like [7].

2.2   Learning in Adversarial Context

The proposed defense mechanisms in the literature widely employed machine learning techniques to counter phishing attacks. However, adversarial sampling attacks can threaten the current defense mechanisms. While there are some general analysis of the vulnerabilities of classification algorithms and the corresponding attacks [14], to the best of our knowledge, there is no other study on adversarial sampling in the context of the phishing attacks. Thus far, researchers studied and formulated these threats in a general manner or in other application contexts like image recognition. In the following, we briefly explore these efforts.

Dalvi *et al.* [9] studied the problem of adversary learning as a game between two active agents: data miner and adversary. The goal of each agent is to minimize its cost and maximize the cost to the other agent. The classifier adapts to the environment and its settings either manually or automatically in this approach. The authors assumed that both sides, including data miner and adversary, have perfect knowledge about a problem. This assumption, however, does not hold in many situations. For example, in the phishing detection system, the adversary does not know the training set or the actual classification algorithm used. In Sect. 3, we modeled the adversary and elaborated why the adversary cannot have perfect knowledge. The attackers may directly or indirectly target the vulnerabilities in the feature selection procedure. Although the attackers might target the trained classification system, it still is an indirect attack on the chosen features.

Xiao *et al.* [15] explored the vulnerabilities of feature selection algorithms under adversarial sampling attacks. They extended a previous framework [16] to investigate the robustness of three well-known feature selection algorithms.

There are few approaches that create more secure machine learning models. Designing a secure learning algorithm is one way to build a more robust classifier against these attacks. Demontis *et al.* [17] investigated a defense method that can improve the security of linear classifier by learning more evenly-distributed feature weights. They presented a secure SVM called Sec-SVM to defend against evasion attacks with feature manipulation. Wang [18] theoretically guaranteed robustness of k-nearest neighbors algorithm in the context of adversarial examples. They introduced a modified version of k-nearest neighbor classifier while k is equal to 1 and theoretically guaranteed its robustness in a large dataset.

Finally, there are some tools for bench-marking and standardizing performance of machine learning classifier against adversarial attacks in the literature. *Cleverhans* [19] is an open-source library that provides an implementation of adversarial sample construction techniques and adversarial training for image datasets. Given the lack of such bench-marking tools for the phishing problem, we tested our approach with our own attack strategies and implementation.

3   Threat Model

In this section, we model the adversarial sampling attack against machine learning based phishing detection approaches. We start with the attacker's *goal*, *knowledge*, and *influence* in general machine learning solutions, and then we explain them in the context of phishing problem. We model the adversarial sample generation for existing phishing instances based on the attacker's abilities and then evaluate the cost that the adversary has to pay for the successful execution of this attack. Finally, we define the vulnerability level for the dataset.

3.1   Attacker's Goal

Biggio *et al.* explored three different goals for the attackers namely *security violation, attack specificity,* and *error specificity* [20]. The goal of an attacker in the *security violation* is to evade well-known security metrics including availability, privacy, and integrity. The attacker may violate the availability of the system by denial-of-service attack. In this case, if the system cannot accomplish the desired task due to the attacker's behavior, the availability of the service would be affected. The attacker needs to obtain sensitive and private information of users with approaches like reverse-engineering to violate the user's privacy.

In the phishing context, the adversary will attack the integrity of the system. The integrity is violated if the attack does not violate the regular system behavior; however, the attacker violates the accuracy of the classifier *e.g.* by luring classifier to label maliciously crafted phishing instances as legitimate to evade the classifier. The attack *specificity* depends on whether an attacker wants to mis-classify a specific set of samples (like phishing) or any given sample. The error *specificity* relates to the attacker's effort to increase a specific type of error in the system and degrade other classifier scores.

In this study, we consider that the adversary desires to attack the *specificity* of the learning model. This leads to the incorrect classification of the adversarial phishing samples as legitimate and thereby, these samples will deceive the end users. Also, with respect to error *specificity*, the adversary wants to decrease the True Positive Rate (TPR).

3.2   Attacker's Knowledge

An attacker may have different levels of knowledge about the machine learning model. An attacker might have detailed knowledge, *i.e., white-box* or *perfect knowledge*, minimal knowledge about the model called *zero knowledge* [15,20] and limited knowledge about the model known as the *gray-box*. If the adversary knows everything about the learning model, parameters, and the training dataset including the classifier parameters, then the attacker has *perfect knowledge*. In the *zero knowledge*, the adversary can probe the model by sending instances and observing the results. The adversary infers information about the model by choosing appropriate data samples. In the *limited knowledge*, it is assumed that adversary knows about features and their representation, and the learning algorithm. However, the adversary does not know about the training set or the algorithm's parameters.

From the dataset point of view, the attacker may have partial or full access to the training dataset. Attacker may also have partial or full knowledge about the feature representation or feature selection algorithm and its criteria. In the worst case scenario, an attacker may know about the subset of selected features. In our study, we assumed that the adversary has *limited knowledge*. The adversary knows about the classifier model and the feature set but does not know about training set, the classifier, or classifier's training parameters.

3.3   Attacker Influence

Two major types of attacker influence have been defined in the literature namely *poisoning* and *evasion* attacks. The *poisoning* attack refers to the case where the adversary injects adversarial instances into the training phase. This injection leads to bypass data later on in the testing phase of the experiment or even at the practical usage of the model. For example, email providers use spam detection services to block emails that include a link to phishing websites. The email system gives the users ability to override the email's label *e.g.* re-labeling a spam email as non-spam to deal in cases of False Positive detection, The system benefits from user's labeling to improve the accuracy by updating the training set. However, in a *poisoning* attack, an attacker with an authorized email account in the system can re-label the correctly detected spam emails as non-spam to poison the training set of the classifier.

While the attacker does not have access to the training set in the *evasion* attack, it tries to intentionally and smartly manipulate features to avoid samples being labeled correctly by the classifier at the testing phase. Similar to the previous example on the spam detection system, a phisher may send an email with intentionally misspelled words to evade the classifier.

In this work, we assume that the attacker has a *limited knowledge* about the learning model, but has unlimited access to the *predict* function of the learning model. The attacker can test as many instances as needed and get the results. With this assumption, an attacker can create a large number of new samples and test them against the classifier to see if they can bypass the model. In the next section, we describe our adversarial sampling approach and outline our method for measuring the effectiveness of the samples in lowering the classifier's accuracy. Section 5 studies how this attack can be effective by showing the degradation of the classifier's score under this attack.

4 Adversarial Sampling for Phishing

We simulate the attacker's approach to generate new adversarial samples based
on the existing phishing instances that are detected by the classifier. The adversary generates new instances based on these current phishing instances to check
whether the generated instances are able to evade the classifier. We assume that
the attacker has full control on the URL and web page content except for the
domain name, which is unique. The attacker has *limited knowledge* about the
classifier and features, as we discussed earlier.

4.1 Defining the Dataset

We use similar notation to that used in [15]. The whole dataset has been gen-
erated by a procedure $\mathcal{P} : \mathcal{X} \mapsto \mathcal{Y}$. In the experiment, we defined two types
of instances: Legitimate (L) and Phishing (P). A learning algorithm trains from
this dataset and will label the new instances. Each instance in the dataset has
d features that are represented as a d-dimensional vector and is labeled as legitimate (L) or phishing (P).

$$x_i = [x_i^1, \cdots x_i^d]^T \in \mathcal{X} \tag{1}$$

Each instance relates to the target label of $y_i \in \mathcal{Y}$, $\mathcal{Y} \in \{0, 1\}$. We denote this
set D with n samples as follows: $\mathcal{D} = \{x_i, y_i\}_{i=1}^n$. The set $\mathcal{T}$ is the subset with t
instances that the adversary can access, $\mathcal{T} \subseteq \mathcal{D}$, $t \leq n$

4.2 Selecting Features for Manipulation

To specify a subset of features, we introduce the notation $\Phi = \{0, 1\}^d$, where
each element denotes whether the corresponding feature has been selected (1)
or not (0). The first step for creating adversarial samples is to select one or
more features for manipulation. Φ^s denote the set of all possible combinations
of s features, $\binom{n}{s}$, that have been selected and π_i^s denotes i^{th} such choice of
features. For example, $\pi_1^3 = (0, 1, 1, 1, 0)$ means that, the first combination from
Φ^s, chooses features 2, 3, and 4, for manipulation. We formalize this in Eq. 2:

$$\pi_i^s \in \Phi^s \ \ where \ i \in \binom{n}{s} \ \ and \ \sum_{i=1}^d \pi_i^s = s \tag{2}$$

Assigning new feature values is the next step after defining the subset of features for manipulation. We assumed that each feature value may be replaced
by values that appeared in existing phishing instances. The intuition is that, if
the value has been found to be assigned to that feature previously for a phishing instance, then the feature is more likely to get that value again in another
phishing instance.

Let T^i denote the set of all values that have appeared for the feature i among
the phishing instances. For example, $T^2 = \{-1, 0, 1\}$ denotes existing phishing
instances have values -1, 0, and 1 in the second feature.

For generating new instances, first, we need to generate all possible feature combinations with different lengths and then, for each combination, we need to permute all possible feature values from T^i. This process is done only for phishing instances that have been predicted correctly by the classifier. Algorithm 1 explains this process for a given phishing instance. It shows how the adversarial instances will be generated based on an original input and desired features for manipulation. There are two inputs for Algorithm 1: an original phishing instance and the selected features to manipulate, and returns as output, the new adversarial instances that have been generated. In lines 2 and 3, the algorithm loops over all of the selected features for manipulation. The algorithm gets all available values for them from the array T that is previously defined. The algorithm adds a series of available feature values to the list L. Line 5 calculates a *Product* function over array L to calculate all possible values for selected features and saves them in L_pr.

Algorithm 1. Generating the Adversarial Samples

Result: New Adversarial Samples
Input : x , *selFeatures*
Output: genSamples

1 Let L and *genSamples* be new array

```
/* Get possible values for selected features              */
```
2 **foreach** *featurePos* in *selFeatures* **do**
3 | L.append $\leftarrow T[featurePos]$

4 **end**
```
/* Product possible values to generate all combination    */
```
5 L_pr $\leftarrow product(L^*)$
```
/* Generate new instances based on new feature values     */
```
6 **foreach** *new_val* in *L_pr* **do**
```
    /* Making a temporary copy of instance                */
```
7 | temp $\leftarrow Copy(x)$
8 | **for** $k \leftarrow 0$ to $Len(new\_val)$ **do**
```
        /* Override the feature value with new value       */
```
9 | | temp$[x[k]] = $ new_val[k]
10 | **end**
```
    /* Adding the new instance to the result array         */
```
11 | genSamples.append (temp)

12 **end**
13 return genSamples

Now, each row in L_pr has the values for all of the selected features for manipulation. We make another loop over L_Pr to assign new values to the original phishing instance x. The algorithm saves them in the result array of *genSamples*.

4.3 Adversary Cost

Attackers have to handle two challenges for generating adversarial instances. From a machine learning point of view, the dataset includes vectors, but the attacker has to change the website in a way that it generates the desired vector similar to adversarial samples. This is not a trivial process, and it has considerable cost for the attacker. Whereas adversarial samples may have a higher chance of evading the classifier, but they may not be visually or functionally similar to the targeted websites. This increases the chance of being detected by the end-user. Thus, the adversary wants to minimize two parameters: the number of manipulated features and the assigned feature values. We consider this as a cost function for the adversary.

In the previous section, we discussed how the attacker controls the number of manipulated features, but it is not the only parameter. If the manipulated feature values are far from the original values, it will increase the chance of evading the classifier. We study this hypothesis in Sect. 5. But, this will also change the visual appearance or behavioral functionality of the website from the targeted website, thereby, increasing the chance of phishing website being detected by the end-user.

In this work, we used the *Euclidean distance* between the original phishing sample and newly generated sample to estimate the cost, a higher distance indicates a larger cost. Consider x_i to be a phishing instance and x_i' a manipulated one based on the original x_i instance. Both are vectors of size n. The *Euclidean distance* between x_i and x_i' will be calculated by Eq. 3:

$$d(x_i, x_i') = \sqrt{\sum_{k=1}^{n} (x_i^k - x_i'^k)^2} \tag{3}$$

If l is the number of manipulated features to generate x_i' from x_i, and d is *Euclidean distance* between them, the total cost c will be derived from this equation: $\mathcal{C}(x_i, x_i') = (l, d)$. This tuple will be used to evaluate the total cost for generating the adversarial instances.

4.4   Vulnerability Level

A phishing instance that has been predicted correctly is vulnerable at the level of l with the cost of d if there is at least one adversarial instance that can bypass with l manipulated features and distance d. We call this instance vulnerable if, by manipulating l features and with the distance of d, it can bypass the classifier. The goal of the attacker here is optimizing the l and d; a multi-objective optimization problem for the attacker. For example, if we have a phishing instance, which has been detected by the classifier, but there is the new instance generated by manipulating 3 features and a *Euclidean distance* of 2.7 bypasses the classifier, the original sample is vulnerable at the level of 3 with a cost of 2.7.

5   Experiments and Results

In this section, we show the effectiveness of our threat model and proposed adversarial sampling attack that degrades the accuracy and efficacy of existing learning models. First, we discuss the datasets utilized and then, we elaborate on three different experiments we have conducted and their results.

5.1 Used Datasets

We obtained four publicly available phishing datasets on the Internet and the details of these datasets are given below.

Dataset 1: DS-1: This set includes 1000 legitimate websites from Alexa.com and 1200 phishing websites from PhishTank.com; 2200 in total. Each instance in this dataset has eight features and all are related to the domain name of the websites. The features used are domain length, presence of non-alphabetic character in the domain name, the ratio of hyperlinks referring to the domain name, the presence of HTTPS protocol, matching domain name with copyright logo, and matching domain name with the page title. With these features, Shirazi *et al.* [7] reported an ACC of 97–98% in the experiments, which is significantly high.

Dataset 2: DS-2: Rami *et al.* [21] created this dataset in 2012 and shared it with UCI machine learning repository [22]. This set includes 30 features and are categorized into five categories: *URL based, abnormal based, HTML-based, JavaScript based,* and *domain-based* features. This dataset includes 4898 legitimate instances from Alexa.com merged with 6158 phishing instances from PhishTank.com; more than 11000 in total making it the most extensive dataset that we have used in this study.

Dataset 3: DS-3: In 2014, Abdelhamid *et al.* [23] shared their dataset on UCI machine learning repository [22]. This dataset includes 651 legitimate websites and 701 phishing websites; 1352 instances in total and includes ten features combination of third-party services and HTML based features for each instance. Authors report an ACC between 90%–95% in their experiments.

Dataset 4: DS-4: This dataset is the most recent dataset publicly available in the literature that we could find and was published in 2018. It has been created by Tan *et al.* [24] and was published on Mendeley[1] dataset library. This set contains 5000 websites from Alexa.com and as well as those obtained by web crawling, labeled as legitimate, and 5000 phishing websites from PhishTank.com and OpenPhish.com. The authors collected this data from January to May 2015 and from May to June 2017. This dataset includes 48 features, a combination of URL-based, and HTML-based features. While this dataset includes the URL length and it may bias the dataset as Shirazi *et al.* [7] explained it, but we kept all features.

Table 1 summarises the number of instances, features, and the portion of legitimate vs phishing instances in each dataset. We have a dataset with a large number of instances, DS-2, and DS-4 with 11000 and 10000 respectively. We also have small dataset DS-3 with 1250 instances. With respect to the number

---

[1] https://data.mendeley.com/.

Table 1. Number of instances, features, and portion of legitimate and phishing websites in each dataset

Dataset	Data shape (#)		Instances (%)	
	Size	Features	Legitimate	Phishing
DS-1	2210	7	44.71	55.29
DS-2	11055	30	55.69	44.31
DS-3	1250	9	43.84	56.16
DS-4	10000	48	50.0	50.0

of features, DS-1, with just seven features is a dataset with a limited number of features and in comparison, DS-4 with 48 features is a large dataset. Also, we used an unbiased dataset like DS-1 and used DS-4 as well though it may be biased concerning some of the features like URL length. Besides, the features in each dataset are selected from different points of view such as URL-based features in DS-2, DS-3, and DS-4, or domain-related features in DS-1, and HTML-Based features in DS-2 and DS-4. These variations validate our hypothesis in a stronger and more general sense. Also, it shows that adversarial sampling is a serious problem that may be happening in different situations and needs to be addressed.

5.2 Exp-1: Evaluation of Datasets

In the first experiment, we tested the performance of each dataset against a wide range of classifiers. The experiment is as follows. We labeled phishing websites in all datasets as +1 and legitimate websites as −1. We used five-fold cross-validation to avoid issues of over-fitting and to test the performance of the learning model against unknown data instance classification. We used six different classifiers namely Decision Tree Decision Tree (DT), Gradient Boosting (GB), Random Forest (RF), K-Nearest Neighbors (KNN), and Support Vector Machine (SVM) with two different kernels: Linear (lin) and Gaussian (rbf) to make the comparison between classifiers. We repeated each experiment 10 times and reported the average and standard deviation of the results. Table 2 explains the achieved results in this experiment.

For DS-1, RF and GB both generate the highest ACC and the TPR in each classifier is almost the same. Also, DS-1 has the best average of TPR among all classifiers. This means that, despite different classifiers, the features are well-defined. RF gives the best TPR (94.25%) and ACC (95.76%) on DS-2. Interestingly, the DT does not generate a good TPR (86.77%).

The experiments on DS-3 dataset did not yield a high TPR or the ACC. Both GB and SVM with Gaussian kernel has the TPR of 87%, which are not that much good. The best ACC, for this dataset, is from GB, with 83%. The experiment on DS-3 gave very good results. Both GB and RF gave a TPR over 97% and accuracy of 97%, which are very high. Also, this dataset has the best average of ACC among different classifiers meaning this dataset performs

Table 2. Evaluation of model against different classifiers with two metrics.

(a) TPR

Cls.	DS-1	DS-2	DS-3	DS-4	Avg.
DT	95.25	86.77	84.97	96.14	**95.25**
GB	96.18	92.25	87.23	97.65	**96.18**
KNN	95.93	90.61	84.95	93.97	**95.93**
RF	96.25	94.25	85.84	97.85	**96.25**
SVM(l)	95	89.62	86.71	94.93	**95**
SVM(r)	93.67	91.88	87.88	95.69	**93.67**
Best	**96.25**	**94.25**	**87.88**	**97.85**	

(b) ACC

Cls.	DS-1	DS-2	DS-3	DS-4	Avg.
DT	94.8	92.1	82.51	95.73	**91.29**
GB	95.49	94.32	83.76	97.52	**92.77**
KNN	94.82	92.21	81.16	93.76	**90.49**
RF	95.35	95.76	82.89	97.8	**92.95**
SVM(l)	93.96	92.4	79.16	94.38	**89.98**
SVM(r)	93.96	94.14	82.4	95.2	**91.43**
Best	**95.49**	**95.76**	**83.76**	**97.8**	

Table 3. The classifier that holds best f1 on each dataset has been selected. TPR and ACC are also reported for comparison

Metric	DS-1	DS-2	DS-3	DS-4
Best classifier	**GB**	**RF**	**GB**	**RF**
Best f1	**95.94**	**95.17**	**85.83**	**97.8**
TPR	96.18	94.25	87.23	97.85
ACC	95.49	95.76	83.76	97.8

very well with different types of classifiers. With six different classifiers, the experiments on both DS-1 and DS-4 show an average ACC of more than 94%, which is significantly high. This confirms that these datasets are well-defined and have a good set of distinguishing features.

We used a single metric of f1 to compare all classifiers and datasets together. Table 3 shows the best f1 score for each dataset with the classifier that has produced that result. It is evident from this table that both GB and RF generate the best results among all of the experiments, so we selected these two classifiers for the next experiments.

5.3 Generating Adversarial Samples

In each dataset, we reserved 200 phishing instances and then trained the model without the 200 reserved phishing instances. The generated adversarial samples need to be similar and valid to the phishing examples; otherwise, those cannot be assumed to be phishing instances. To assign new values to the features and generate new instances, we just used previously seen values in the phishing instances. With this strategy, it is guaranteed that the newly assigned value is valid and has already been seen in other phishing instances in the dataset. We discussed this process earlier in Sect. 4. We randomly selected combination of features, up to four different features, and changed the values of each feature with all possible feature values.

After creating each new sample, we tested our new sample against the selected classifier and checked whether it could bypass the classifier or not. If it did, we

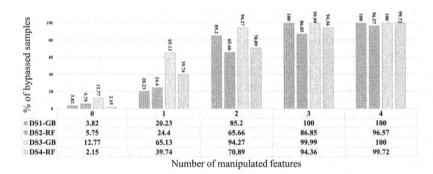

	0	1	2	3	4
DS1-GB	3.82	20.23	85.2	100	100
DS2-RF	5.75	24.4	65.66	86.85	96.57
DS3-GB	12.77	65.13	94.27	99.99	100
DS4-RF	2.15	39.74	70.89	94.36	99.72

Number of manipulated features

Fig. 1. Robustness of datasets against adversarial samples

consider the original phishing instance to be a vulnerable instance. We calculate the distance between the new instance and the original one to find the closest instance that can bypass the classifier.

5.4   Exp-2: Robustness of Learning Model

This experiment studies the robustness of datasets and learning models against generated adversarial samples. We selected one classifier that performs best for each dataset based on the f1 score from Table 3. For the datasets DS-1 and DS-3, we selected GB and RF for DS-2 and DS-4.

In this experiment, we counted the number of reserved phishing instances that are vulnerable. This means that, there should be at least one optimized manipulated instance based on the original sample that can bypass the classifier. With small perturbation on these instances, they can bypass the classifier and elude the users to release their critical information. Based on our hypothesis, these are vulnerable instances and can be assumed as a threat to the learning model. We repeated each experiment ten times and reported the average of the results.

Figure 1 shows the results of Exp-2. While the x-axis shows the number of manipulated features, zero manipulated feature means that the test happened with the original phishing instances without any perturbation. The trend of results reveals that by increasing the number of perturbation, the number of evaded samples increase proportionally. We continued increasing the perturbed features for up to four different features at a time. We observed that with four features, almost all phishing instances bypass the classifier model.

For example, Fig. 1 shows that less than 4% of phishing instances in DS-1 can bypass the classifier without any perturbation. With only one manipulated feature, more than 20% of phishing instances can bypass the classifier. With two manipulated features, almost all of instances can bypass the GB. The results are almost the same for other datasets. In another case, while just 12% of original phishing instances (the instances without any changes) have been misclassified in DS-3, the results significantly go up to 65% with only one perturbed feature.

This experiment shows how vulnerable the machine learning models are for the phishing problem. Small perturbation on features can bypass the classifier and degrade the accuracy significantly.

5.5 Exp-3: Dataset Vulnerability Level

In this experiment, we studied the cost that an adversary has to pay to bypass a classifier. From an adversary point of view, it is not inexpensive to manipulate an instance with new feature values and bypass the classifier. In Sect. 4.3, we assessed the cost and in Sect. 4.4, we defined the term *vulnerability level* for one instance. Similar to previous experiment 5.4, we reserved 200 phishing instances from each dataset and chose the classifier for each dataset based on Table 3. For datasets DS-1 and DS-3, we chose GB while we chose RF for both DS-2 and DS-4 datasets. Averaging the *vulnerability level* for each of the 200 selected instances and repeating the experiment ten times, we assessed the vulnerability level for the whole dataset.

Figure 2 presents the results of this experiment for all datasets for two parameters: the number of manipulated features and the average cost of adversarial instances. It is evident that, by increasing the number of manipulated features, the cost also increases steadily. For example, for the dataset DS-1, the average cost, for adversarial samples, with one manipulated feature is 0.95 and with four manipulated features the cost is 3.93.

Furthermore, it is clear that the average cost for some datasets is more than that of other datasets. For example, in the DS-4, the adversary has to pay more cost especially when the number of features increases to three and four in comparison to the other datasets. This shows that this dataset is more robust against these attacks and has a lower vulnerability level. Also, it is clear that with one single feature manipulation with a small cost, it is possible to bypass a classifier. This needs to be considered when a dataset and features are designed.

5.6 Comparing the Results with Previous Experiments

In this section, we compare our approach with some of the previous researches in this field. Table 4 compared nine different approaches in the literature. We summarized the pros and cons of each approach and show the dataset size and best accuracy results of each approach. We studied a wide range of previous efforts by focusing on machine learning techniques. Some of the techniques solely focused on the URL itself [11,13] but others look at both URL and the content of the page [7,25]. The use of third-party services is another difference between approaches. While using third-party services like search or DNS inquires leverage the feature set and make the feature set more reliable it also endangers the privacy of the users. Third-party enquiries to fetch the feature value reveal the browsing history of the end-users The previous studies have been done on variable sizes of datasets. While some of the datasets have less than 5 thousand records [7,25], there are also datasets with millions of instances [5,13]. Also, for approaches analyzing just the URL without the webpage content, creating

Table 4. Comparisons of different approaches in the literature including our proposed approach

Author	Description	Size	ACC
Niakanlahiji et al. [4]	-Scalable feature-rich framework with a series of new and existing features -Not using third-party services, Language agnostic	22.3K	95%
Sahinguz et al. [11]	-Real-time detection mechanism based on NLP of URLs, Language independent -Tested on a large dataset, Not using third-party service	73K	97%
Verma et al. [12]	-Features based on lexical-, distance-, and length-related features of the URL -Using four large datasets	115K	99.3%
Jiang et al. [5]	-Combined the URL and DNS information, Used a deep neural network with the help of NLP, Automatically extracts hidden features	7M	96%
Tian et al. [13]	-Studied five types of domain squatting, Using dataset of over 224 million registered domains, Using visual and OCR analysis, Found new phishing instances that evaded common blacklist	234M	N/A
Pereira et al. [6]	-Detecting algorithmically generated domain, Graph-based algorithm to extract the dictionaries that are being used to generate algoritmically domains	80K	99%
Shirazi et al. [7]	-Studying limitation current approaches: large number of features and bias in the datasets, Focused on the domain name, Running at the client-side -Not using third-party services	2.2 K	97–98%
Li et al. [8]	-Extract the features from both URL and HTML of the page -Not using third-party services	50K	97%
Bulakh et al. [25]	-Companies can define their phishing detection mechanism and protect the customers -Can be used as an complimentary service besides other detection approaches	1.3K	96.34%
Our work	-Evaluate the performance of existing datasets including [7,21,23,24] -Using multiple classifiers and comparing the results	2–10K	81–95%
Our work	-Proposing adversarial sampling attack against the learning model, Showing the feasibility of the attack, Prove the vulnerability of current model, Modeling the vulnerability level and cost	2–10K	0%

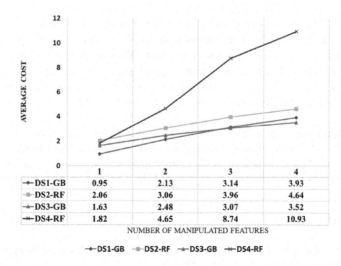

	1	2	3	4
DS1-GB	0.95	2.13	3.14	3.93
DS2-RF	2.06	3.06	3.96	4.64
DS3-GB	1.63	2.48	3.07	3.52
DS4-RF	1.82	4.65	8.74	10.93

NUMBER OF MANIPULATED FEATURES

DS1-GB   DS2-RF   DS3-GB   DS4-RF

Fig. 2. The manipulation cost for adversarial samples based on number of manipulated features

massive datasets are easier. Most of the approaches achieved high accuracy of over the 95%. Both [6,12] achieved accuracy of 99%, which is significantly high. Tian *et al.* [13] found new phishing samples that were not detected by common phishing detection mechanisms even after one month. We also added the results of this study to Table 4. We trained the classifier on the four public datasets and achieved very high accuracy. When we added the manipulated features in the testing phase, the accuracy degraded significantly and finally became zero. These experiments prove that our proposed attack is sufficient to evade existing classifiers for phishing detection.

6   Conclusion and Future Work

In this work, we explained the limitation of machine learning techniques when adversarial samples are taken into consideration. We introduced the notion of vulnerability level for data instances and datasets based on the adversarial attacks and quantified it. We achieved high accuracy in the absence of this attack using seven different well-studied classifiers in the literature: more than 95% for all classifiers except one that had 82%. However, when we evaluated the best-performing classifier against the adversarial samples, the performance of the classifier degraded significantly. With only one feature perturbation, the TPR falls from 82–97% to 79%-45% and, increasing the number of perturbed features to four, the TPR fell to 0%, meaning that all of the phishing instances were able to bypass the classifier. We continued our experiments by considering the adversary cost in the experiment. We showed that both the number of manipulated features and the total manipulation cost, which can be derived from the difference between original phishing sample and the adversarial sample, are essential. This means that from an attacker point of view, not only changing the minimum number of instances is desired, but also it is important that the adversarial sample has the minimum cost. This is an impressive result and shows the weakness of well-known defense mechanisms against phishing attacks. In future, we want to design robust machine learning models that are immune to adversarial learning attacks.

Acknowledgements. This work is supported in part by funds from NSF Awards CNS 1650573, CNS 1822118 and funding from CableLabs, Furuno Electric Company, SecureNok, and AFRL. Research findings and opinions expressed are solely those of the authors and in no way reflect the opinions of the NSF or any other federal agencies.

References

1. Zhang, Y., Xiao, Y., Ghaboosi, K., Zhang, J., Deng, H.: A survey of cyber crimes. Secur. Commun. Netw. **5**, 422–437 (2012)
2. Felix, J., Hauck, C.: System security: a hacker's perspective. Interex Proc. **1**, 6–6 (1987)
3. APWG: Phishing attack trends report - 3q 2018 (2018). Accessed 24 Jan 2019
4. Niakanlahiji, A., Chu, B.-T., Al-Shaer, E., PhishMon: a machine learning framework for detecting phishing webpages. In: Intelligence and Security Informatics, pp. 220–225 (2018)
5. Jiang, J., et al.: A deep learning based online malicious URL and DNS detection scheme. In: Security and Privacy in Communication Systems, pp. 438–448 (2017)
6. Pereira, M., Coleman, S., Yu, B., DeCock, M., Nascimento, A.: Dictionary extraction and detection of algorithmically generated domain names in passive DNS traffic. In: Bailey, M., Holz, T., Stamatogiannakis, M., Ioannidis, S. (eds.) RAID 2018. LNCS, vol. 11050, pp. 295–314. Springer, Cham (2018). https://doi.org/10.1007/978-3-030-00470-5_14
7. Shirazi, H., Bezawada, B., Ray, I.: "Kn0w Thy Doma1n Name": unbiased phishing detection using domain name based features. In: Access Control Models and Technologies, pp. 69–75 (2018)
8. Li, Y., Yang, Z., Chen, X., Yuan, H., Liu, W.: A stacking model using URL and HTML features for phishing webpage detection. Future Gener. Comput. Syst. **94**, 27–39 (2019)
9. Dalvi, N., Domingos, P., Sanghai, S., Verma, D., et al.: Adversarial classification. In: International Conference on Knowledge Discovery and Data Mining, pp. 99–108 (2004)
10. ISTR Internet Security Threat Report. Technical report, vol. 23
11. Sahingoz, O.K., Buber, E., Demir, O., Diri, B.: Machine learning based phishing detection from URLs. Expert Syst. Appl. **117**, 345–357 (2019)
12. Verma, R., Dyer, K.: On the character of phishing URLs: accurate and robust statistical learning classifiers. In: Data and Application Security and Privacy, pp. 111–122 (2015)
13. Tian, K., Jan, S.T.K., Hu, H., Yao, D., Wang, G.: Needle in a haystack: tracking down elite phishing domains in the wild. In: Internet Measurement Conference, pp. 429–442 (2018)

14. Huang, L., Joseph, A.D., Nelson, B., Rubinstein, B.I.P., Tygar, J.D.: Adversarial machine learning. In: ACM Workshop on Security and Artificial Intelligence, pp. 43–58 (2011)
15. Xiao, H., Biggio, B., Brown, G., Fumera, G., Eckert, C., Roli, F.: Is feature selection secure against training data poisoning? In: International Conference on Machine Learning, pp. 1689–1698 (2015)
16. Biggio, B., Fumera, G., Roli, F.: Security evaluation of pattern classifiers under attack. IEEE Trans. Knowl. Data Eng. **26**, 984–996 (2014)
17. Demontis, A., et al.: Yes, machine learning can be more secure! a case study on android malware detection. Depend. Secure Comput. (2017)
18. Wang, Y., Jha, S., Chaudhuri, K.: Analyzing the robustness of nearest neighbors to adversarial examples. In: International Conference on Machine Learning, pp. 5120–5129 (2018)
19. Papernot, N., Goodfellow, I., Sheatsley, R., Feinman, R., McDaniel, P.: cleverhans v1. 0.0: an adversarial machine learning library. arXiv preprint arXiv:1610.00768, October 2016
20. Biggio, B., Roli, F.: Wild patterns: ten years after the rise of adversarial machine learning. arXiv preprint arXiv:1712.03141 (2017)
21. Mohammad, R.M., Thabtah, F., McCluskey, L.: An assessment of features related to phishing websites using an automated technique. In: Internet Technology and Secured Transactions, pp. 492–497 (2012)
22. Dheeru, D., Taniskidou, E.K.: UCI machine learning repository (2017)
23. Abdelhamid, N., Ayesh, A., Thabtah, F.: Phishing detection based associative classification data mining. Expert Syst. Appl. **41**(13), 5948–5959 (2014)
24. Tan, C.L.: Phishing dataset for machine learning: feature evaluation (2018)
25. Bulakh, V., Gupta, M.: Countering phishing from brands' vantage point. In: International Workshop on Security and Privacy Analytics, pp. 17–24 (2016)

Is My Phone Listening in? On the Feasibility and Detectability of Mobile Eavesdropping

Jacob Leon Kröger[1,2(✉)] and Philip Raschke[1]

[1] Technische Universität Berlin, Berlin, Germany
{kroeger, philip. raschke}@tu-berlin. de
[2] Weizenbaum Institute for the Networked Society, Berlin, Germany

Abstract. Besides various other privacy concerns with mobile devices, many people suspect their smartphones to be secretly eavesdropping on them. In particular, a large number of reports has emerged in recent years claiming that private conversations conducted in the presence of smartphones seemingly resulted in targeted online advertisements. These rumors have not only attracted media attention, but also the attention of regulatory authorities. With regard to explaining the phenomenon, opinions are divided both in public debate and in research. While one side dismisses the eavesdropping suspicions as unrealistic or even paranoid, many others are fully convinced of the allegations or at least consider them plausible. To help structure the ongoing controversy and dispel misconceptions that may have arisen, this paper provides a holistic overview of the issue, reviewing and analyzing existing arguments and explanatory approaches from both sides. Based on previous research and our own analysis, we challenge the widespread assumption that the spying fears have already been disproved. While confirming a lack of empirical evidence, we cannot rule out the possibility of sophisticated large-scale eavesdropping attacks being successful and remaining undetected. Taking into account existing access control mechanisms, detection methods, and other technical aspects, we point out remaining vulnerabilities and research gaps.

Keywords: Privacy · Smartphone · Eavesdropping · Spying · Listening · Microphone · Conversation · Advertisement

1 Introduction

Smartphones are powerful tools that make our lives easier in many ways. Since they are equipped with a variety of sensors, store large amounts of personal data and are carried throughout the day by many people, including in highly intimate places and situations, they also raise various privacy concerns.

One widespread fear is that smartphones could be turned into remote bugging devices. For years, countless reports have been circulating on the Internet from people who claim that things they talked about within earshot of their phone later appeared in targeted online advertisements, leading many to believe that their private conversations must have been secretly recorded and analyzed.

S. N. Foley (Ed.): DBSec 2019, LNCS 11559, pp. 102–120, 2019.
https://doi.org/10.1007/978-3-030-22479-0_6

The reported suspicious ads range across many product and service categories, including clothing, consumer electronics, foods and beverages, cars, medicines, holiday destinations, sports equipment, pet care products, cosmetics, and home appliances – and while some of these ads were described as matching an overall discussion topic, others allegedly promoted a brand or even a very specific product mentioned in a preceding face-to-face conversation [6, 12]. Some people claim to have experienced the phenomenon frequently and that they have successfully reproduced it in private experiments. Interestingly, many of the purported witnesses emphasize that the advertised product or service seems not related to places they have visited, terms they have searched for online, or things they have mentioned in text messages, emails or social media [6, 40]. Furthermore, some reports explicitly rate it as unlikely that the respective advertisements were selected by conventional targeting algorithms, as they lay notably outside the range of advertising normally received and did sometimes not even appear to match the person's consumer profile (e.g. in terms of interests, activities, age, gender, or relationship status) [6, 41].

Numerous popular media outlets have reported on these alleged eavesdropping attacks [3]. In a Forbes article, for instance, the US-based market research company Forrester reports that at least 20 employees in its own workforce have experienced the phenomenon for themselves [40]. The same holds true for one in five Australians, according to a recent survey [38]. Even the US House Committee on Energy and Commerce has started to investigate the issue by sending letters to Google and Apple inquiring about the ways in which iOS and Android devices record private conversations [77].

Many commentators, including tech bloggers, researchers and business leaders, on the other hand, view the fear that private companies could target their ads based on eavesdropped conversations as baseless and paranoid. The reputational risk, it is argued, would be far too high to make this a viable option [76]. With regard to CPU, battery and data storage limitations, former Facebook product manager Antonio García Martínez even considers the alleged eavesdropping scenario to be economically and technically unfeasible [51]. As an alternative explanation for suspiciously relevant ads, he points to the many established and well-documented methods that companies successfully use to track, profile and micro-target potential customers. Yet another possible explanation states that the frequently reported phenomenon is merely a product of chance, potentially paired with some form of confirmation bias [41]. Finally, some commentators also suggest that topics of private conversations are sometimes inspired by unconsciously processed advertisements, which may later cause the perception of being spied upon when the respective ad is encountered again [28].

Many views, theories and arguments have been put forward in attempt to explain the curious phenomenon, including experimental results and positions from the research community. However, a consensus has not yet been reached, not even regarding the fundamental technical feasibility of the alleged eavesdropping attacks. Therefore, this paper reviews, verifies and compares existing arguments from both sides of the discourse. Apart from providing a structured overview of the matter, conclusions about the feasibility and detectability of smartphone-based eavesdropping are drawn based on existing research and our own analysis.

In accordance with the reports found on the phenomenon, this paper will focus on smartphones – specifically, iOS and Android devices. Since smartphones are the most widespread consumer electronics device, and since iOS and Android together clearly dominate the mobile OS market [70], this choice seems justified to us. However, most of the considerations in this paper are applicable to other types of mobile devices and other operating systems as well.

The remainder of this paper is structured as follows. In Sect. 2, we describe the underlying threat model, distinguishing between three possible adversaries. Section 3 examines the possibility of using smartphone microphones for stealthy eavesdropping, expanding on aspects of security permissions and user notifications. Similarly, Sect. 4 considers smartphone motion sensors as a potential eavesdropping channel, taking into account sampling frequency limits enforced by mobile operating systems. Section 5 then looks into the effectiveness of existing mitigation and detection techniques developed by Google, Apple, and the global research community. In Sect. 6, the ecosystem providers themselves are considered as potential adversaries. Section 7 evaluates the technical and economic feasibility of large-scale eavesdropping attacks. After that, Sect. 8 examines ways in which governmental and criminal hackers can compromise the speech privacy of smartphone users. Finally, Sect. 9 provides a discussion of analysis results, followed by a conclusion in Sect. 10.

2   Threat Model

To target advertisements based on smartphone eavesdropping, an organization A, who is responsible for selecting the audience for certain online ads (either the advertiser itself or a contractor entrusted with this task, such as an advertising network[1]), needs to somehow gain access to sensor data[2] from the corresponding mobile device, or to information derived from the sensor data.

Initially, speech is recorded through the smartphone by an actor B, which could be either (1) the operating system provider itself, e.g. Apple or Google, (2) non-system apps installed on the device, or (3) third-party libraries[3] included in these apps. Potentially after some processing and filtering, which can happen locally on the device or on remote servers, actor B shares relevant information extracted from the recording – directly or through intermediaries – with organization A (unless A and B are one and the same actor, which is also possible).

Organization A then uses the received information to identify the smartphone owner as a suitable target for specific ads and sends a corresponding broadcast request to an ad publisher (organization A could also publish the ads itself if it has access to ad distribution channels). Finally, the publisher displays the ads on websites or apps – either on the smartphone through which the speech was recorded or on other devices

---

[1] Advertising networks are companies that match demand and supply of online ad space by connecting advertisers to ad publishers. They often hold extensive amounts of data on individual internet users to enable targeted advertising [17].

[2] "sensor data" can refer to either audio recordings or motion sensor data (see Sects. 3, 4).

[3] The role and significance of third-party apps will be further explained in Sect. 3.1.

that can be linked[4] to the smartphone owner, for example through logins, browsing behavior, or IP address matching. The websites and apps on which the advertisements appear do not reveal who recorded the smartphone owner's speech. Not even organization A necessarily understands how and by whom the received profiling information was initially collected. For illustration, Fig. 1 presents a simplified overview of the threat model.

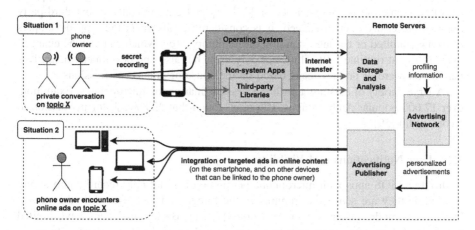

Fig. 1. A schematic and simplified overview of the threat model.

3 Microphone-Based Eavesdropping

Modern smartphones have the capability to tape any sort of ambient sound through built-in microphones, including private conversations, and to transmit sensitive data, such as the recording itself or information extracted from recorded speech, to remote servers over the Internet. Mobile apps installed on a phone could exploit these capabilities for secret eavesdropping. Aspects concerning app permissions and user notifications that could affect the feasibility and visibility of such an attack are examined in the following two subsections.

3.1 Microphone Access Permission

Before an app can access microphones in Android and iOS devices, permission has to be granted by the user. However, people tend to accept such requests blindly if they are interested in an app's functionality [10]. A survey of 308 Android users found that only 17% of respondents paid attention to permissions during app installation, and no more than 3% of the participants correctly answered the related comprehension questions [24].

---

[4] For more information on cross-device tracking, refer to [65].

Encouraging app development at the expense of user privacy, current permission systems are much less strict than they were in early smartphones and have been criticized as "coarse grained and incomplete" [59]. Also, once a permission is granted, it is usually not transparent for users when and for which particular purpose data is being collected and to which servers it is being sent [62].

To include analytics and advertising capabilities, apps commonly make use of third-party libraries, i.e., code written by other companies. These libraries share multimedia permissions, such as microphone access, with their corresponding host app and are often granted direct Internet access [39]. Apart from the concern that third-party libraries are easily over-privileged, it is considered problematic that app developers often have limited or no understanding of the library code, which can also be changed dynamically at runtime [59]. Thus, not only users but also app developers themselves may be unaware of privacy leaks based on the abuse of granted permissions.

A large variety of existing apps has access to smartphone microphones. Examining over 17.000 popular Android apps, Pan et al. found that 43.8% ask for permission to record audio [59].

3.2 User Notifications and Visibility

Android and iOS apps with microphone permission can not only record audio at any time while they are active, i.e. running in the foreground, but also while they are in background mode, under certain conditions [7, 31]. Background apps have limited privileges and are often suspended to conserve the device's limited resources. In cases, however, where they request the system to stay alive and continue recording while not in the foreground, there are ways to indicate this to the user.

In iOS, the status bar will automatically turn bright red when recording takes place in the background, allowing the user to immediately detect potentially unwanted microphone activity [8].

While the latest release of Android (version 9 Pie) implements similar measures [31], some older versions produce no visible indication when background apps access the microphone [10]. In this context, it might be worth noting that Android has been widely criticized for its slow update cycle, with hundreds of millions of devices running on massively outdated versions [56]. Also, quite obviously, notifications in the graphical user interface are only visible as long as the device's screen is not turned off. And finally, some experimenters have already succeeded in circumventing the notification requirements for smartphone media recordings [69].

4 Motion Sensor-Based Eavesdropping

Adversaries might be able to eavesdrop on conversations through cell phones without accessing the microphone. Studies have shown that smartphone motion sensors – more specifically, accelerometers and gyroscopes – can be sensitive enough to pick up sound vibrations and possibly even reconstruct speech signals [36, 54, 79].

4.1 Experimental Research Findings

There are opposing views on whether non-acoustic smartphone sensors capture sounds at normal conversational loudness. While Anand and Saxena did not notice an apparent effect of live human speech on motion sensors in several test devices [3], other studies report very small but measurable effects of machine-rendered speech, significant enough to reconstruct spoken words or phrases [54, 79].

Using only smartphone gyroscopes, researchers from Israel's defense technology group Rafael and Stanford University were able to capture acoustic signals rich enough to identify a speaker's gender, distinguish between different speakers and, to some extent, track what was being said [54]. In a similar experiment, Zhang et al. demonstrated the feasibility of inferring spoken words from smartphone accelerometer readings in real-time, even in the presence of ambient noise and user mobility [79]. According to their evaluation, the achieved accuracies were comparable to microphone-based hotword detection applications such as Samsung S Voice and Google Now.

Both [79] and [54] have notable limitations. First of all, their algorithms were only able to detect a small set of predefined keywords instead of performing full speech recognition. Also, the speech in both experiments was produced by loudspeakers or phone speakers, which may result in acoustic properties different from live human speech. In [54], the playback device and the recording smartphone even shared a common surface, leading critics to suggest that the observed effect on sensor readings was not caused by aerial sound waves, but rather by direct surface vibrations [3]. Also, in contrast to Zhang et al., this approach only achieved low recognition accuracies, particularly for speaker-independent hotword detection. By their own admission, however, the authors of [54] are "security experts, not speech recognition experts" [32]. Therefore, the study should be regarded as an initial exploration rather than a perfect simulation of state-of-the-art spying techniques. With regard to the effectiveness of their approach, the researchers pointed out several possible directions for future improvement.

It might also be noteworthy that patents have already been filed for methods to capture acoustic signals through motion sensors, including a "method of detecting a user's voice activity using an accelerometer" [21] and a "system that uses an accelerometer in a mobile device to detect hotwords" [55].

4.2 Sampling Frequency Limits

In order to limit energy consumption and because typical applications of smartphone motion sensors do not require highly sampled data, current mobile operating systems impose a cap on the sampling frequency of motion sensors, such as a maximum of 200 Hz for accelerometer readings in Android [3] and 100 Hz for gyroscopes in iOS [32]. For comparison, the fundamental frequency of the human speaking voice typically lies between 85 Hz and 155 Hz for men and 165 Hz and 255 Hz for women [79]. Thus, if at all, non-acoustic smartphone sensors can only capture a limited range of speech sounds, which presents a challenge to speech reconstruction attacks.

With the help of the aliasing effect explained in [54], however, it is possible to indirectly capture tones above the enforced frequency limits. Furthermore, experiments show that motion sensor signals from multiple co-located devices can be merged to obtain a signal with increased sampling frequency, significantly improving the effectiveness of speech reconstruction attacks [36]. Two or more smartphones that are located in proximity to each other and whose sensor readings are shared – directly or indirectly – with the same actor may therefore pose an increased threat to speech privacy.

It should also be noted that motion sensors in smartphones are usually capable of delivering much higher sampling frequencies (often up to 8 kHz) than the upper bounds prescribed by mobile operating systems [3]. Researchers already expressed concern that adversaries might be able to override and thereby exceed the software-based limits through patching applications or kernel drivers in mobile devices [3, 54].

4.3 Sensor Access Permissions and Energy Efficiency

While certain hardware components, such as camera, microphone and the GPS chip, are typically protected by permission mechanisms in mobile operating systems, motion sensors can be directly accessed by third-party apps in iOS and Android without any prior notification or request to the user [32, 45]. Thus, there is usually no way for smartphone owners to monitor, let alone control when and for what purposes data from built-in accelerometers and gyroscopes is collected. Even visited websites can often access smartphone motion sensors [32]. Exploiting accelerometers and gyroscopes to intrude user privacy is also much more energy-efficient and thus less conspicuous than recording via microphone [79].

5 Existing Mitigation and Detection Techniques

Many methods are applied by ecosystem providers and security researchers to screen mobile apps for vulnerabilities and malicious behavior. The following two subsections examine existing efforts with regard to their potential impact on the feasibility and detectability of mobile eavesdropping attacks.

5.1 App Inspections Conducted by Ecosystem Providers

Both iOS and Android apply a combination of static, dynamic and manual analysis to scan new and existing apps on their respective app market for potential security threats and to ensure that they operate as advertised [78]. Clearly, as the misbehavior of third-party apps can ultimately damage their own reputation, the platforms have strong incentives to detect and prevent abuse attempts.

Nevertheless, countless examples of initially undetected malware and privacy leaks have shown that the security screenings provided by Google and Apple are not always successful [19]. Google Play's app inspection process has even been described as "fundamentally vulnerable" [29]. In a typical cat-and-mouse game, malicious apps evolve quickly to bypass newly implemented security measures [63], sometimes by using "unbearably simple techniques" [29]. In Android devices from uncertified manufacturers, malware may even be pre-installed before shipment [14]. Significant vulnerabilities have also been found in official built-in apps. Apple's FaceTime app, for example, allowed potential attackers to gain unauthorized access to iPhone cameras and microphones without any requirement of advanced hacking skills [15].

Leaving security loopholes aside, the existing security mechanisms do not guarantee privacy protection in terms of data minimization and transparency. Many mobile apps collect personal data with no apparent relevance to the advertised functionality [18, 62]. Even well-known apps like Uber have not been prevented from collecting sensitive user data that is not required for the service they offer [46].

There are also many documented cases of mobile apps using their microphone access in unexpected ways. An example that has received a lot of media attention recently is the use of so-called "ultrasonic beacons", i.e. high-pitched Morse-style audio signals inaudible to the human ear which are secretly played in stores or embedded in TV commercials and other broadcast content in order to be able to unobtrusively track the location, activities and media consumption habits of consumers [10]. For this to work, the data subject needs to carry a receiving device that records and scans ambient sound for relevant ultrasonic signals and sends them back to the tracking network for automated comparison. A constantly growing number of mobile apps – several hundred already, some of them very popular – are using their microphone permission for exactly that purpose, often without properly informing the user about it [10, 47]. These apps, some of which are targeted at children and would not require audio recording for their core functionality, may even detect sounds while the phone is locked and carried in a pocket [47]. Even in cases where users are aware that their phone listens in, it is not clear to them what the audio stream is filtered for exactly and what information is being exfiltrated. Thus, the example of ultrasonic beacons shows how apps that have been approved into Apple's App Store and Google Play can exploit their permissions for dubious and potentially unexpected tracking purposes.

Finally, it should not be overlooked that smartphone apps can also be obtained from various non-official sources, circumventing Apple's and Google's permission systems and auditing processes [62]. In Android, users are free in choosing the source of their applications [78]. Following a more restrictive policy, iOS only allows users to install apps downloaded from the official Apple App Store. However, kernel patches can be used to gain root access and remove software restrictions in iOS ("iOS jailbreaking"), which enables users to install apps from uncertified publishers [62].

5.2   App Inspections Conducted by the Research Community

In addition to the checks conducted by Google and Apple, mobile apps are being reviewed by a broad community of security and privacy researchers. A wide and constantly expanding range of manual and automated methods is applied for this purpose.

Pan et al., for instance, scanned 17,260 popular Android apps from different app markets for potential privacy leaks [59]. Through examining their media permissions, privacy policies and outgoing network flows, the researchers tried to identify apps that upload audio recordings to the Internet without explicitly informing the user about it.

While unveiling other serious forms of privacy violations, they found no evidence of such behavior. Based on these findings, the widely held suspicion of companies secretly eavesdropping on smartphone users was already portrayed as refuted in news headlines [34, 80].

However, the study comes with numerous limitations: Apart from considering only a small fraction of the over 2 million available Android apps, the researchers did not examine media exfiltration from app background activity, did not consider the use of privileged APIs, only tested a limited amount of each app's functionalities for a short amount of time, used a controlled test environment with no real human interactions, did not consider iOS apps at all, and were not able to detect media that was intentionally obfuscated, encrypted at the application-layer, or sent over the network in non-standard encoding formats. Perhaps most importantly, Pan et al. were not able to rule out the scenario of apps transforming audio recordings into less detectable text transcripts or audio fingerprints before sending the information out. This would be a very realistic attack scenario. In fact, various popular apps are known to compress recorded audio in such a way [10, 33]. While all the choices that Pan et al. made regarding their experimental setup and methodology are completely understandable and were communicated transparently, the limitations do limit the significance of their findings. All in all, their approach would only uncover highly unsophisticated eavesdropping attempts.

Of course, many other researchers have also tried to detect privacy leaks in iOS and Android apps [62]. Besides analyzing decompiled code, permission requests and generated network traffic, other factors, such as battery power consumption and device memory usage, can also be monitored to detect suspicious app behavior [67]. Although some experts claim to have observed certain mobile apps recording and sending out audio with no apparent justification [58], the scientific community has not yet produced any hard evidence for large-scale eavesdropping through smartphone microphones.

Like the above-cited work by Pan et al., however, other existing methods to identify privacy threats in mobile devices also come with considerable limitations. Due to its closed-source nature, there is generally a lack of scalable tools for detecting malicious apps within iOS [19]. While, on the other hand, numerous efficient methods have been proposed for automatically scanning Android apps, none of these approaches is totally effective at detecting privacy leaks [59]. As with security checks of the official app stores (see Sect. 5.1), there is a wide range of possible obfuscation techniques and covert channels to circumvent detection mechanisms developed by the scientific community [10, 67]. Furthermore, many of the existing approaches do not indicate if detected data exfiltration activities are justified with regard to an app's advertised functionality [62]. Yerukhimovich et al. even suggest that apps classified as safe or non-malicious are more likely to leak private information than typical "malware" [78].

Therefore, the fact that no evidence for large-scale mobile eavesdropping has been found so far should not be interpreted as an all-clear. It could only mean that it is difficult – under current circumstances perhaps even impossible – to detect such attacks effectively.

6 Ecosystem Providers as Potential Adversaries

Not only third-party apps but also mobile operating systems themselves can access privacy-sensitive smartphone data and transfer it over the Internet. It has been known for years that both, iOS and Android, do so extensively [5]. Examining the amount of data sent back to Google's and Apple's servers from test devices, a recent study found that iPhones – on average – received four requests per hour from their manufacturer during idle periods, and eighteen requests during periods of heavy use [68]. Leaving these numbers far behind, Android phones received forty hourly requests from Google when in idle state and ninety requests during heavy use. Of course, the number of requests per hour has only limited informational value. Data is often collected much more frequently, such as on a secondly basis or even constantly, to be later aggregated, compressed and sent out in data bundles [5].

While the establishment of network connections can be monitored, many aspects of data collection and processing in smartphones remain opaque. The source code of iOS is not made publicly available, and while Android is based on code from the Android Open Source Project, several of Google's proprietary apps and system components are closed-source as well [2]. Due to the resulting lack of transparency, it cannot be reliably ruled out that sensitive data is collected and processed without the will or knowledge of the smartphone owner – although, naturally, this would represent a considerable legal and reputational risk for the corresponding platform provider.

As an intermediary between applications and hardware resources, operating systems control the access to smartphone sensors, including microphones, accelerometers and gyroscopes, and can also decide whether or not sensor activity is indicated to the user on the device's screen. Other than with third-party apps, there is no superior authority in the system supervising the actions and decisions of iOS and Android. While external security experts can carry out inspections using similar methods as outlined in Sect. 5.2, they also face similar limitations. There is no reason to assume that operating systems refrain from using sophisticated obfuscation techniques to conceal their data collection practices. Additionally, being in control of the whole system, iOS and Android can access data on different levels of their respective software stack, which gives them more options for stealthy data exfiltration and could possibly impede detection.

7 Technical and Economic Feasibility

Even where adversaries manage to get around security measures and evade detection, it remains questionable whether a continuous and large-scale eavesdropping operation for the purpose of ad targeting would be technically feasible and economically viable. Based on estimations of CPU, battery, network transfer and data storage requirements, some commentators already stated their conclusion that such an operation would be far too expensive [51, 76] and may "strain even the resources of the NSA" [71]. Taking into account their underlying assumptions, these estimates appear valid. However, there are several ways in which smartphone-based eavesdropping could be made much more efficient and scalable, including:

- **Low quality audio recording.** To reduce the required data storage, processing power and energy consumption, adversaries could record audio at low bitrates. Speech signals do not even have to be intelligible to the human ear to be recognized and transcribed into text by algorithms [54].
- **Local pre-processing.** Some steps in the processing of recordings (e.g. transcription, extraction of audio features, data filtering, keyword matching, compression) can be performed locally on the device in order to transmit only the most relevant data to remote servers and thus reduce network traffic and required cloud storage.
- **Keyword detection instead of full speech recognition.** The amounts of processing power required for automatic speech recognition can be prohibitively high for local execution on mobile devices. A less CPU-intensive alternative to full speech recognition is keyword detection, where only a pre-defined vocabulary of spoken words is recognized. Such systems can even run on devices with much lower computational power than smartphones, such as 16-bit microcontrollers [25]. It has been argued that it would still be too taxing for mobile devices to listen out for the "millions or perhaps billions" of targetable keywords that could potentially be dropped in private conversations [51]. However, instead of listening for specific product and brand names, audio recordings can simply be scanned for trigger words that indicate a person's interest, such as "love", "enjoyed", or "great", in order to identify relevant snippets of the recording, which can then be analyzed in more depth. In fact, this very audio analysis method has already been patented, with the specific declared purpose of informing "targeted advertising and product recommendations" [22].
- **Selective recording.** Instead of recording continuously, an adversary could only record at selected moments using wake words or triggers based on time, location, user activity, sound level, and other context variables. This could significantly reduce the amount of required storage and network traffic [67].

Mobile apps that use all or some of the above techniques can be light enough to run smoothly on smartphones, as numerous commercial apps and research projects show [9, 10, 33, 58, 67].

But even if it is possible for companies to listen in on private conversations, some argue that this information might not be of much value to advertisers, since they would need to know a conversation's context and speaker personalities very well in order to accurately infer personal preferences and purchase intentions from spoken phrases [51]. This argument is reasonable, but can equally be applied to many other profiling methods, including online tracking and location tracking, which are widely used nonetheless. Of course, where contextual information is sparse, such methods may lead to wrong conclusions about the respective data subject, possibly resulting in poor and inefficient ad targeting. However, this would not conflict with the above-mentioned reports of suspected eavesdropping: While the ads were perceived as inspired by topics raised in private conversations, they did not always reflect the purported witnesses' actual needs and wants [6, 12].

From an outside perspective, it cannot be precisely determined how profitable certain types of personal data are for advertisers. It is therefore difficult, if not impossible, to draw up a meaningful cost-benefit calculation. However, it can generally be assumed that private conversations contain a lot of valuable profiling information, especially when speakers express their interest in certain products or services. It is also worth mentioning that some of the world's largest companies earn a significant portion of their revenue through advertising – for Google and Facebook, this portion amounted to 85% and 98% in 2018, respectively [1, 23]. Profits from advertising can be considerably increased through effective targeting, which requires the collection of detailed personal information [68]. There is no doubt that smartphone sensor data can be very useful for this purpose. A recently filed patent describes, for example, how "local signals" from a mobile device, including motion sensor data and audio data from the microphone, can be analyzed to personalize a user's Facebook news feed [50].

8 Unauthorized Access to Smartphones

Although this is most likely no explanation for suspicious ad placement, it should be noted that there are many ways in which skilled computer experts or "hackers" can gain unauthorized access to mobile devices. The widespread use of smartphones makes them a particularly attractive hacking target [4].

Not only cyber criminals, but also law enforcement agencies and secret services invest heavily in their capabilities to exploit software flaws and other security vulnerabilities in consumer electronics [73]. It has been known for some time that intelligence agencies, such as NSA, GCHQ, and CIA, are equipped with tools to secretly compromise devices running iOS, Android and other mobile operating systems, enabling them "to move inside a system freely as if they owned it" [66, 75].

In addition to accessing sensitive data, such as geo-location, passwords, personal notes, contacts, and text messages, this includes the ability to turn on a phone's microphone without a user's consent or awareness [11]. With the help of specialized tools, smartphone microphones can even be tapped when the device is (or seems) switched off [73]. Such attacks can also be successful in high-security environments. In a recent case, for example, more than 100 Israeli servicemen had their phones infected with spyware that allowed unknown adversaries to control built-in cameras and microphones [57].

Besides the United States and some European nations, other developed countries, such as Russia, Israel and China, also have highly sophisticated spying technology at their disposal [75]. Less developed countries and other actors can buy digital eavesdropping tools from a flourishing industry of surveillance contractors at comparatively low prices [60]. That not only secret services but also law enforcement agencies in the US can be authorized to convert smartphones into "roving bugs" to listen in on private conversations has been confirmed in a 2012 court ruling [16]. Eavesdropping capabilities of criminal organizations should not be underestimated, either. According to a report by McAfee and the Center for Strategic and International Studies (CSIS), there are 20 to 30 cybercrime groups with "nation-state level" capacity in countries of the former Soviet Union alone [52].

9 Discussion

So far, despite significant research efforts, no evidence has been found to confirm the widespread suspicion that firms are secretly eavesdropping on smartphone users to inform ads. To the best of our knowledge, however, the opposite has not been proven either. While some threat scenarios (e.g. the constant transfer of uncompressed audio recordings into the cloud) can be ruled out based on existing security measures and considerations regarding an attack's visibility, cost and technical feasibility, there are still many security vulnerabilities and a fundamental lack of transparency that potentially leave room for more sophisticated attacks to be successful and remain undetected.

In comparison with the researchers cited in this paper, it can be assumed that certain companies have significantly more financial resources, more training data, and more technical expertise in areas such as signal processing, data compression, covert channels, and automatic speech recognition. This is – besides unresolved contradictions between cited studies and large remaining research gaps – another reason why existing work should not be seen as final and conclusive, but rather as an initial exploration of the issue.

While this paper focuses on smartphones, it should be noted that microphones and motion sensors are also present in a variety of other Internet-connected devices, including not only VR headsets, wearable fitness trackers and smartwatches, but also baby monitors, toys, remote controls, cars, household appliances, laptops, and smart speakers. Some of these devices may have weaker privacy safeguards than smartphones. For instance, they may not ask for user permission before turning on the microphone or may not impose a limit on sensor sampling frequencies. Numerous devices, including smart TVs [13], smart speakers [27], and connected toys [26], have already been suspected to spy on private conversations of their users. Certain smart home devices, such as home security alarms, may even contain a hidden microphone without disclosing it in the product specifications [44]. For these reasons, it is essential to also thoroughly examine non-smartphone devices when investigating suspicions of eavesdropping.

It is quite possible, at the same time, that the fears of advertising companies eavesdropping on private conversations are unfounded. Besides the widespread attribution to chance, one alternative approach to explaining strangely accurate advertisements points to all the established tracking technologies commonly employed by advertisers that do not depend on any phone sensors or microphones [51].

Drawing from credit card networks, healthcare providers, insurers, employers, public records, websites, mobile apps, and many other sources, certain multi-national corporations already hold billions of individual data points on consumers' location histories, browsing behaviors, religious and political affiliations, occupations, socioeconomic backgrounds, health conditions, personality traits, product preferences, and so on [17, 64]. Although their own search engines, social networks, email services, route planners, instant messengers, and media platforms already give them intimate insight into the lives of billions of people, advertising giants like Facebook and Google also intensively track user behavior on foreign websites and apps. Of the 17.260 apps examined in [59], for example, 48.22% share user data with Facebook in the background. Through their analytics services and like buttons, Google and Facebook can track clicks and scrolls of Internet users on a vast number of websites [17].

The deep and potentially unexpected insights that result from such ubiquitous surveillance can be used for micro-targeted advertising and might thereby create an illusion of being eavesdropped upon, especially if the data subject is ill-informed about the pervasiveness and impressive possibilities of data linkage.

Even without being used for audio snooping, smartphones (in their current configuration) allow a large variety of actors to track private citizen in a much more efficient and detailed way than would ever have been possible in even the most repressive regimes and police states of the 20th century. At the bottom line, whether sensitive information is extracted from private conversations or collected from other sources does not make much difference to the possibilities of data exploitation and the entailing consequences for the data subject. Therefore, whether justified or not, the suspicions examined in this paper eventually lead to a very fundamental question: What degree of surveillance should be considered acceptable for commercial purposes like targeted advertising? Although this paper cannot offer an answer to this political question, it should not be forgotten that constant surveillance is by no means a technical necessity and that, by definition, democracies should design and regulate technology to primarily reflect the values of the public, not commercial interests.

Certainly, the fear of eavesdropping smartphones should never be portrayed as completely unfounded, as various criminal and governmental actors can gain unauthorized access to consumer electronics. Although such attacks are unlikely to result in targeted advertisement, they equally deprive the user of control over his or her privacy and might lead to other unpredictable harms and consequences. For example, digital spying tools have been used to infiltrate the smartphones of journalists [49] and human rights activists [60] for repressive purposes.

Finally, it should be recognized that – apart from the linguistic contents of speech – microphones and motion sensors may unexpectedly transmit a wealth of other sensitive information. Through the lens of advanced analytics, a voice recording can reveal a speaker's identity [53], physical and mental health state [20, 37], and personality traits [61], for example. Accelerometer data from mobile devices may implicitly contain information about a user's location [35], daily activities [48], eating, drinking and smoking habits [72, 74], degree of intoxication [30], gender, age, body features and emotional state [43] and can also be used to re-construct sequences of text entered into a device, including passwords [42].

10 Conclusion

After online advertisements seemingly adapted to topics raised in private face-to-face conversations, many people suspect companies to secretly listen in through their smartphones. This paper reviewed and analyzed existing approaches to explaining the phenomenon and examined the general feasibility and detectability of mobile eavesdropping attacks. While it is possible, on the one hand, that the strangely accurate ads were just a product of chance or conventional profiling methods, the spying fears were not disproved so far, neither by device manufacturers and ecosystem providers nor by the research community.

In our threat model, we considered non-system mobile apps, third-party libraries, and ecosystem providers themselves as potential adversaries. Smartphone microphones and motion sensors were investigated as possible eavesdropping channels. Taking into account permission requirements, user notifications, sensor sampling frequencies, limited device resources, and existing security checks, we conclude that – under the current levels of data collection transparency in iOS and Android – sophisticated eavesdropping operations could potentially be run by either of the above-mentioned adversaries without being detected. At this time, no estimate can be made as to the probability and economic viability of such attacks.

References

1. Alphabet Inc.: Alphabet Announces Fourth Quarter and Fiscal Year 2018 Results (2019). https://abc.xyz/investor/static/pdf/2018Q4_alphabet_earnings_release.pdf?cache=adc3b38
2. Amadeo, R.: Google's iron grip on Android: Controlling open source by any means necessary (2018). https://arstechnica.com/gadgets/2018/07/googles-iron-grip-on-android-controlling-open-source-by-any-means-necessary/
3. Anand, S.A., Saxena, N.: Speechless: analyzing the threat to speech privacy from smartphone motion sensors. In: 2018 IEEE Symposium on Security and Privacy, San Francisco, CA, pp. 1000–1017. IEEE (2018). https://doi.org/10.1109/SP.2018.00004
4. Aneja, L., Babbar, S.: Research trends in malware detection on Android devices. In: Panda, B., Sharma, S., Roy, N. (eds.) Data Science and Analytics. Communications in Computer and Information Science, vol. 799, pp. 629–642. Springer, Singapore (2018). https://doi.org/10.1007/978-981-10-8527-7_53
5. Angwin, J., Valentino-DeVries, J.: Apple, Google Collect User Data (2011). https://www.wsj.com/articles/SB10001424052748703983704576277101723453610
6. Anonymous: YouTube user demonstrating how Facebook listens to conversations to serve ads (2017). https://www.reddit.com/r/videos/comments/79i4cj/youtube_user_demonstrating_how_facebook_listens/
7. Apple: Background Execution. https://developer.apple.com/library/archive/documentation/iPhone/Conceptual/iPhoneOSProgrammingGuide/BackgroundExecution/BackgroundExecution.html
8. Apple: Record - iPhone User Guide. https://help.apple.com/iphone/11/?lang=en#/iph4d2a39a3b
9. Arcas, B.A., et al.: Now playing: continuous low-power music recognition. arXiv Comput. Res. Repos. abs/1711.10958 (2017). http://arxiv.org/abs/1711.10958
10. Arp, D., et al.: Privacy threats through ultrasonic side channels on mobile devices. In: 2017 IEEE European Symposium on Security and Privacy (EuroS&P), Paris, France, pp. 35–47. IEEE (2017). https://doi.org/10.1109/EuroSP.2017.33
11. Ball, J.: Angry Birds and "leaky" phone apps targeted by NSA and GCHQ for user data (2014). https://www.theguardian.com/world/2014/jan/27/nsa-gchq-smartphone-app-angry-birds-personal-data
12. BBC News Services: Is your phone listening in? Your stories (2017). https://www.bbc.com/news/technology-41802282

13. Beres, D.: How To Stop Your Smart TV From Eavesdropping On You (2015). https://www.huffpost.com/entry/your-samsung-tv-is-spying-on-you_n_6647762
14. Bocek, V., Chrysaidos, N.: Android devices ship with pre-installed malware (2018). https://blog.avast.com/android-devices-ship-with-pre-installed-malware
15. Bogost, I.: FaceTime Is Eroding Trust in Tech (2019). https://www.theatlantic.com/technology/archive/2019/01/apple-facetime-bug-you-cant-escape/581554/
16. Brown, A.J.: United States v. Oliva (United States Court of Appeals, D.C. No. 3:07-cr-00050-BR-1) (2012)
17. Christl, W.: Corporate Surveillance in Everyday Life. Cracked Labs, Vienna (2017)
18. Christl, W., Spiekermann, S.: Networks of Control: A Report on Corporate Surveillance, Digital Tracking, Big Data & Privacy. Facultas, Vienna (2016)
19. Cimitile, A., et al.: Machine learning meets iOS malware: identifying malicious applications on Apple environment. In: Proceedings of the 3rd International Conference on Information Systems Security and Privacy, Porto, Portugal, pp. 487–492. SciTePress (2017). https://doi.org/10.5220/0006217304870492
20. Cummins, N., et al.: Speech analysis for health: current state-of-the-art and the increasing impact of deep learning. Methods (2018). https://doi.org/10.1016/j.ymeth.2018.07.007
21. Dusan, S.V., et al.: System and Method of Detecting a User's Voice Activity Using an Accelerometer (Patent No.: US9438985B2) (2014). https://patents.google.com/patent/US9438985B2/en
22. Edara, K.K.: Keyword Determinations from Voice Data (Patent No.: US20140337131A1) (2014). https://patents.google.com/patent/US20140337131A1/en
23. Facebook: Facebook Reports Fourth Quarter and Full Year 2018 Results. https://s21.q4cdn.com/399680738/files/doc_financials/2018/Q4/Q4-2018-Earnings-Release.pdf
24. Felt, A.P., et al.: Android permissions: user attention, comprehension, and behavior. In: Proceedings of the Eighth Symposium on Usable Privacy and Security (SOUPS 2012), Washington, D.C. ACM Press (2012). https://doi.org/10.1145/2335356.2335360
25. Fourniols, J.-Y., et al.: An overview of basics speech recognition and autonomous approach for smart home IOT low power devices. J. Signal Inf. Process. 9, 239–257. https://doi.org/10.4236/jsip.2018.94015
26. de Freytas-Tamura, K.: The Bright-Eyed Talking Doll That Just Might Be a Spy (2018). https://www.nytimes.com/2017/02/17/technology/cayla-talking-doll-hackers.html
27. Fussell, S.: Behind Every Robot Is a Human (2019). https://www.theatlantic.com/technology/archive/2019/04/amazon-workers-eavesdrop-amazon-echo-clips/587110/
28. Ganjoo, S.: Is Facebook secretly listening your conversations? New report says yes, security experts say no proof (2018). https://www.indiatoday.in/technology/features/story/is-facebook-secretly-listening-your-conversations-new-report-says-yes-security-experts-say-no-proof-1255870-2018-06-09
29. Gao, G., Chow, M.: Android Applications, Can You Trust Google Play on These. Tufts University (2016)
30. Gharani, P., et al.: An Artificial Neural Network for Gait Analysis to Estimate Blood Alcohol Content Level. arXiv Comput. Res. Repos. abs/1712.01691 (2017). https://arxiv.org/abs/1712.01691
31. Google: Android 9 Pie. https://www.android.com/versions/pie-9-0/
32. Greenberg, A.: The Gyroscopes in Your Phone Could Let Apps Eavesdrop on Conversations (2014). https://www.wired.com/2014/08/gyroscope-listening-hack/
33. Grosche, P., et al.: Audio content-based music retrieval. In: Müller, M., et al. (eds.) Multimodal Music Processing. Dagstuhl Follow-Ups. Dagstuhl Publishing, Wadern (2012)

34. Hale, J.L.: Does Your Smartphone Listen To You? A New Study Debunked This Common Conspiracy (2018). https://www.bustle.com/p/does-your-smartphone-listen-to-you-a-new-study-debunked-this-common-conspiracy-9682413

35. Han, J., et al.: ACComplice: location inference using accelerometers on smartphones. In: 2012 Fourth International Conference on Communication Systems and Networks (COMSNETS), pp. 1–9 (2012). https://doi.org/10.1109/COMSNETS.2012.6151305

36. Han, J., et al.: PitchIn: eavesdropping via intelligible speech reconstruction using non-acoustic sensor fusion. In: Proceedings of the 16th ACM/IEEE International Conference on Information Processing in Sensor Networks (IPSN), pp. 181–192. ACM Press, Pittsburgh (2017). https://doi.org/10.1145/3055031.3055088

37. Hashim, N.W., et al.: Evaluation of voice acoustics as predictors of clinical depression scores. J. Voice **31**(2), 256.e1–256.e6 (2017). https://doi.org/10.1016/j.jvoice.2016.06.006

38. Hassan, B.: 1 in 5 Aussies convinced their smartphone is spying on them (2018). https://www.finder.com.au/press-release-july-2018-1-in-5-aussies-convinced-their-smartphone-is-spying-on-them

39. He, Y., et al.: Dynamic privacy leakage analysis of Android third-party libraries. In: 1st International Conference on Data Intelligence and Security (ICDIS), pp. 275–280 (2018). https://doi.org/10.1109/ICDIS.2018.00051

40. Khatibloo, F.: Is Facebook Listening (And So What If They Are)? (2017). https://www.forbes.com/sites/forrester/2017/03/17/is-facebook-listening-and-so-what-if-they-are/

41. Kleinman, Z.: Is your smartphone listening to you? (2016). https://www.bbc.com/news/technology-35639549

42. Kröger, J.: Unexpected inferences from sensor data: a hidden privacy threat in the internet of things. In: Strous, L., Cerf, V.G. (eds.) Internet of Things. Information Processing in an Increasingly Connected World. IFIP Advances in Information and Communication Technology, vol. 548, pp. 147–159. Springer, Cham (2019). https://doi.org/10.1007/978-3-030-15651-0_13

43. Kröger, J.L., et al.: Privacy implications of accelerometer data: a review of possible inferences. In: Proceedings of the 3rd International Conference on Cryptography, Security and Privacy (ICCSP). ACM, New York (2019). https://doi.org/10.1145/3309074.3309076

44. Lee, D.: Google admits error over hidden microphone (2019). https://www.bbc.com/news/technology-47303077

45. Liu, X., et al.: Discovering and understanding Android sensor usage behaviors with data flow analysis. World Wide Web **21**(1), 105–126 (2018). https://doi.org/10.1007/s11280-017-0446-0

46. Lomas, N.: Uber to end controversial post-trip tracking as part of privacy drive (2017). http://social.techcrunch.com/2017/08/29/uber-to-end-controversial-post-trip-tracking-as-part-of-privacy-drive/

47. Maheshwari, S.: That Game on Your Phone May Be Tracking What You're Watching on TV (2017). https://www.nytimes.com/2017/12/28/business/media/alphonso-app-tracking.html

48. Mannini, A., et al.: Activity recognition using a single accelerometer placed at the wrist or ankle. Med. Sci. Sports Exerc. **45**(11), 2193–2203 (2013). https://doi.org/10.1249/MSS.0b013e31829736d6

49. Marczak, B., et al.: Hacking Team and the Targeting of Ethiopian Journalists (2014). https://citizenlab.ca/2014/02/hacking-team-targeting-ethiopian-journalists/

50. Marra, C.J., et al.: Ranking of News Feed in a Mobile Device Based on Local Signals (Pub. No.: US20170351675A1) (2017). https://patents.google.com/patent/US20170351675A1/en

51. Martínez, A.G.: Facebook's Not Listening Through Your Phone. It Doesn't Have To (2017). https://www.wired.com/story/facebooks-listening-smartphone-microphone/

52. McAfee: Net Losses: Estimating the Global Cost of Cybercrime. Center for Strategic and International Studies (CSIS), Washington, D.C. (2014)
53. McLaren, M., et al.: The 2016 speakers in the wild speaker recognition evaluation. In: Proceedings of the 16th Annual Conference of the International Speech Communication Association (INTERSPEECH), pp. 823–827 (2016). https://doi.org/10.21437/Interspeech.2016-1137
54. Michalevsky, Y., et al.: Gyrophone: recognizing speech from gyroscope signals. In: Proceedings of the 23rd USENIX Security Symposium, pp. 1053–1067 (2014)
55. Mohapatra, P., et al.: Energy-efficient, Accelerometer-based Hotword Detection to Launch a Voice-control System. (Patent No.: US20170316779A1) (2017). https://patents.google.com/patent/US20170316779A1/en
56. Morris, I.: Android Is Still Failing Where Apple's iOS Is Winning (2018). https://www.forbes.com/sites/ianmorris/2018/04/13/android-is-still-failing-where-apples-ios-is-winning/
57. Naor, I.: Breaking The Weakest Link Of The Strongest Chain (2017). https://securelist.com/breaking-the-weakest-link-of-the-strongest-chain/77562/
58. Nichols, S., Morgans, J.: Your Phone Is Listening and it's Not Paranoia (2018). https://www.vice.com/en_uk/article/wjbzzy/your-phone-is-listening-and-its-not-paranoia
59. Pan, E., et al.: Panoptispy: Characterizing Audio and Video Exfiltration from Android Applications. Proc. Priv. Enhanc. Technol. **2018**(4), 33–50 (2018). https://doi.org/10.1515/popets-2018-0030
60. Perlroth, N.: Governments Turn to Commercial Spyware to Intimidate Dissidents (2017). https://www.nytimes.com/2016/05/30/technology/governments-turn-to-commercial-spyware-to-intimidate-dissidents.html
61. Polzehl, T.: Personality in Speech. Springer, Cham (2015). https://doi.org/10.1007/978-3-319-09516-5
62. Quattrone, A.: Inferring Sensitive Information from Seemingly Innocuous Smartphone Data. The University of Melbourne (2016)
63. Rahman, M., et al.: Search rank fraud and malware detection in Google Play. IEEE Trans. Knowl. Data Eng. **29**(6), 1329–1342 (2017). https://doi.org/10.1109/TKDE.2017.2667658
64. Ramirez, E., et al.: Data Brokers. A Call for Transparency and Accountability. Federal Trade Commission, Washington, D.C. (2014)
65. Ramirez, R., et al.: Cross-Device Tracking: An FTC Staff Report. Federal Trade Commission, Washington, D.C. (2017)
66. Rosenbach, M., et al.: iSpy: How the NSA Accesses Smartphone Data (2013). http://www.spiegel.de/international/world/how-the-nsa-spies-on-smartphones-including-the-blackberry-a-921161.html
67. Schlegel, R., et al.: Soundcomber: a stealthy and context-aware sound trojan for smartphones. In: Proceedings of the Network and Distributed System Security Symposium (NDSS) (2011)
68. Schmidt, D.C.: Google Data Collection. Digital Content Next, New York (2018)
69. Sidor, S.: Exploring limits of covert data collection on Android: apps can take photos with your phone without you knowing (2014). http://www.ez.ai/2014/05/exploring-limits-of-covert-data.html)
70. Statista: Global mobile OS market share in sales to end users from 1st quarter 2009 to 2nd quarter 2018. https://www.statista.com/statistics/266136/global-market-share-held-by-smartphone-operating-systems/
71. Stern, J.: Facebook Really Is Spying on You, Just Not Through Your Phone's Mic (2018). https://www.wsj.com/articles/facebook-really-is-spying-on-you-just-not-through-your-phones-mic-1520448644

72. Tang, Q., et al.: Automated detection of puffing and smoking with wrist accelerometers. In: Proceedings of the 8th International Conference on Pervasive Computing Technologies for Healthcare. pp. 80–87 (2014)

73. Taylor, P.: Edward Snowden interview: "Smartphones can be taken over" (2015). https://www.bbc.com/news/uk-34444233

74. Thomaz, E., et al.: A practical approach for recognizing eating moments with wrist-mounted inertial sensing. In: Proceedings of the ACM International Conference on Ubiquitous Computing, pp. 1029–1040. ACM Press (2015). https://doi.org/10.1145/2750858.2807545

75. Timberg, C., et al.: WikiLeaks: The CIA is using popular TVs, smartphones and cars to spy on their owners (2017). https://www.washingtonpost.com/news/the-switch/wp/2017/03/07/why-the-cia-is-using-your-tvs-smartphones-and-cars-for-spying/?noredirect=on&utm_term=.c16 2373021c3

76. Triggs, R.: No, your smartphone is not always listening to you (2018). https://www.androidauthority.com/your-phone-is-not-listening-to-you-884028/

77. Tsukayama, H., Romm, T.: Lawmakers press Apple and Google to explain how they track and listen to users (2018). https://www.washingtonpost.com/technology/2018/07/09/lawmakers-press-apple-google-explain-how-they-track-listen-users/

78. Yerukhimovich, A., et al.: Can smartphones and privacy coexist? Assessing technologies and regulations protecting personal data on Android and iOS devices. MIT Lincoln Laboratory, Lexington, MA (2016). https://doi.org/10.7249/RR1393

79. Zhang, L., et al.: AccelWord: energy efficient hotword detection through accelerometer. In: Proceedings of the 13th Annual International Conference on Mobile Systems, Applications, and Services (MobiSys), pp. 301–315. ACM Press (2015). https://doi.org/10.1145/2742647.2742658

80. No, Phones Aren't Listening to Your Conversations, but May Be Recording In-App Videos: Study (2018). https://www.justandroid.net/2018/07/05/no-phones-arent-listening-to-your-conversations-but-may-be-recording-in-app-videos-study/

Droids in Disarray: Detecting *Frame Confusion* in Hybrid Android Apps

Davide Caputo, Luca Verderame, Simone Aonzo, and Alessio Merlo[(⊠)]

DIBRIS, University of Genoa, Viale F. Causa, 13, 16145 Genoa, Italy
{davide.caputo,luca.verderame,simone.aonzo,alessio.merlo}@unige.it

Abstract. Frame Confusion is a vulnerability affecting hybrid applications which allows circumventing the isolation granted by the Same-Origin Policy. The detection of such vulnerability is still carried out manually by application developers, but the process is error-prone and often underestimated. In this paper, we propose a sound and complete methodology to detect the Frame Confusion on Android as well as a publicly-released tool (i.e., FCDroid) which implements such methodology and allows to detect the Frame Confusion in hybrid applications, automatically. We also discuss an empirical assessment carried out on a set of 50K applications using FCDroid, which revealed that a lot of hybrid applications suffer from Frame Confusion. Finally, we show how to exploit Frame Confusion on a news application to steal the user's credentials.

Keywords: Frame Confusion · Android security · Static analysis · Dynamic analysis

1 Introduction

Nowadays, the landscape of mobile devices is mostly divided between Android and iOS, with a market share of 74% and 23%, respectively[1]. From a technical standpoint, Android and iOS have remarkable differences both in terms of OS architecture and Software Development Kit (SDK). Such heterogeneity negatively impacts the application (hereafter, app) development process, as companies must rely on different developer teams (be them internal or outsourced) for each platform, thereby increasing the costs of both app development and maintenance. A promising way to overcome the limitation posed by such multi-platform development process is a *cross-platform* framework, which allows to implement an app using a unique programming language and automatically generate a corresponding Android and iOS version. Cross-platform frameworks based on web technologies (i.e., HTML, CSS, and JavaScript), like Cordova [11] or Phone-Gap [26], allow for the development of the so-called *hybrid applications*, which combine elements of both *native* (i.e., OS-specific) and *web* apps.

---

[1] http://gs.statcounter.com/os-market-share/mobile/worldwide.

© IFIP International Federation for Information Processing 2019
Published by Springer Nature Switzerland AG 2019
S. N. Foley (Ed.): DBSec 2019, LNCS 11559, pp. 121–139, 2019.
https://doi.org/10.1007/978-3-030-22479-0_7

Hybrid apps allow developers to write code based on platform-neutral web technologies and wrap them into a single native app that can render HTML/CSS content and execute JavaScript – like a standard web browser – through a component called *WebView* on Android, and *WKWebView* in iOS. Such component acts as a bridge between the web (i.e., the JavaScript code) and the native world (i.e., the Java or Swift code) through the definition of ad-hoc interfaces. Such interfaces (called *JavaScriptInterfaces* in Android or *WKScriptMessageHandlers* in iOS) allow the developer to define a set of function calls that can be mutually invoked by the two worlds using asynchronous callbacks. As a result, they grant access to the complete set of OS functionality to hybrid apps, thereby making them equivalent to native apps.

However, from a security standpoint, the interaction between the native and the web worlds – which rely on different security models and requirements – can expose hybrid apps to ad-hoc and complex vulnerabilities, like those described in [9,15,20,22,23]. Among them, the *Frame Confusion* vulnerability [22] in hybrid apps has been discovered some years ago and it has been fixed on iOS[2] but not on Android (neither in the latest version, i.e., Android Pie 9.0). To this regard, we argue that a lot of hybrid apps still suffer from such vulnerability and that there is still a lack of (i) an extensive analysis of Frame Confusion, (ii) a methodology to automatically detect Frame Confusion in hybrid apps, and (iii) a reliable solution to mitigate the problem.

Frame Confusion is basically due to the ability of JavaScript code to invoke Android native code through web pages containing at least an *Iframe* element. Such element allows loading external contents (e.g., advertisements, video and payment systems) from domains which differ from the domain of the hybrid app. For this reason, any *Iframe* is in charge of *containerizing* the rendered sub-page, and should execute content only within the scope of its own domain, as prescribed by the Same-Origin Policy (SOP). However, in case of web pages with multiple Iframes, the WebView is unable to identify the Iframe that invokes a function in the native code, and thus the result of the invocation is always executed in the main app page, thereby inducing the confusion problem. Such misbehavior occurs as the *JavaScriptInterface* is bound by the OS to the entire WebView element, without any distinction among the domains (and thus the Iframes) that invoke the function calls. Therefore, the Frame Confusion vulnerability allows to bypass the isolation granted by the Iframe security model and to build a communication channel between web pages belonging to different domains, (i.e., the main app page and the inner Iframes). As a consequence, such vulnerability can affect the confidentiality and the integrity of hybrid apps: a malicious Iframe can, for instance, force the main app to expose private information (like session cookies or internal app files) or mount sophisticated phishing attacks.

Contribution of the Paper. In this work, we focus on the Frame Confusion vulnerability on Android. Therefore, hereafter we refer specifically to the Android OS.

---

[2] https://cordova.apache.org/docs/en/latest/guide/appdev/security/index.html# iframes-and-the-callback-id-mechanism.

Our contribution is three-fold. First, we propose a methodology for systematically detecting the Frame Confusion vulnerability in hybrid apps on Android. Then, we present *FCDroid*, a tool that implements such methodology to automatically identify hybrid apps on Android that suffer from the Frame Confusion vulnerability. FCDroid combines static and dynamic analysis techniques in order to reduce false positive and false negative rates. Finally, we discuss the results of an extensive analysis carried out through FCDroid on a set of 50,000 apps downloaded from the Google Play Store. The experimental results indicate that the 49.35% of the analyzed apps are hybrid, as they use the WebView component and enable JavaScript execution, while about 6.63% of them (i.e., 1637 apps) were found to be vulnerable to Frame Confusion for a total of more than 250.000.000 app installations worldwide. To further validate the proposed methodology, we have manually analyzed some of these vulnerable apps to find out possible attacks exploiting the Frame Confusion vulnerability. To this regard, we were able to exploit Frame Confusion in an Asian news application that has more than 1M users worldwide; such attack allows to steal the user's credential of the primary social media website.

Organization of the Paper. The rest of the paper is organized as follows: Sect. 2 introduces some technical background on hybrid apps and the Frame Confusion, while Sect. 3 discusses the detection methodology. Section 4 presents FCDroid, while Sect. 5 shows the experimental results. Section 6 discusses the exploitation of the Frame Confusion on an actual news app, and Sect. 7 presents some related work. Finally, Sect. 8 concludes the paper.

2   Technical Background

Landscape of Mobile Apps. Mobile apps can be divided into three categories, namely, (i) *native*, (ii) *web*, and (iii) *hybrid* apps.

Native apps are binary, platform-specific files which are installed on the device and execute by interacting with a set of API calls exposed by the mobile OS. As a consequence, they must be developed in the OS-specific language (i.e., Java/Kotlin for Android and Objective-C/Swift for iOS), and they have full and direct access to the OS API. On one hand, native apps exhibit the best performance for CPU-intensive workloads, while, on the other hand, they need to be re-implemented to execute on a different mobile OS. As this is a daunting task mostly for small-medium enterprises, there is a growing trend towards web or hybrid apps.

Web apps render HTML5 and execute Javascript code within the device browser (which is a native app). For this reason, they are highly portable and platform-independent, but the interaction with the underlying OS is limited to the API accessible by the browser itself. Consequently, they have restricted functionalities and, in general, limited performance.

Hybrid apps have been proposed to overcome the limitations of both native and web apps, namely granting (i) portability over platforms, (ii) access to the whole OS API and, (iii) reasonable performance. Hybrid apps are programmed once in cross-platform web technologies (i.e., HTML, CSS, and JavaScript) as web apps, and then wrapped into a platform-specific native container, i.e., the WebView. The WebView may also allow the interaction between the web and the native part, acting like a *bridge* between the web code and the host OS API, thereby allowing to render HTML/CSS content, execute JavaScript code, and get access to the full OS API.

WebView. The WebView is an Android app component which embeds a mini-browser for rendering HTML/web pages and execute JavaScript code in mobile apps.

The WebView allows defining ad-hoc interfaces, called `Javascript-Interfaces`, that enable to invoke Java methods from the JavaScript code. This feature allows cross-platform frameworks (e.g., Cordova, PhoneGap) to design a set of plugins that can be embedded in apps and offer platform-specific functionality, such as the API for the file-system or the GPS location. To enable JavaScript interfaces, the developer needs to bind a set of Java methods to a WebView component using the `addJavascriptInterface` method. The communication between the JavaScript and the Java code is handled by the WebView using *asynchronous* callbacks. In detail, when some JavaScript code invokes Java code through an interface bounded to the WebView, it does not wait for the result: instead, when the result is ready, the Java code outside the WebView invokes a JavaScript callback function, passing the result back to the web page. This mechanism provides improved app performance and responsiveness, particularly in the case of long-running operations that would block the UI.

WebView Security Mechanisms. As the WebView deals with web content that can include untrusted HTML and JavaScript code, it can suffer from well-known web security vulnerabilities such as cross-site scripting [6,7,17] or file-based cross-zone scripting [9]. As countermeasures, the Android OS includes a set of mechanisms aimed at limiting the capability of the WebView to the minimum functionality required by hybrid apps. By default, the WebView does not execute JavaScript, thus requiring developers to enable this feature using the `setJavascriptEnabled` method. Besides, the application can either enable or disable the access of the WebView to specific resources like files, databases or geolocation [30] through the `WebSettings` object.

Since API 17, the Java methods – which are exposed through a JavaScript interface – need to be explicitly annotated with the `@JavascriptInterface` [16]. The aim is to restrict the access to the OS API, in order to prevent the invocation of any public Java method through code reflection [1].

To further increase the resilience of the WebView component against untrusted contents, since API level 21 the Android OS implements the Web-View as an independent app, thus offering a centralized update mechanism that relieves the developer from the burden of manually updating each hybrid app [31].

Moreover, since API Level 26 the WebView renderer executes in a separate process [33]. Finally, since Android 8, the WebView incorporates Google's Safe Browsing protections to detect and warn users about potentially dangerous websites. Unfortunately, this option needs to be explicitly enabled by the developer through a specific tag in the Android Manifest [32].

2.1 Frame Confusion

Frame Confusion is a vulnerability affecting hybrid apps that allows malicious interactions among the main web page (hereafter, *main frame*) and different web domains hosted in inner Iframe elements (hereafter, *child frames*) through the asynchronous bridge between the Java and the JavaScript code, granted by the WebView. To this aim, the WebView maintains a map that links the Web-View instance with a list of function calls in the native code registered to the *JavascriptInterface*. However, such a map does not include any restriction on web domains (and thus web pages) that can access the attached interfaces. Thus, if the main frame contains multiple child frames, each of them can independently and asynchronously access all the interfaces bound to the WebView component, in order to interact with the Java code. For this reason, the WebView is forced to return the results of each native method invocation to the main frame and not to the actual caller, be it the main frame or a child frame, thereby causing potentially unintended interactions between different frames, i.e., the *Frame Confusion*.

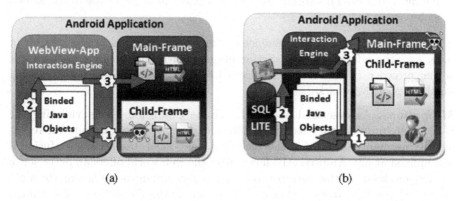

(a)   (b)

Fig. 1. Exploitation of Frame Confusion from (a) the child frame and (b) the main frame.

Such interaction between the native and the web worlds allows bypassing the Same Origin Policy (SOP), which - in a standard web browser - completely isolates the contents of the main frame from the child frames, since they belong to different domains.

Attacking and Exploiting the Frame Confusion. The Frame Confusion vulnerability can be exploited either by using a compromised child frame or the main frame, as shown in Fig. 1 (taken from [22]). In detail, if an attacker is able to compromise a child frame (Fig. 1a), he triggers the invocation of native function calls through the WebView App (step 1), which computes the result (step 2) and sends the callback to the main frame (step 3). For instance, a malicious advertisement campaign - embedded in a child frame - can exploit this attack and affects the main frame, by, e.g., inducing an unwanted phone call or force the sending of an SMS.

On the other hand, in case of a compromised main frame (Fig. 1b), the attacker is able to intercept all the callbacks triggered by the child frames, thus leading to possible information leaks. As an example, a benign child frame could inadvertently expose sensitive information like, e.g., the GPS location or the result of a SQL query, to the main frame in control of the attacker, through a native method invocation.

The exploitation of the Frame Confusion vulnerability requires the attacker to affect any web domain in the main or a child Iframes that has access to the JavaScript interfaces. This condition is achieved through:

- *The direct control of a web page.* In such a scenario, the attacker can be able either to take control over an existing web domain or to create an ad-hoc website, e.g., a malicious advertisement campaign.
- *The injection of malicious code in an existing web page.* In this case, the attacker can exploit a weakness in the communication protocol of the hybrid app, e.g., a clear-text communication or a misconfigured SSL connection, to mount a Man-In-The-Middle attack[3] and inject malicious code in the loaded web pages.

It is worth noticing that the presence of other vulnerabilities in the JavaScript code, e.g., the adoption of JavaScript libraries with known vulnerabilities [25] or the presence of XSS vulnerabilities [4,6,28], further boosts the exploiting capabilities of the attacker.

Mitigations. As described above, the Frame Confusion allows violating the SOP by circumventing the sandbox of Iframes. Unfortunately, despite the recent security mechanisms added in the WebView component, the Frame Confusion is still unfixed at any Android API level. Still, the web world offers an extra set of security mechanisms that are able to restrict the communication among the main frame and the child frames, thus preventing the Frame Confusion vulnerability, i.e.:

- the Iframe `sandbox` attribute [29], which enables a set of extra restrictions on any content hosted by an Iframe and, among them, it allows blocking the execution of JavaScript code. Although effective in principle, this mechanism completely prevents the execution of any JavaScript code, thus limiting the functionalities of the web page.
- the Content Security Policy (CSP) [10] that allows for the definition of fine-grained restrictions on the execution of JavaScript code, including the possibility to define a set of trusted domains that are able to execute JavaScript,

---

[3] https://www.owasp.org/index.php/Man-in-the-middle_attack.

in a white-listing fashion. Although effective against the loading of an undesired web domain, the CSP cannot prevent the injection of the malicious code in a white-listed domain, thereby resulting ineffective against the Frame Confusion.

Furthermore, previous security mechanisms are not enabled by default, thus leaving the burden of their configuration to the developer. All in all, at the current state of the art, none of the existing security mechanisms are able to effectively prevent the Frame Confusion.

3   A Frame Confusion Detection Methodology

The lack of a solution for preventing the Frame Confusion asks for – at least – a methodology to automatically detect such vulnerability. Unfortunately, at the current state of the art, the only way to detect Frame Confusion is through manual source-code inspection, mostly carried out by app developers. Such activity is error-prone and requires good skills in security analysis by the developing team. Furthermore, the complexity of Frame Confusion leads developers to false positives/negatives or, in the worst case, to underestimate or ignore the problem. To overcome this limitation, we propose a novel methodology for the automatic identification of the Frame Confusion in Android. To achieve such result, we first define a blueprint of the Frame Confusion vulnerability, and then we build an analysis flow that is able to detect it automatically, by exploiting a fruitful combination of static and dynamic analysis techniques.

3.1   Vulnerability Blueprint

The design of an automatic and rigorous analysis flow for the Frame Confusion vulnerability demands for the selection of a set of features that *enable* the vulnerability. To this aim, we argue that a minimal set of such features is the following:

1. the app requires the Internet permission in order to access web domains using a WebView component;
2. the app uses at least a WebView (W) that is configured to execute JavaScript code;
3. W sets at least a `JavascriptInterface` (JI);
4. JI injects at least a public Java method (m) that can be accessed from the JavaScript code;
5. in case of an app targeted to API level 17 or higher, m needs to be further annotated with the `@javascriptinterface` tag;
6. W loads at least a web page (WP) that contains one or more Iframe elements;
7. WP does not enforce any mitigation technique among those described in the previous section.

3.2   Detection Algorithm

The Frame Confusion detection methodology can be summarized by the pseudocode listed in Algorithm 1. Given a generic Android app in *.apk* format, the algorithm begins by retrieving a list of the Android permissions used by the app (row 1). If the list does not include the Internet permission, then the app cannot use the WebView component, and therefore it is marked as *not vulnerable* (rows 2–4). Otherwise, the algorithm computes the list of all the invoked methods of the app (row 5) in order to locate the presence of setJavaScriptEnabled, and addJavascriptInterface APIs.

If a setJavaScriptEnabled invocation (row 9) is recognized, the algorithm further investigates the flag parameter of the call (rows 10–14). A True value indicates that the WebView enables the execution of JavaScript and thus its object reference is retrieved (row 12) and included in the list of those that enable JavaScript (row 13).

Instead, the presence of a addJavascriptInterface indicates that a WebView component is configured to expose a bridge between Java and JavaScript. If this is the case, the algorithm extracts *(i)* the WebView object from which the addJavascriptInterface method is invoked (row 17), and *(ii)* the Java object injected in the JavascriptInterface (row 18). After that, the algorithm needs to detect if the Java object injected in the interface contains public methods that can potentially be accessed from JavaScript code (rows 19–27). Moreover, in case of apps targeted to API level 17 or above, the public methods of the object need to be further annotated with the @javascriptinterface tag (rows 19–22). If the injected Java object contains methods accessible from JavaScript, then the corresponding WebView instance can be added to the list of those that expose potentially vulnerable interfaces (row 21 or row 25).

Next, if the analysis is not able to find at least a WebView - with JavaScript enabled - that contains a JavaScript interface with exposed Java methods, then the app is marked as not vulnerable (rows 29–34). Otherwise, the analysis collects from the Website collector module all the website pages accessed by the WebView that are (i) included in the resources of the *.apk* package (row 35), (ii) statically invoked by loadURL methods (row 36), and (iii) dynamically reached during the execution of the app (row 37).

Thereafter, the algorithm collects every website that uses at least an Iframe element that loads an external page (either embedded in HTML pages or generated by JavaScript) and that does not enforce any of the mitigation techniques discussed in the previous section (rows 40–48). Finally, if the app loads at least one vulnerable website, it is marked as *vulnerable*. On the contrary, if the app uses the appropriate security mechanisms or does not use any Iframe is marked as *non vulnerable*.

Algorithm 1. Frame Confusion Detection

Input : APK Package
Output: vulnerable, notVulnerable

1 listPermissions = `getPermissionFromApk`(app);
2 **if** *"android.permission.INTERNET"* **not in** listPermissions **then**
3 | **return** notVulnerable;
4 **end**

5 methodsList = `getAllInvMet`(app);
6 JSWebView = list();
7 IWebView = list();

8 **foreach** *method* in methodsList **do**
9 | **if** *method.getName == "setJavaScriptEnabled"* **then**
10 | flagParam = `getFlagParam`(*method*);
11 | **if** flagParam == *True* **then**
12 | webViewObj = `getInvObj`(*method*);
13 | JSWebView.add (webViewObj);
14 | **end**
15 | **end**
16 | **else if** *method.getName == "addJavascriptInterface"* **then**
17 | webViewObj = `getInvObj`(*method*);
18 | interface = `getInterfaceObj`(*method*);
19 | **if** `getSDK`(app) > 17 **then**
20 | **if** `containAnnotatedPubMet` *(interface)* **then**
21 | | IWebView.add (webViewObj);
22 | **end**
23 | **end**
24 | **else if** `containPubMet`(interface) **then**
25 | | IWebView.add (webViewObj);
26 | **end**
27 | **end**
28 **end**

29 **if** `len` *(JSWebView)* == 0 **or** `len` *(IWebView)* == 0 **then**
30 | **return** notVulnerable;
31 **end**
32 **if** `len` *(IWebView ∩ JSWebView)* == **0 then**
33 | **return** notVulnerable;
34 **end**

35 resourceFiles = `getAllResourceApk`(app);
36 dumpWebStat = `getStaticUrl`(methodsList);
37 dumpWebDyn = `getDynamicUrl`(app);
38 filesToCheck = dumpWebDyn **union** resourceFiles **union** dumpWebStat;
39 vulnerablePages = list();

40 **foreach** *file* in filesToCheck **do**
41 | **if** `isHTMLfile`(*file*) **or** `isJSfile`(*file*) **then**
42 | **if** `containIframe`(*file*) **then**
43 | **if not** `containCSP`(*file*) **and not** `containSandboxAtt`(*file*) **then**
44 | | vulnerablePages.add (file);
45 | **end**
46 | **end**
47 | **end**
48 **end**

49 **if** `len` *(vulnerablePages)* > 0 **then**
50 | **return** vulnerable;
51 **end**
52 **return** notVulnerable;

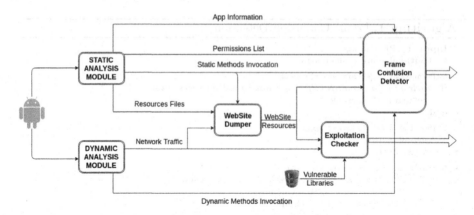

Fig. 2. The FCDroid Architecture.

4   The FCDroid tool

FCDroid[4] implements the proposed detection methodology to automatically identify the presence of the Frame Confusion vulnerability in Android apps.

The rest of the section discusses *(i)* the implementation challenges addressed by FCDroid and *(ii)* its architecture, emphasizing the underlying tools and technologies.

4.1   Implementation Challenges

The Frame Confusion detection methodology poses several challenges in terms of implementation. Indeed, an automatic detection tool needs to:

1. achieve maximum coverage, i.e., by detecting all possible app execution paths that may lead to the vulnerability;
2. recognize the actual configuration of WebView components, which may *dynamically* enable JavaScript or define new interfaces;
3. analyze all the web pages loaded inside some potentially vulnerable Web-Views, by also considering those loaded according to (i) the user's input, and (ii) the value of runtime variables.

To address such challenges, an automatic tool can rely on static and dynamic analysis techniques. Static analysis techniques can examine all possible execution paths and variable values, not just those invoked during execution. However, static approaches can *(i)* introduce false positives and *(ii)* be unable to detect complex scenario, like, e.g., values provided by the user or resources loaded at runtime. On the other hand, dynamic analysis techniques allow to detect the actual behavior of the app, but it is limited by *(i)* the coverage of the analysis and *(ii)* the time required for the analysis, thus producing potential false negatives.

---

[4] FCDroid is available at https://www.fcdroid.com.

To this aim, FCDroid combines static and dynamic analysis techniques to overcome the limitations of both techniques and achieve more accurate detection results.

4.2 FCDroid Architecture

The FCDroid architecture, depicted in Fig. 2, is composed by five main building blocks: the Static Analysis Module (SAM), the Dynamic Analysis Module (DAM), the WebSite Dumper (WD), the Frame Confusion Detector (FCD), and the Exploitation Checker (EC).

Static Analysis Module (SAM). The Static Analysis Module relies on *Apktool* [34] to disassemble the app package (in the *.apk* format) and translate the Dalvik bytecode contained in the app into Smali [14] language. In addition to that, SAM brings the resources contained in the app back to their original form, e.g., from binary compiled XML files into textual XML files. Then, the module extracts the list of permissions requested by the app and the target Android API level according to the content of the `AndroidManifest.xml` file. Finally, the SAM inspects each extracted Smali file in order to locate all the API invocations related to the WebView component. In detail, the module detects:

- `setJavaScriptEnabled` that enables the JavaScript code in a WebView object. If found, the SAM also extracts the variable containing the boolean flag passed as an argument;
- `addJavascriptInterface`, that creates a JavaScript interface object. In this case, the SAM retrieves the Java class of the injected object and the name assigned to the interface;
- `loadUrl` and `evaluateJavaScript`, that allows the loading of specific URLs or JavaScript code inside the WebView. In case, the module also extracts the URL address or the script code, if statically defined;

The collected pieces of information are then sent to the WebSite Dumper and the Frame Confusion Detector to continue the analysis.

Dynamic Analysis Module (DAM). The Dynamic Analysis Module is in charge of executing the app into a controlled testing environment in order to monitor the stimulation of the WebView components at runtime. To this aim, it installs the app into an Android Emulator and stimulates the app automatically, trying to explore its possible execution states. This allows the DAM to *(i)* monitor the invocations of WebView-related API along with their execution parameters, and *(ii)* intercept all the network traffic generated by the app. In order to stimulate the app automatically, the DAM relies on DroidBot [21], an open-source tool that can automatically explore the app UI and mimic the interaction with a user. Unlike many existing input generators that rely on static analysis and instrumentation of the app to generate inputs, DroidBot works in black-box mode, i.e., it does not need to know in advance the structure of the app, and it is resilient to obfuscation techniques. In order to keep track of API invocations, the DAM module provides the Android emulator with an ApiMonitor module. ApiMonitor, based on the Xposed[5] framework, allows the DAM to intercept and collect each method executed by the app during the analysis, saving its invocation and the value of parameters on a JSON file. Furthermore, the DAM module intercepts and stores all the network traffic generated by the app using the HTTP/HTTPs proxy *mitmproxy* [12].

WebSite Dumper (WD). The WebSite Dumper module aims at extracting and retrieving all the websites invoked by the WebView components. To do that, it retrieves from the SAM and DAM modules the list of URLs accessed by app WebView components.

For each identified URL, the WebSite Dumper determines whether it refers to a local or a remote resource. In the first case, it collects and stores the static resource obtained by the app package. In the latter case, the WD module dumps the content of the remote website by downloading the web pages recursively, up to a maximum of 3 levels deep, by using the *wget* tool[6]. Finally, the module polishes the results and maintains only HTML and JavaScript files that will be inspected by both the FCD and the EC module.

Frame Confusion Detector (FCD). The Frame Confusion Detector module implements the core logic of FCDroid for the detection of the vulnerability. At first, FCD collects from the SAM the list of permissions of the app and verifies that the app requires the Internet permission. If so, the module analyzes the list of invoked APIs (both those statically extracted by the SAM module and those evaluated at runtime by the DAM) to verify the existence of at least a WebView instance that enables JavaScript and exposes a JavaScript interface. Furthermore, if an exposed interface is found, the FCD parses the class of the injected Java object to determine the existence of methods that can be accessed from JavaScript.

Finally, the FCD also needs to detect the amount of potentially-vulnerable webpages. To this aim, the module collects the websites dumped by the WD and checks whether a page contains at least an Iframe element and does not enforce any mitigation techniques (i.e., the `Content-Security-Policy` meta tag in the HTML header or the sandbox attribute). At the end of the analysis, the FCD module marks the application as *vulnerable* or *not vulnerable*.

Exploitation Checker (EC). The Exploitation Checker is the module responsible for the detection of app configurations that can boost the exploitation of the Frame Confusion Vulnerability. In details, the EC can identify:

- *The adoption of unencrypted communication channels*, by analyzing the network traffic generated by the DAM module and by extracting the list of URLs that are accessed in plain HTTP.
- *The presence of buggy/vulnerable Javascript libraries* by relying on the RetireJS [13] tool, which allows obtaining a list of known-to-be-vulnerable JavaScript libraries that are executed within the WebView.

---

[5] https://repo.xposed.info/.
[6] https://www.gnu.org/software/wget/.

Table 1. Statistics on the Frame Confusion blueprint.

	Percentage	Ratio
Internet Permission	96.45%	48226/50k
Use WebView	49.35%	24675/50k
JavaScript Enabled	49.35%	24675/50k
JavaScript Interface	44.84%	22420/50k

Table 2. Statistics on the web pages accessed by hybrid apps.

	Percentage	Ratio
Web pages with Iframes	1.2%	87k/6.7M
Web pages with Iframes and CSP	27.96%	24108/87k
Web pages with Iframes and sandbox attribute	0%	0/87k

Table 3. Statistics on the exploiting conditions of vulnerable apps.

	Percentage	Ratio
Insecure connections	59.98%	982/1637
XSS vulnerabilities	27.48%	450/1637
Vulnerable JS libraries	79.96%	1309/1637

- *The presence of JavaScript code vulnerable to DOM-XSS attacks*[7], by including a customized implementation of JSPrime [24]. Such tool inspects the JavaScript code in order to detect unsanitized input variables that could allow an attacker to execute arbitrary JavaScript code in the victim's WebView.

5   Experimental Results

We empirically assessed the reliability of the proposed methodology, by systematically analyzing 50.000 apps with FCDroid[8]. Such apps have been downloaded from the Google Play Store in December 2018, and they are the top free Android apps ranked by the number of installations and average ratings according to Androidrank [2]. Our experiments were conducted using an Intel Xeon 3106@1.70 GHz, with 32 GB RAM, running Ubuntu 18.04.

Frame Confusion Identification. The FCDroid tool discovered that 49.35% of apps (i.e., 24675 out of 50000) are hybrid, since they use at least a WebView component, thereby highlighting the wide adoption of such component in the Android ecosystem.

As shown in Table 1, all the apps with at least one WebView component enable the execution of JavaScript, while 44.84% also attach (at least) a

---

[7] https://www.owasp.org/index.php/DOM_Based_XSS.
[8] The complete list of experimental results is available at https://www.fcdroid.com/results.

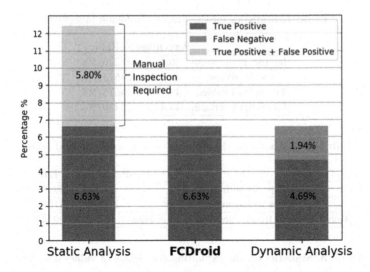

Fig. 3. FCDroid vs static and dynamic analysis

JavaScript interface, which contains properly annotated methods. Such methods can be invoked from the websites loaded inside the apps.

Furthermore, FCDroid inspected all the websites accessed by the hybrid apps obtaining the results described in Table 2. In detail, 1.2% (87k/6.7M) of websites contain at least an Iframe element; among those pages, the 27.96% include CPS policies while none of the visited pages enforces the sandbox attribute. Therefore, such findings indicate that most of the web pages that use Iframe elements are potentially vulnerable to Frame Confusion. Finally, *our analysis identifies that 6.63% (i.e., 1637) of hybrid apps are potentially vulnerable to Frame Confusion.* To estimate the impact of such results on the Android users' community, we cross-referenced our findings with the Google Play Store meta-data, obtaining that the total sum of official installations for vulnerable apps is greater than 250.000.000.

Exploitation Conditions. We further inspected the vulnerable apps with FCDroid, in order to detect the presence of other vulnerabilities in the app configuration that can make easier the exploitation of the Frame Confusion. As shown in Table 3, 59.98% of vulnerable apps access websites using an insecure connection, i.e., plain HTTP, while 27.48% contain code vulnerable to XSS attacks. Finally, 79.96% of vulnerable apps include JavaScript libraries with known security vulnerabilities.

Advantages and Limitations of FCDroid. FCDroid combines static and dynamic analysis techniques to maximize the detection accuracy. To prove that, we compared the analysis results obtained by the static and the dynamic analysis with the hybrid approach of FCDroid, as shown in Fig. 3. In detail, the static analysis allows to identify 12.43% of apps as potentially vulnerable to Frame Confusion. Unluckily, such amount contains both true and false positives. To

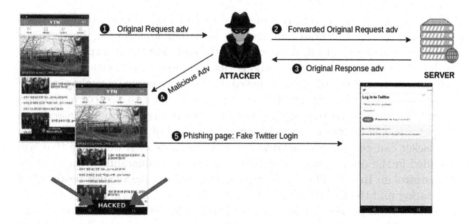

Fig. 4. YTN news: flow of the attack.

discriminate, each app should be manually inspected, through a time-consuming and error-prone process. On the contrary, the dynamic analysis is not exhaustive, i.e., it may be unable to reach statically defined web pages, like those hardcoded in the app package, that are loaded in WebView components. Therefore, only 4.69% of apps are successfully detected as vulnerable (true positives) through dynamic analysis.

To overcome such limitations, the mixed approach (i.e., static and dynamic analysis) adopted by FCDroid detected 6.63% of (true positive) vulnerable apps automatically. Indeed, FCDroid allowed to automatically validate more than half of the potentially positive results detected by the static analysis. Furthermore, FCDroid increased the detection rate of the dynamic analysis by 1.94%.

Nonetheless, the experimental results also unveiled some limitations of the current FCDroid implementation. First, the dynamic analysis is limited to the public surface of the app (i.e., the one that does not require any user authentication) and executes in a predefined time-frame (i.e., 60 s). Furthermore, the implementation of FCDroid does not detect dynamically-generated Iframe elements, like, i.e., those created at runtime by the JavaScript code.

6   Attacking and Exploiting the Frame Confusion

In this section, we discuss the impact of a successful exploitation of Frame Confusion by attacking a news app (i.e., YTN News[9]) which has been found vulnerable by FCDroid. At the moment of writing, YTN is available on the Google Play Store and has more than 1M downloads.

We manually reverse-engineered and analyzed the app: it uses the WebView component to load a home page with several Iframes. The Iframe at the bottom

---

[9] https://play.google.com/store/apps/details?id=com.estsoft.android.ytn

of the web page loads an advertisement (step 1 in Fig. 4). Our manual investigation confirms the FCDroid findings: the WebView uses HTTP, JavaScript is enabled, and there is an interface exposed through addJavascriptInterface. The interface exposes different methods that are able to get some information about the device. One of these exposed methods is named liveLogin. This method has three parameters of type *string*, the first two are converted into integers and used to customize the WebView, while the last one is passed as a parameter to the loadUrl method without any kind of sanitization. Therefore, an attacker can easily inject arbitrary JavaScript code or a web page that will be loaded in the main frame. In order to exploit the vulnerability, the attacker must control an Iframe. There exist two approaches to achieve such result: (1) if the attacker and the mobile device belong to the same network, the attacker can carry out a Man-in-the-Middle (MitM) attack, otherwise, (2) the attacker can create an ad-hoc advertising campaign. In our use case, we carried out a MitM attack, and we were able to control the advertisement (steps 2, 3 and 4 in Fig. 4). For the sake of precision, since the WebView uses HTTP, the attacker can also target the main frame; however, in this example, we focused on a child frame, since we only aim to prove the exploitability of Frame Confusion. Thus, given the absence of any security mechanisms, we can access the exposed interface and exploit the Frame Confusion by invoking the liveLogin method with a URL pointing to our malicious web page (step 5 in Fig. 4) containing a fake Twitter login page.

It is worth pointing out that the WebView is a promising vector attack for phishing, as there are no GUI components that prompt the actual URL and the transport protocol (e.g., HTTP/HTTPS), thereby making hard to distinguish between the legitimate Twitter website and a well-crafted phishing site [3].

As a final remark, it is worth noticing that this is just one among a set of potential exploitation of Frame Confusion under such specific app settings. For instance, it is possible to download a large file (since the app has the WRITE-_EXTERNAL_STORAGE permission) or continue to load the same pages within the WebView to carry out simple Denial-of-Service attacks.

7   Related Work

The steady growth of hybrid apps has attracted the attention of both academic and industrial security research communities. The main approaches for the security analysis of hybrid apps can be divided into static and dynamic. In static analysis methodologies, the hybrid app is analyzed according to its source (or binary) code without being executed. For instance, Lee et al. proposed HybriDroid [18], a static analysis framework that examines the inter-communication between the native and the web counterpart of the app to identify development bugs or potential leaks of sensitive information. Other works, like [8,27], and [35] propose some detection methodologies for *code injection attacks* based on app-instrumentation or machine learning techniques. However, any of the proposed static analysis techniques suffer from the over-approximation of the app execution paths which drastically reduce the accuracy due to a high rate of false positives [19]. On the other hand, dynamic analysis techniques aim at analyzing the security of the app runtime behavior in a controlled environment. The sole work based on dynamic analysis techniques for hybrid apps is BridgeTaint, proposed by Bai et al. [5]. BridgeTaint tracks sensitive data exchanged through the bridge and uses a cross-language taint mapping method to perform the taint analysis in both domains. Although dealing with the dynamic monitoring of the bridge between the Java and the JavaScript worlds, BridgeTaint is only focused on data analysis aimed at the identification of data leaks. Anyway, none of the approaches mentioned above is either able to identify the Frame Confusion vulnerability. The work proposed by Luo et al. [22] is the only research paper that explicitly discusses this vulnerability. Indeed, the authors – who also coined the term *"Frame Confusion"* – have also studied the security implications of the two-way interaction between the native and the web code in hybrid apps. Anyway, they focus only on detecting the security weaknesses of the WebView component and the JavaScript interfaces, as well as some statistics on the usage of the WebView API and JavaScript interfaces. Furthermore, their analysis is manual, and it has been carried out on a reduced dataset made by only 132 apps.

To the best of our knowledge, our methodology is the first approach allowing us to systematically detect the Frame Confusion vulnerability. Furthermore, the adoption of both static and dynamic analysis techniques allows overcoming the limitations of both approaches.

8   Conclusion

In this work, we have proposed a methodology for systematically detecting the Frame Confusion vulnerability in hybrid Android apps. Then, we have implemented a tool, FCDroid, based on our methodology, which combines static and dynamic analysis techniques to reduce false positive and false negative rates. The results obtained with FCDroid show that Frame Confusion is a concrete problem: among the top 50.000 apps by installations on the Google Play Store, 24675 use the WebView component and we find that 1637 apps (i.e., about the 6.63% among the ones with at least a WebView component) are vulnerable. Although we took into consideration only the top apps, the Frame Confusion already affects more than 250.000.000 installations. Moreover, we have also discovered that about 59.98% of such vulnerable apps load the page within the WebView using a clear-text connection thereby easing phishing attacks.

Future extension of this research will be the study of proper remediations that could prevent the Frame Confusion without disabling the execution of JavaScript inside Iframes in hybrid apps. As a final remark, our attack to the YTN news app[10] suggests that the WebView is a promising vector for phishing attacks, as the user has no way to discriminate whether she is interacting with the legitimate website or a well-crafted phishing one.

---

[10] We responsibly disclosed our finding to the app developers in January 2019.

References

1. Thomas, D.R., Beresford, A.R., Coudray, T., Sutcliffe, T., Taylor, A.: The lifetime of Android API vulnerabilities: case study on the JavaScript-to-Java interface. In: Christianson, B., Švenda, P., Matyáš, V., Malcolm, J., Stajano, F., Anderson, J. (eds.) Security Protocols 2015. LNCS, vol. 9379, pp. 126–138. Springer, Cham (2015). https://doi.org/10.1007/978-3-319-26096-9_13
2. AndroidRank: Androidrank market data (2018). https://www.androidrank.org/
3. Aonzo, S., Merlo, A., Tavella, G., Fratantonio, Y.: Phishing attacks on modern android. In: Proceedings of the ACM Conference on Computer and Communications Security (CCS), Toronto, Canada, October 2018
4. Backes, M., Gerling, S., Styprekowsky, P.V.: A Local Cross-Site Scripting Attack against Android Phones, pp. 1–6. Saarland University (2011). http://www.infsec.cs.uni-saarland.de/projects/android-vuln/
5. Bai, J., Wang, W., Qin, Y., Zhang, S., Wang, J., Pan, Y.: BridgeTaint: a bi-directional dynamic taint tracking method for JavaScript bridges in Android hybrid applications. IEEE Trans. Inf. Forensics Secur. **14**(3), 677–692 (2019). https://doi.org/10.1109/TIFS.2018.2855650
6. Bao, W., Yao, W., Zong, M., Wang, D.: Cross-site scripting attacks on android hybrid applications. In: Proceedings of the 2017 International Conference on Cryptography, Security and Privacy - ICCSP 2017, pp. 56–61. ACM Press, New York (2017). https://doi.org/10.1145/3058060.3058076, http://dblp.uni-trier.de/db/conf/iccsp/iccsp2017.html
7. Bhavani, A.B.: Cross-site scripting attacks on Android WebView. Int. J. Comput. Sci. Netw. **2**(2), 1–5 (2013)
8. Chen, Y.L., Lee, H.M., Jeng, A.B., Wei, T.E.: DroidCIA: a novel detection method of code injection attacks on HTML5-based mobile apps. In: 2015 Proceedings of 14th IEEE International Conference on Trust, Security and Privacy in Computing and Communications, vol. 1, pp. 1014–1021 (2015). https://doi.org/10.1109/Trustcom.2015.477
9. Chin, E., Wagner, D.: Bifocals: analyzing WebView vulnerabilities in Android applications. In: Kim, Y., Lee, H., Perrig, A. (eds.) WISA 2013. LNCS, vol. 8267, pp. 138–159. Springer, Cham (2014). https://doi.org/10.1007/978-3-319-05149-9_9
10. Content Security Policy: Content security policy (2016). http://content-security-policy.com, https://developers.google.com/web/fundamentals/security/csp/
11. Apache Cordova: (2018). https://cordova.apache.org/
12. Cortesi, A., Hils, M., Kriechbaumer, T., Contributors: mitmproxy: a free and open source interactive HTTPS proxy (Version 4.0) (2010). https://mitmproxy.org/
13. Erlend, O.: RetireJS - Scanner detecting the use of JavaScript libraries with known vulnerabilities (2019). https://retirejs.github.io/retire.js/
14. Gruver, B.: Smali - Assembler/Disassembler for the dex format (2019). http://github.com/JesusFreke/smali/
15. Hu, J.: A tale of two cities : how WebView induces bugs to Android applications, vol. 1, pp. 702–713 (2018). https://doi.org/10.1145/3238147.3238180
16. JavascriptInterface: (2019). https://developer.android.com/reference/android/webkit/JavascriptInterface
17. Jin, X., Hu, X., Ying, K., Du, W., Yin, H., Peri, G.N.: Code injection attacks on HTML5-based mobile apps. In: Proceedings of 2014 ACM SIGSAC Conference on Computer and Communications Security - CCS 2014, pp. 66–77 (2014). https://doi.org/10.1145/2660267.2660275

18. Lee, S., Dolby, J., Ryu, S.: HybriDroid: static analysis framework for Android hybrid applications. In: Proceedings of 31st IEEE/ACM Int. Conference on Automated Software Engineering - ASE 2016, pp. 250–261 (2016). http://dl.acm.org/citation.cfm?doid=2970276.2970368

19. Li, L., et al.: Static analysis of android apps: a systematic literature review. Inf. Softw. Technol. **88**, 67–95 (2017). https://doi.org/10.1016/j.infsof.2017.04.001

20. Li, T., et al.: Unleashing the walking dead : understanding cross-app remote infections on mobile WebViews. In: CCS, pp. 829–844 (2017). https://doi.org/10.1145/3133956.3134021, https://acmccs.github.io/papers/p829-liA.pdf

21. Li, Y., Yang, Z., Guo, Y., Chen, X.: DroidBot: a lightweight UI-guided test input generator for android. In: Proceedings of 2017 IEEE/ACM 39th International Conference on Software Engineering Companion, ICSE-C 2017, pp. 23–26 (2017). https://doi.org/10.1109/ICSE-C.2017.8

22. Luo, T., Hao, H., Du, W., Wang, Y., Yin, H.: Attacks on WebView in the Android system. In: Proceedings of the 27th Annual Computer Security Applications Conference on - ACSAC 2011, p. 343 (2011). https://doi.org/10.1145/2076732.2076781

23. Neugschwandtner, M., Lindorfer, M., Platzer, C.: A view to a kill: WebView exploitation. In: LEET (2013). http://publik.tuwien.ac.at/files/PubDat_223415.pdf

24. Das Patnaik, N., Sabyasachi Sahoo, S.: JSPrime (2013). https://dpnishant.github.io/jsprime/

25. OWASP: using components with known vulnerabilities (2017). https://www.owasp.org/index.php/Top_10-2017_A9-Using_Components_with_Known_Vulnerabilities

26. Adobe PhoneGap: (2018). https://phonegap.com/

27. Rizzo, C., Cavallaro, L., Kinder, J.: BabelView: evaluating the impact of code injection attacks in mobile Webviews (2017). http://arxiv.org/abs/1709.05690

28. Sedol, S., Johari, R.: Survey of cross-site scripting attack in Android Apps. Int. J. Inf. Comput. Technol. **4**(11), 1079–1084 (2014)

29. w3: Sandbox attribute (2018). https://www.w3.org/wiki/Html/Elements/iframe

30. WebSetting: (2019). https://developer.android.com/reference/android/webkit/websettings

31. WebView: (2019). https://play.google.com/store/apps/details?id=com.google.android.webview&hl=en

32. WebViewSafeBrowsing: (2018). https://developer.android.com/guide/webapps/managing-webview

33. WebViewSecurity: (2017). https://android-developers.googleblog.com/2017/06/whats-new-in-webview-security.html

34. Wiśniewski, R., Tumbleson, C.: Apktool A tool for reverse engineering Android apk files (2018). http://ibotpeaches.github.io/Apktool/

35. Yan, R., Xiao, X., Hu, G., Peng, S., Jiang, Y.: New deep learning method to detect code injection attacks on hybrid applications. J. Syst. Softw. **137**, 67–77 (2018). https://doi.org/10.1016/j.jss.2017.11.001

Privacy

Geo-Graph-Indistinguishability: Protecting Location Privacy for LBS over Road Networks

Shun Takagi$^{(\boxtimes)}$ (ID), Yang Cao (ID), Yasuhito Asano (ID), and Masatoshi Yoshikawa (ID)

Graduate School of Informatics, Department of Social Informatics, Kyoto University,
Kyoto, Japan
`shun.0721@hotmail.co.jp`

Abstract. In recent years, Geo-Indistinguishability (GeoI) has been increasingly explored for protecting location privacy in location-based services (LBSs). GeoI is considered a theoretically rigorous location privacy notion since it extends differential privacy to the setting of location privacy. However, GeoI does not consider the road network, which may cause insufficiencies in terms of both privacy and utility for LBSs over a road network. In this paper, we first empirically evaluate the privacy guarantee and the utility loss of GeoI for LBSs over road networks. We identify an extra privacy loss when adversaries have the knowledge of road networks and the degradation of LBS quality of service. Second, we propose a new privacy notion, Geo-Graph-Indistinguishability (GeoGI), for protecting location privacy for LBSs over a road network and design a Graph-Exponential mechanism (GEM) satisfying GeoGI. We also show the relationship between GeoI and GeoGI to explain theoretically why GeoGI is a more suitable privacy notion over road networks. Finally, we evaluate the empirical privacy and utility of the proposed mechanism in real-world road networks. Our experiments confirm that GEM achieves higher utility for LBSs over a road network than the planar Laplace mechanism for GeoI under the same empirical privacy level.

Keywords: Location privacy · Geo-Indistinguishability ·
Road network · Differential privacy

1 Introduction

In recent years, the spread of smartphones and the improvement of GPS has led to a growing use of location-based services (LBSs). While such services have provided enormous benefits to individuals and society, the exposure of the user's location raises privacy issues. By using the location information, it is easy to identify sensitive personal information, such as that pertaining to home and

This work was supported by JSPS KAKENHI Grant Numbers (S) No. 17H06099, (A) No. 18H04093, (C) No. 18K11314.

© IFIP International Federation for Information Processing 2019
Published by Springer Nature Switzerland AG 2019
S. N. Foley (Ed.): DBSec 2019, LNCS 11559, pp. 143–163, 2019.
https://doi.org/10.1007/978-3-030-22479-0_8

family. Many methods for protecting location information have been proposed in the past decade. Most of these methods perturb the true location by using a location privacy-preserving mechanism before sending it to an LBS provider or sharing it with a third party. A mechanism takes a true location as input and outputs a perturbed location that follows a probability distribution over a location domain.

Andrés et al. [10] defined a formal notion of location privacy based on the well-known concept of differential privacy. The definition is called geo-indistingui-shability (GeoI), and an output of a mechanism achieving it guarantees indistinguishability of the true location from other locations, that is, strong privacy protection. This notion is derived from differential privacy [5], which provides a rigorous guarantee of indistinguishability of two neighboring databases. One of the most appealing features of GeoI, inherited from differential privacy, is the guarantee of privacy protection against any attacker to a certain degree.

However, since the proposal of GeoI, many studies [3,14,25] have identified its weaknesses. Yu et al. [25] showed that GeoI did not protect location privacy against the optimal inference attack [16], and they proposed the framework that adapted GeoI to the expected inference error [17], a complementary notion of GeoI. Chatzikokolakis et al. [3] focused on the fact that GeoI did not consider privacy for semantic information (such as population density) and proposed a mechanism that guarantees the protection of this privacy by using a graph whose weight of edges contains such information. However, this graph does not consider a road network. Oya et al. [14] quantified the privacy against an adversary who knows that the true location is one of two locations. The researchers showed that an output of the mechanism achieving GeoI results in a worse quality of service to protect the user's location privacy.

In this study, we find that GeoI provides inadequate privacy guarantee and insufficient utility for some LBSs over road networks, such as the k-nearest neighbor search (e.g., searching for k restaurants nearest to the user location). In such LBSs, the quality of service can be improved by taking advantage of the road network instead of the Euclidean space. This is because objects can usually move only on a predefined set of trajectories as specified by the road network and it is natural for these LBSs to use the distance on the road network [4,9,15]. GeoI may not be practical for LBSs over a road network because it assumes that (1) the perturbed location can be any location in a continuous plane and (2) the distance between locations is measured by the Euclidean distance.

Due to assumption (1), GeoI may result in unexpected privacy loss. For example, as shown in Fig. 1, if a user's perturbed location is unreasonable, such as a position on the sea, an adversary can realize that such a location is impossible, which may cause unexpected privacy leakage. Next, due to assumption (2), GeoI may offer inadequate utility for LBSs over a road network. Taking k-nearest neighbor search using shortest path for example, as shown in Fig. 2, a user of LBS searching the nearest restaurant expects that restaurant 2 will be returned in a higher probability than restaurant 1. This is because the path to restaurant 1 is farther than to restaurant 1 because of the river. However, if the user uses a

Fig. 1. Perturbation of the location to the unreasonable location.

Fig. 2. Difference of the Euclidean distance and the shortest path length on a road network.

GeoI mechanism, the probabilities outputting restaurants 1 and 2 are the same since the two Euclidean distances between restaurants and the user are the same.

In this paper, we study how to protect location privacy while preserving high utility for LBS over a road network. Our contributions are threefold. First, we identify two insufficiencies of GeoI by two empirical evaluations. We quantitatively analyze the change of the inference error w.r.t. adversaries having the knowledge of road network or not when the location is protected by the planar Laplace mechanism that Andrés et al. [10] proposed, and show its privacy leakage. Additionally, we propose the formulation of the utility of the mechanism for LBSs over a road network. We compute its utility of the planar Laplace mechanism on a graph (PLG), which is our extension of PLM for LBSs over a road network, on two real-world road networks, which shows that it performs poorly w.r.t. this formulation. Second, we propose a new privacy notion based on differential privacy, called Geo-Graph-Indistinguishability (GeoGI) that takes a road network into consideration so that we can construct a more suitable mechanism than GeoI mechanism for protecting location privacy in LBSs over a road network. We design a Graph-Exponential Mechanism (GEM) satisfying GeoGI. We also show the relationship between GeoI and GeoGI to explain theoretically why GeoGI is more suitable privacy notion for LBSs over a road network. Third, to better understand the proposed mechanism, we empirically evaluate privacy and utility of the approach in the case of two real-world road networks in Japan; the results verify that the proposed GEM for GeoGI achieves higher utility than PLM for GeoI when both mechanisms have the same empirical privacy level.

2 Preliminary and Problem Setting

2.1 Geo-Indistinguishability [10]

In this section, we describe the definition of Geo-Indistinguishability (GeoI). Let $\mathcal{X}$ be a set of locations and let $\mathcal{Z}$ be a set of query outputs. Intuitively, a mechanism K achieving GeoI guarantees that $K(x)$ and $K(x')$ are similar to a certain degree for any two locations $x, x' \in \mathcal{X}$. This means that even if an adversary obtains an output of the mechanism, he cannot distinguish the true location from other locations to a certain degree.

The multiplicative distance $d_{\mathcal{P}}$ that expresses the distance between two probability distributions σ_1 and σ_2 on $\mathcal{S}$ is defined as $d_{\mathcal{P}}(\sigma_1, \sigma_2) = \sup_{S \in \mathcal{S}} |\ln \frac{\sigma_1(S)}{\sigma_2(S)}|$ with the convention that $|\ln \frac{\sigma_1(S)}{\sigma_2(S)}| = 0$ if both σ_1 and σ_2 are zero and ∞ if only one of them is zero. $d(x, x')$ represents the Euclidean distance between x and x'. Given $\epsilon \in \mathbb{R}^+$, ϵ-GeoI is defined as follows.

Definition 1 (ϵ-geo-indistinguishability). *The mechanism satisfies ϵ-GeoI iff $\forall x, x' \in \mathcal{X}, Z \subseteq \mathcal{Z}$*

$$d_{\mathcal{P}}(\Pr(K(x) \in Z), \Pr(K(x') \in Z)) \le \epsilon d(x, x') \tag{1}$$

Mechanism Satisfying ϵ-GeoI. The authors of [10] introduced a mechanism called planar Laplace mechanism (PLM) for achieving ϵ-GeoI. The probability distribution that PLM generates is called the planar Laplace distribution and, as its name suggests, is derived from a two-dimensional version of the Laplace distribution as follows: $\Pr(PLM_\epsilon(x) = z) = \frac{\epsilon^2}{2\pi} e^{-\epsilon d(x,z)}$.

2.2   Problem Statement

As we described in Sect. 1, some LBSs improve their services by using a road network. In this paper, we assume these LBSs, where the road network G is defined as a weighted undirected graph (V, E). Let V be a set of vertices which represent points on the road network, each of which has a coordinate on the Euclidean plane, and let E be a set of edges. The weight $w(a, b)$ of edges between connected vertices $a \in V$ and $b \in V$ represents the shortest distance of the road on the Euclidean plane connecting the two vertices a and b, which leads to $w(a, b) \ge d(a, b)$. Then, $d_s(v, v')$ represents the shortest path length between $v \in V$ and $v' \in V$, and following inequality holds. This is because $d(v, v')$ stands for the shortest distance as the crow flies while $d_s(v, v')$ stands for the shortest distance on the road network.

$$d_s(v, v') \ge d(v, v') \tag{2}$$

In these LBSs, a user who wants to receive the service sends a vertex that represents his location to an untrusted LBS provider, and the LBS provider performs the service's computations w.r.t. the vertex (using the road network) and provides the service. Furthermore, we assume that there is no trusted server. Hence, a user needs to protect privacy on his device by himself.

Quantification of Utility and Privacy Guarantee. When a user uses a mechanism to protect his privacy, the quality of the service the user receives degrades. Shokri et al. [16] generally quantified this quality loss, referred to as service quality loss (SQL), in LBSs when a user uses mechanism K:

$$SQL(\pi_u, K, d_q) = \sum_{r,r'} \pi_u(r) \Pr(K(r) = r') d_q(r, r') \tag{3}$$

Table 1. Summary of notation

Symbol	Meaning
$LBSs$	Location-based Services.
u, a	A user and an adversary.
$\mathcal{X}$	Set of locations of users on the Euclidean plane.
$\mathcal{Z}$	Set of query outcomes that represent the users' perturbed locations
G	Weighted undirected graph (V, E) that represents a road network.
V	Set of vertices on the Euclidean plane.
E	Set of edges. A weight is the shortest distance on a road connecting two vertices.
$\pi_u(r)$	The probability of being at location r when accessing the LBS.
$\pi_a(r)$	The adversary's knowledge about user's location that represents the probability of being at r.
K	A mechanism. Given a location, K outputs a perturbed location.
$d(x, x')$	Euclidean distance between x and x'.
$d_s(v, v')$	The shortest path length on the graph between v and v'.

Here, π_u is the probability distribution representing the probability of user's location, called a prior of the user. $d_q(r, r')$ represents the metric of a degree of dissimilarity, which depends on the LBS. Thus, this means the expected value of a degree of dissimilarity between the actual location of the user and the location obfuscated by mechanism K. Shokri et al. used the Euclidean distance from a natural idea that the longer the Euclidean distance between the true location and the obfuscated location is, the worse the service becomes. We call this SQL_e.

Shokri et al. also quantified the degree of a privacy protection by a mechanism. The researchers translated location privacy into adversarial error (AE) by measuring how accurately an adversary could infer the user's true location. Formally, AE can be formulated as follows:

$$AE(\pi_a, K, h, d_q) = \sum_{\hat{r}, r', r} \pi_a(r) \Pr(K(r) = r') \Pr(h(r') = \hat{r}) d_q(\hat{r}, r) \qquad (4)$$

Here, π_a is the probability distribution representing the adversarial knowledge about the user's actual location, called a prior of the adversary. An inference mechanism h outputs an inferred point by drawing a point according to the probability distribution when given an obfuscated point, which stands for the inference of the adversary. Thus, $\Pr(h(r') = \hat{r})$ means the probability of estimating $\hat{r}$ as the actual location of the user when the adversary observes r'. Therefore, AE represents the expected value of a degree of dissimilarity $d_q(\hat{r}, r)$ between the user's true location r and the location $\hat{r}$ the adversary infers. As in the case of SQL, the researchers used the Euclidean distance as d_q.

The major notations in this paper are summarized in Table 1.

3   Evaluating Privacy and Utility of Geo-Indistinguishability

In this section, we empirically show two insufficiencies of GeoI caused by considering a road network. First, we describe the model of an adversary [17]. Then, we show the insufficiency of the privacy guarantee by modeling an adversary who considers the road network and quantifying the accuracy of the attack of this adversary. Next, we describe the formulation of the utility of an output of the mechanism [16]. Then, we propose the way of applying it for LBSs over a road network, and we show by experimentations with two real-world road networks that PLM performs poorly for this formulation.

3.1   Empirical Privacy Evaluation

We assume that the adversary also uses a road network to infer the user's actual location because a road network is publicly available. In the paper [10], the authors did not consider such an adversary, and we empirically show that if the adversary considers a road network, this may lead to privacy leakage even if the mechanism satisfies GeoI, which is referred to as an insufficiency of the privacy protection.

Adversarial Model. First, we describe the model of the adversary who tries to infer the user's actual location. Shokri et al. [17] modeled the adversary who knows the prior π_a and the mechanism that the user uses and can solve problems with any computational complexity. Although this assumption is advantageous for the adversary, showing the protection against this adversary will guarantee strong privacy. When the adversary obtains the user's obfuscated location r', he tries to infer the user's true location by the optimal inference attack. In this attack, an adversary solves the following mathematical optimization problem and obtains the optimal probability distribution and constructs the optimal inference mechanism h w.r.t. his knowledge; by using this mechanism with input r', he estimates the user's true location.

$$\underset{h}{\text{minimize}} \sum_{\hat{r},r',r} \pi_a(r) \Pr(K(r) = r') \Pr(h(r') = \hat{r}) d_p(r, \hat{r})$$

$$\text{subject to} \sum_{\hat{r}} h(r')(\hat{r}) = 1, \forall r' \tag{5}$$

$$h(r')(\hat{r}) \geq 0, \forall r', \hat{r}$$

We model an adversary who knows a road network in this way. If an adversary knows a road network, the domain of his prior π_a is V, and d_p is d_s because we assume that the adversary also tries to improve his inference w.r.t. the shortest distance. In this setting, this is a linear programming problem because $\Pr(h(r') = \hat{r})$ represents a variable and the other terms are constant so that the objective

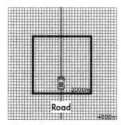

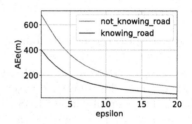

Fig. 3. A synthetic map. The dimensions are 4000 m * 4000 m, and each lattice point has a coordinate. The red line indicates the road, and a user is located inside the black frame. (Color figure online)

Fig. 4. Adversarial error (AE) in the scenarios of the adversary knowing or not knowing the road network. (Color figure online)

function and the constraints are linear. We solve this problem using CBC (coin-or branch and cut)[1] solver of the PuLP library of Python.

Experiment. In the following paragraph, we show that the adversary who knows a road network can attack with higher accuracy than can the adversary who does not know it. To make this easy to understand, we use a simple synthetic data illustrated in Fig. 3.

This map consists of 1600 squares with the side length of 100 m; that is, the area dimensions are 4000 m * 4000 m, and each lattice point has a coordinate. The red line through the center represents the road where the user is considered to be, and the other area represents locations where the user must not be, such as on the sea.

Thus, we assume that the user's location is determined according to a uniform distribution on the red line and inside the black frame to make an output of the mechanism planarly spread; the adversary who does not consider uses the prior given by the uniform distribution inside the black frame, and the adversary who considers the road network uses the prior given by the uniform distribution on the red line inside the black frame. Then, Fig. 4 shows AE_e of each adversary w.r.t. the privacy parameter ϵ of the mechanism. Comparing AE_e of both adversaries, it is clear that the adversary with the prior considering the road network can estimate the user's true location more accurately.

Remark 1. This results from that the PLM could obfuscate the user's location to a place where the user must not be; in the case of Fig. 3, anywhere except the red line. Thus, determining this place contributes to the accuracy of the attack. Figure 4 shows that an adversarial error could become approximately half in this particular case, so this is an insufficiency of the privacy protection of GeoI.

---

[1] https://projects.coin-or.org/Cbc.

3.2   Utility

If a user uses LBSs over a road network and uses a mechanism on the Euclidean plane, such as PLM, the user or the LBS provider needs to map the perturbed location to a vertex of the road network because the LBS provider presumes that the user is located at a vertex of the graph to take advantage of a road network. For example, in the LBS that searches for the nearest restaurants, the LBS provider needs to compute the shortest path length between vertices where the user and restaurants are located. If the user is located outside of the graph, the shortest path length cannot be computed.

Then, it is worth noting that if the user (rather than the service provider) performed the mapping to the vertex before the user sent the perturbed location, it would prevent the perturbed location from being the location where the user must not be and would improve the privacy protection. In this view, we propose the mechanism on the graph, which is defined as the algorithm that outputs the perturbed vertex when given a vertex. Then, we can formulate SQL on a road network as SQL_s using the shortest path length as the metric.

$$SQL_s(\pi_u, K) = SQL(\pi_u, K, d_s) = \sum_{v,v'} \pi_u(v) \Pr(K(v) = v') d_s(v, v') \qquad (6)$$

In this formulation, we anticipate the lower utility of an output of the mechanism achieving GeoI because it cannot consider SQL_s due to the definition using the Euclidean distance. For example, if there is a river with no bridges so that the user cannot cross it, the opposite riverside is far away and obfuscating to the opposite riverside results in the lower utility (see Fig. 2). However, GeoI may consider the opposite riverside to be close.

Then, we can also formulate the adversarial error over a road network that we call AE_s as follows:

$$AE_s(\pi_a, K, h) = AE(\pi_a, K, h, d_s)$$
$$= \sum_{\hat{r}, r', r} \pi_u(r) \Pr(K(r) = r') \Pr(h(r') = \hat{r}) d_s(\hat{r}, r) \qquad (7)$$

This formula expresses the expected value of the shortest path length between the actual location and the location that an adversary with the prior of π_a infers with inference mechanism h when a user uses mechanism K. Intuitively, this represents the adversarial error on a road network because we use the shortest path length d_s as the metric. Thus, when the adversary can infer the true location on the road network with a high accuracy, the formula (7) will have a small value. It can be stated that if AE_s is small, the privacy protection level of the mechanism is low.

Experiments. Here, we empirically show that the mechanism satisfying GeoI may perform worse w.r.t. SQL_s than we expect, since the definition of GeoI

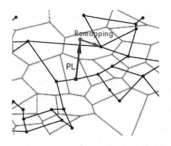

Fig. 5. Output of PLG.

Fig. 6. Tokyo (left) and Akita (right) road networks.

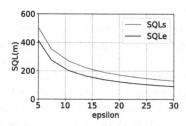

Fig. 7. SQL_s and SQL_e of PLG vs. ϵ on the Tokyo graph.

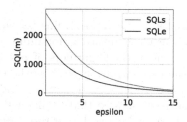

Fig. 8. SQL_s and SQL_e of PLG vs. ϵ on the Akita graph.

considers SQL as SQL_e. To illustrate this, we compare SQL_s and SQL_e of the same GeoI mechanism. As we stated, we use a mechanism on a graph for LBSs over a road network. Then, we propose a natural and straightforward way of converting PLM to a planar Laplace mechanism on a graph (PLG). First, a user perturbs the location using PLM. Next, the user maps the perturbed location on the Euclidean plane to the nearest vertex on the road network. Figure 5 is an example of an output of PLG. Formally, we formulate this mechanism as follows:

$$Pr(PLG_\epsilon(v) = w) = \int_{S_w} Pr(PLM_\epsilon(x_v) = z)dz \qquad (8)$$

Here, x_v is the coordinate of vertex v, and let S_w be a Voronoi cell of vertex w when the Voronoi diagram created from the graph on the Euclidean plane is given.

Theorem 1. *PLG_ϵ satisfies ϵ-GeoI on the graph.*

We refer the reader to the appendix for the proof. Thus, it is assumed that this is a straightforward way of GeoI mechanism in our setting where a user needs to output a vertex as we stated in Chap. 2.2, and because we cannot use PL, it is reasonable to use this mechanism instead of PL. We compute SQL_s and SQL_e

on two road networks. We used OpenStreetMap[2] to retrieve two maps of areas of two cities, Tokyo and Akita, in Japan with dimensions of $4000\,\mathrm{m} * 4000\,\mathrm{m}$ as in Fig. 6.

We plot the results in Figs. 7 and 8 around the range where SQL is reasonable. If the user uses the same mechanism (i.e., the same ϵ), it is observed that the utility for the LBS over a road network is worse. This outcome is caused by the difference of the Euclidean distance and the shortest path length between two vertices. Additionally, it is observed that the difference between SQL_s and SQL_e on the Akita graph is larger than that on the Tokyo graph at the same SQL_e because the difference between the two distances on the Akita graph is larger than that on the Tokyo graph.

Remark 2. GeoI constrains the mechanism to use the Euclidean distance so that the mechanism cannot improve its utility of the output for LBSs using a road network. Regardless of how hard we try to improve the utility of the mechanism output, as long as there is the constraint of GeoI, the mechanism cannot consider a road network, and we cannot improve the mechanism. Additionally, we showed that SQL_s of PLG is worse on two graphs; however, there may be a road network that results in much worse utility than we showed. Unless the mechanism considers a road network, the mechanism cannot guarantee high utility.

4   Geo-Graph-Indistinguishability

In this section, we propose a new definition of location privacy called Geo-Graph-Indistinguishability (GeoGI) for LBSs using a road network, which is tolerant to the weaknesses of GeoI. We first formally define GeoGI, and then propose a mechanism satisfying GeoGI which is called Graph-Exponential Mechanism (GEM). Finally, we clarify the relationship between GeoI and GeoGI, and describe validity of the definition of GeoGI.

4.1   Definition

Given a graph $G = (V, E)$ representing a road network, let $\mathcal{W}$ be a set of vertices that a mechanism outputs. Then, mechanism K on the graph returns the random vertex $w \in \mathcal{W}$ according to a probability distribution when given a vertex $v \in V$. Then, given $\epsilon \in \mathbb{R}^+$, ϵ-Geo-Graph-Indistinguishability is defined as follows.

Definition 2 (ϵ-*Geo-Graph-Indistinguishability*). *A mechanism K on a road network $G = (V, E)$ satisfies ϵ-Geo-Graph-Indistinguishability iff $\forall v, v' \in V, \forall W \subseteq \mathcal{W}$,*

$$d_{\mathcal{P}}(\Pr(K(v) \subseteq W), \Pr(K(v') \subseteq W)) \leq \epsilon d_s(v, v') \qquad (9)$$

---

[2] https://openstreetmap.jp/.

The definition can be also formulated as $\forall v, v' \in V, \forall W \subseteq \mathcal{W}, \frac{\Pr(K(v) \subseteq W)}{\Pr(K(v') \subseteq W)} \leq$ $e^{\epsilon d_s(v,v')}$. This formulation implies that GeoGI is an instance of $d_{\mathcal{X}}$-privacy [7] proposed by Chatzikokolakis et al. as are GeoI and differential privacy. The authors showed that an instance of $d_{\mathcal{X}}$-privacy guaranteed strong privacy. We refer the reader to the appendix for further details. Intuitively, this definition guarantees that for any $v, v' \in V$, the closer to v' a vertex v is w.r.t. the shortest path length, the more similar $K(v)$ and $K(v')$ are.

It is worth noting that the definition of GeoGI includes a given graph representing a road network, and this results in the privacy protection level and utility varying depending on the road network even if the privacy parameter ϵ remains the same.

4.2   Graph Exponential Mechanism

In this section, we propose a mechanism that achieves ϵ-GeoGI. Given parameter $\epsilon \in \mathbb{R}^+$ and graph $G = (V, E)$, we define GEM_ϵ for any user's location $v \in V$ and perturbed location $w \in \mathcal{W}$ as follows.

Definition 3. GEM_ϵ *takes v as an input and outputs w with the following probability.*

$$\Pr(GEM_\epsilon(v) = w) = \alpha(v)e^{-\frac{\epsilon}{2}d_s(v,w)} \tag{10}$$

where α is a normalization factor and $\alpha(v) = \frac{1}{\sum_{w \in V} e^{-\frac{\epsilon}{2}d_s(v,w)}}$.

The pseudocode of GEM is described in Appendix (Sect. 8.3) due to space limitation. This mechanism employs the idea of exponential mechanism [12] that is one of the general mechanisms for achieving differential privacy.

Theorem 2. GEM_ϵ *satisfies ϵ-GeoGI.*

We refer the reader to the appendix for the proof. This mechanism considers a road network so that high utility for LBSs over a road network can be expected. Moreover, since this mechanism satisfies GeoGI, strong privacy based on differential privacy is guaranteed.

Creating the Probability Distribution and Drawing a Random Point.

Because we assume that the LBS provider is untrusted and there is no trusted server, a user needs to create this distribution by himself and choose the perturbed vertex according to the distribution. In this section, we describe a method to do this and its issues caused by the number of vertices.

To create the probability distribution, (i) the user gets shortest path lengths to all vertices from the vertex where the user is located. (ii) Then the user computes $e^{-\frac{\epsilon}{2}d_s}$ and based on this distribution, (iii) chooses a point.

Phase (i) is acceptable if the server which has enough computing power computes the all shortest lengths and sends users it in advance. This is because the shortest path length can be computed by Dijkstra's algorithm; this computational complexity of this operation depends on the data structure. if we use a naive method, it is $O(|E| + |V|^2)$, and it can be improved by using Fibonacci heap to $O(|E| + |V| \log |V|)$, where $|V|$ and $|E|$ represent the counts of edges and vertices. However, There is a problem if the user needs to compute it and the size of vertices is large because the user uses a mobile phone with limited computing power. So we have to consider the better algorithm. On a road network, a fast algorithm computing the shortest path length has been studied; we refer the reader to [1] that may be applied to our algorithm. Phase (ii) is no problem because it computational complexity is $O(|V|)$. For phase (iii), when the number of vertices is much larger than we expected, we may not be able to effectively sample the vertices according to the distribution. This problem has also been studied and is known as consistent weighted sampling (CWS): we refer the reader to [11,23]. We believe that these studies can be applied to our algorithm and it can be computed even if the size of vertices is somewhat large.

4.3   Analyzing the Relationship Between GeoI and GeoGI

In this section, we describe the relationship between GeoI and GeoGI. There are two major differences: one is the domain, and the other is the distance metric.

First, we state the difference of the location domain. We can design a GeoI mechanism on the Euclidean plane; however, the same cannot be done for GeoGI because GeoGI constrains a mechanism to use a vertex on the graph due to the definition using the shortest path length. Since we exploit the mechanism for LBSs using a road network, the constraint does not pose a problem. Moreover, this constraint prevents the perturbation to a location where the user must not be located and improve the privacy guarantees, as we stated in Sect. 3.1. We refer the mechanism which meets this constraint as a mechanism on the graph.

The other difference is the used metric. In this part, we assume the mechanism on a graph; otherwise, we cannot design a GeoGI mechanism. Then, the following theorem holds due to the used metric of GeoGI.

Theorem 3. *If a mechanism on the graph satisfies ϵ-GeoI, it satisfies ϵ-GeoGI.*

This is due to the definition of a graph that represents the road network. Using Inequality (2), we derive the following inequality:

$$d_{\mathcal{P}}(\Pr(K(v) \subseteq (W)), \Pr(K(v') \subseteq (W))) \leq \epsilon d(v, v') \leq \epsilon d_s(v, v') \qquad (11)$$

This inequation shows that the GeoI mechanism is also the GeoGI mechanism, but the reverse is not always true. For example, PLG satisfies both GeoI and GeoGI. This means that GeoGI relaxes the restriction of GeoI. Thus, we can design a more suitable mechanism which improves the utility for a road network.

It is worth noting whether this relaxing of the definition leads to weakening of the guarantees of privacy protection. In short, GeoGI has no guarantees of privacy protection w.r.t. the Euclidean distance so that if a user uses a mechanism that satisfies GeoGI to protect the location, the adversary may easily distinguish the user's location in terms of the Euclidean distance. In what follows, we show

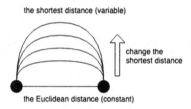

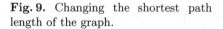

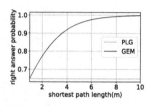

Fig. 9. Changing the shortest path length of the graph.

Fig. 10. Change of TP according to GEM and PLG.

this fact using the notion of the true probability (TP). The probability that an adversary can distinguish user's location is represented as

$$TP(\pi_u, K, h) = \sum_{\hat{v}, v, w} \pi_u(v) \Pr(K(v) = w) \Pr(h(w) = \hat{v}) \delta(\hat{v}, v) \qquad (12)$$

Here $\delta(\hat{v}, v)$ is a function that returns 1 if $\hat{v} = v$ holds and otherwise returns 0. TP means the expected value of a probability that an adversary can remap the obfuscated location to the true location.

We assume a set of graphs, each of which has only two vertices. The Euclidean distance between the vertices is the same for all the graphs, but the weight of the edge between them is different for each graph (Fig. 9). Next, we assume that each prior the user and the adversary have is a uniform distribution on two vertices of this graph, and we compute TP of PLG and GEM. Figure 10 shows the change of TP when the weight, that is, the shortest path length, changes. Due to the guarantee of the Euclidean distance of GeoI, PLG does not degrade TP even if the shortest path length changes, however, since GeoGI does not have a guarantee of the Euclidean distance, GEM significantly degrades TP, which means that the adversary can know the user's true location.

GeoGI can achieve better utility than can GeoI by guaranteeing privacy protection in terms of the shortest path length instead of the Euclidean distance. This idea comes from the interpretation of privacy; in this paper, we assume that privacy and the utility can be interpreted as the shortest path length on the graph and that it should be acceptable for LBSs on a road network. Therefore, GeoGI may not be suitable for protecting location privacy if the privacy should be interpreted as the Euclidean distance, e.g., querying the weather conditions where we need to protect a wide range of locations.

4.4 Discussion

We assume that a graph representing the road network is given in GeoGI. We note that setting a different graph (road network) in the definition of GeoGI implies a different privacy model. Thus, GeoGI can be personalized. For example, a conservative user may want to use a global graph in GeoGI that covers all possible locations on Earth, while a liberal user may use a smaller graph in

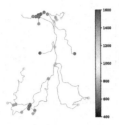 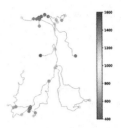

Fig. 11. SQLr$_s$ of each vertex on Akita graph.

Fig. 12. AEr$_s$ of each vertex on Akita graph.

GeoGI that only covers the city of her residence. The graph may also depend on application scenarios in practice. For example, if the application is vehicle navigation, the graph should cover all highway road network instead of pedestrian lanes. In summary, the privacy level and utility depend on the shape of the graph, such as its density and size, and its relationships should be shown because this has the potential to lead to improvement of the privacy protection and utility. This topic is left to future research.

Additionally, the utility and the privacy protection level depend on the vertex where the user uses the GeoGI mechanism because, as opposed to the Euclidean plane that spreads uniformly, each vertex relates differently with other vertices and because to satisfy GeoGI, the mechanism needs to vary the probability distribution depending on the vertex's relationship. This complexity obscures the mechanism performance for a user, and a user will not know how to adjust the privacy parameter. Then, we propose a way of measuring the performance of the mechanism used by a user located at a certain vertex. We formulate the mechanism's utility as SQLr$_s$ and its privacy protection level as AEr$_s$ as follows:

$$SQLr_s(v, K) = \sum_{v'} \Pr(K(v) = v')d_s(v, v') \tag{13}$$

$$AEr_s(v, K, h) = \sum_{v', \hat{v}} \Pr(K(r) = r') \Pr(h(v') = \hat{v})d_s(\hat{v}, v) \tag{14}$$

When a user is located at vertex v, SQLr$_s$ represents the expected value of the shortest path length between v and the perturbed vertex v'. AEr$_s$ represents the expected value of the shortest path length between v and vertex $\hat{v}$ inferred by the adversary with inference mechanism h (in this case, we assume optimal inference attack). We show SQLr$_s$ and AEr$_s$ of the Akita graph using GEM with $\epsilon = 0.002$ in Figs. 11 and 12, respectively. As we can see, the utility loss (i.e., SQLr$_s$) and privacy (i.e., AEr$_s$) differ on different locations in spite of the same privacy parameter. We can develop a tool to visualize the privacy and utility of the mechanism under different privacy parameters, which may help users to determine a proper privacy parameter. We defer this to a long version of this work.

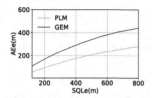
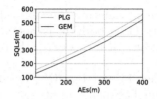
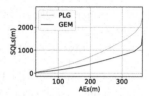

Fig. 13. AE_e when using GEM and PLM against the adversary who knows the road network.

Fig. 14. SQL_s of PLG and GEM w.r.t. AE_s on the Tokyo graph.

Fig. 15. SQL_s of PLG and GEM w.r.t. AE_s on the Akita graph.

5 Experiments

In this section, we show that GEM outperforms the GeoI mechanism in terms of utility and privacy protection for LBSs on a road network. To demonstrate this conclusion, we performed two experiments as follows. First, since GEM, in contrast to PL, may perturb the input location to a location that is out of the road network, such as on the sea (as stated in Sect. 3.1), it is assumed that GEM achieves better privacy guarantee when the adversaries have the knowledge of road networks. To show this, we computed AE_e of GEM on the synthetic graph we used in Sect. 3.1. Next, because GEM, in contrast to PLG, considers a road network (as stated in Sect. 3.2), it is assumed that the output quality of GEM is higher than PLG. To show this, we computed SQL_s of GEM for two cities we used in Sect. 3.2. Additionally, we compared the results with those of PL and PLG.

5.1 Privacy Protection Level of GEM

We computed AE_e of GEM on the graph of Fig. 3. As in Sect. 3.1, we assume that the adversary knew the road network so that his prior was a uniform distribution on the red line inside the black frame. Since GEM, in contrast to PL, outputs only the locations on the red line, it is assumed that AE_e of GEM is higher than that of PLM. To fairly compare AE_e of each mechanism, we performed the comparison under the same utility SQL_e.

As is shown in Fig. 13, AE_e of GEM is higher than that of PL in case of the adversary who knows a road network. This means that GE can protect user privacy more strongly than can PL because GE guarantees that the output is on the road network.

5.2 Utility of GEM

We computed SQL_s of GEM on two graphs of Fig. 6. Since GEM considers the road network, it is assumed that SQL_s of GEM is higher than that of PLG. To fairly compare SQL_s of each mechanism, we performed the comparison under the same AE_s. Then, we assumed that both the priors that a user and an adversary have are uniform distributions on the graph with a range of 2000 m from the centers of maps.

As we can see from Figs. 14 and 15, SQL_s of GEM is lower than that of PLG. Thus, a GEM output has higher utility for LBSs using a road network than does a PLG output. Additionally, it can be said that the difference of the SQL between PLG and GEM is larger on the Akita graph than on the Tokyo graph. The reason is that the difference between the Euclidean distance and the shortest path length is larger for vertices of the Akita graph.

6   Related Work

6.1   Location Privacy on a Road Network

To the best of our knowledge, this is the first study to propose the perturbation with the differential privacy approach over the road network. However, several studies explored location privacy on a road network.

Tyagi et al. [20] studied location privacy over a road network for VANET users, and they show that there are no comprehensive privacy-preserving techniques or frameworks that cover all privacy requirements or issues to maintain the desired level of location privacy.

Wang et al. [21] and Wen et al. [22] proposed the method of privacy protection for the user who wishes to receive location-based services and travels over roads. The authors use k-anonymity as the protection method and take advantage of the road network constraints.

A series of key features distinguish our solution from these studies: (a) we use the differential privacy approach so that our solution has a guarantee of privacy protection against any attacker and (b) we assume that there is no trusted server. We highlight these two points as advantages of our proposed method.

6.2   State-of-the-Art Privacy Models

Since GeoI was published, many related applications have been proposed. To et al. [18] developed an online framework of privacy-preserving spatial crowdsourcing service using GeoI. Tong et al. [19] proposed a framework of privacy-preserving ridesharing service based on GeoI and differential privacy approach. It may be possible to improve these applications by using GeoGI instead of GeoI. Additionally, Bordenabe et al. [13] proposed an optimized mechanism satisfying GeoI; it may be possible to apply this method to GeoGI.

According to [10], if a user uses the GeoI mechanism multiple times, this causes privacy degradation due to correlations in the data; this scenario also applies to GeoGI. This issue remains a difficult and intensely investigated problem in the field of differential privacy. There are two kinds of approaches attempting to solve this problem. The first is to develop a mechanism for multiple perturbations that satisfies existing notion, such as differential privacy and GeoI [6,8].

Kairouz et al. [8] studied the composition theorem and proposed a mechanism that upgrades the privacy guarantee. Chatzikokolakis et al. [6] proposed a method of controlling privacy using GeoI when locations are correlated. The second approach is to propose a new privacy notion for correlated data [2, 24]. Xiao et al. [24] proposed δ-location set privacy to protect each location in a trajectory when a moving user sends locations. Cao et al. [2] proposed PriSTE, a framework for protecting spatiotemporal event privacy. We believe that these studies can be applied to our work.

7   Conclusion and Future Work

In this paper, by evaluating privacy and utility of PL, we have shown that the definition of GeoI is insufficient for LBSs over a road network to protect privacy and output the useful perturbed location. The core of our proposal is a new notion of privacy that we call GeoGI, which takes the place of GeoI for such LBSs, and a mechanism GEM, based on the exponential mechanism, to perturb the user location. We have shown how GeoGI relates to GeoI and that GeoGI is a more suitable privacy definition for such LBSs w.r.t. privacy protection and utility. We also have shown the effectiveness of our proposed approach by comparing GEM with PLG in the example of two cities in Japan.

In the future, we aim to extend the privacy model to several graphs. Although in this paper, we represented a road network as an undirected graph, it should be represented as a directed graph because of the existence of one-way roads, and this may degrade the utility. Additionally, we need to consider the movement mode such as walking, driving, and flying. Finally, we need to pay attention to the fact that multiple perturbations of correlated data such as trajectory data may degrade the level of protection even if the mechanism satisfies GeoGI as in case of GeoI and differential privacy. This topic has been intensely studied, and we believe that it can be applied to GeoGI. We plan to solve these problems in future research.

8   Appendix

8.1   Proofs

Theorem 1. PLG_ϵ *on graph* $G = (V, E)$ *satisfies* ϵ*-GeoI on graph* G.

Proof. This proposition can be formulated as follows for all vertices $v, v' \in V, W \subseteq \mathcal{W}$,

$$d_p(\Pr(PLG(v) \subseteq W), \Pr(PLG(v') \subseteq W)) \leq \epsilon d(v, v') \tag{15}$$

Furthermore, we derive

$$\forall v, v' \in V, w \in W, \Pr(PLG(v) = w) \leq e^{\epsilon d(v, v')} \Pr(PLG(v') = w) \tag{16}$$

Since PL_ϵ satisfies ϵ-GeoI, for all $z \in \mathcal{Z}$, we derive

$$\Pr(PL_\epsilon(x_v) = z) \le e^{\epsilon d(x_v, x_{v'})} \Pr(PL_\epsilon(x_{v'}) = z) \tag{17}$$

By the theorem of integral inequality we obtain

$$\int_{S_w} \Pr(PL_\epsilon(x_v) = z)dz \le \int_{S_w} e^{\epsilon d(x_v, x_{v'})} \Pr(PL_\epsilon(x_{v'}) = z)dz$$
$$= e^{\epsilon d(x_v, x_{v'})} \int_{S_w} \Pr(PL_\epsilon(x_{v'}) = z)dz \tag{18}$$

Using (8) and (18), we obtain

$$\Pr(PLG(v) = w) \le e^{\epsilon d(v, v')} \Pr(PLG(v') = w) \tag{19}$$

This concludes the proof.

Theorem 2. GEM_ϵ satisfies ϵ-GeoGI.

Proof. This proposition can be formulated for all vertices $v, v' \in V, W \subseteq \mathcal{W}$:

$$d_p(\Pr(GEM(v) \subseteq W), \Pr(GEM(v') \subseteq W)) \le \epsilon d_s(v, v') \tag{20}$$

The ratio of $\Pr(GEM(v) = w)$ and $\Pr(GEM(v') = w)$ is expressed as follows:

$$\frac{\Pr(GEM(v) = w)}{\Pr(GEM(v') = w)} = \frac{\alpha(v) e^{-\frac{\epsilon}{2} d_s(v,w)}}{\alpha(v') e^{-\frac{\epsilon}{2} d_s(v',w)}} = \frac{a(v)}{a(v')} e^{\frac{\epsilon}{2}(d_s(v',w) - d_s(v,w))} \tag{21}$$

When $-d_s(v, w) + d_s(v', w)$ has the maximum value for $w \in W$, (21) reaches the maximum value too. Due to the triangle inequality, the inequality $\forall w \in W, -d_s(v, w) + d_s(v', w) \le -d_s(v, w) + d_s(v', v) + d_s(v, w) = d_s(v, v')$ holds, and the following inequality is derived:

$$\frac{\Pr(GEM(v) = w)}{\Pr(GEM(v') = w)} \le \frac{\alpha(v)}{\alpha(v')} e^{\frac{\epsilon}{2} d_s(v, v')} \tag{22}$$

Next, we show that the following inequality holds:

$$\frac{\alpha(v)}{\alpha(v')} < e^{\frac{\epsilon}{2} d_s(v, v')} \tag{23}$$

The inequality 23 is expressed as follows for any $v, v' \in V, w \in W$ of any graph G:

$$\sum_{w \in V} e^{-\frac{\epsilon}{2} d_s(v', w)} - e^{\frac{\epsilon}{2} d_s(v, v')} \sum_{w \in V} e^{-\frac{\epsilon}{2} d_s(v, w)} < 0 \tag{24}$$

Using the triangle inequality, we have $\forall w \in W, d(v, w) - d(v, v') \le d_s(v', w)$, $e^{-\frac{\epsilon}{2} d_s(v', w)} \le e^{-\frac{\epsilon}{2}(d_s(v,w) - d_s(v,v'))}$. Therefore,

$$\sum_{w \in V} (e^{-\frac{\epsilon}{2} d_s(v', w)} - e^{-\frac{\epsilon}{2}(d_s(v,w) - d_s(v,v'))}) \le \sum_{V \setminus v} (e^{-\frac{\epsilon}{2} d_s(v', V)} - e^{-\frac{\epsilon}{2} d_s(v', V)}) < 0$$

Using (22) and (23), we obtain

$$\frac{\Pr(GEM(v) = w)}{\Pr(GEM(v') = w)} < e^{\frac{\epsilon}{2}d_s(v,v')}e^{\frac{\epsilon}{2}d_s(v,v')} = e^{\epsilon d_s(v,v')} \tag{25}$$

8.2 d_χ-privacy

As we stated in Sect. 4, GeoGI is an instance of d_χ-privacy: due to this characterization, we can give two characterizations of GeoGI that mathematically show the guarantee of strong privacy protection. In this section, we stated the characterizations of GeoGI.

Hiding Function. The first characterization uses the concept of a hiding function $\phi : V \to V$. For any hiding function and a secret location $v \in V$, when an attacker who has a prior distribution that expresses the user's location information obtains each output $w \sim K(v), w' \sim K(\phi(v))$ of a mechanism satisfying ϵ-GeoGI, the following inequality holds for the multiplicative distance between its two posterior distributions:

$$d_\mathcal{P}(p(v|w), p(v|w')) \leq 2\epsilon d_s(\phi) \tag{26}$$

Let $d_s(\phi(v)) = sup_{v \in V} d_s(v, \phi(v))$ be the maximum distance between an actual vertex and its hidden version. This inequality guarantees that the adversary's conclusions are the same (up to $2\epsilon d_\chi(\phi)$) regardless of whether ϕ has been applied or not.

Informed Attacker. The other characterization is shown by the multiplicative distance between the prior distribution and its posterior distribution that is derived by obtaining an output of the mechanism. By measuring its distance, we can determine how much the adversary has learned about the secret. We assume that an adversary (informed attacker) knows that the vertex v where the user is located in N. When the adversary obtains an output of the mechanism. The following inequality holds for the multiplicative distance between his prior distribution $\pi_{|N}(v) = \pi(v|N)$ and its posterior distribution $p_{|N}(v|w) = p(v|w, N)$:

$$d_\mathcal{P}(\pi_{|N}, p_{|N}(v|w)) \leq \epsilon d_s(N) \tag{27}$$

Let $d_s(N) = max_{v,v' \in N} d_s(v, v')$ be the maximum distance between vertices in N. This inequality guarantees that when $d_s(N)$ is small, the adversary's prior distribution and its posterior distribution are similar. In other words, the more the adversary knows about the actual location, the less he cannot learn about the location from an output of the mechanism.

8.3   Pseudocode of GEM

Algorithm 1. Graph Exponential Mechanism (GEM).

Input: v, G, ϵ.
Output: Sanitized location w of input v.
1 initialization;
2 Compute shortest distances to all other vertices from v by Dijkstra's
   algorithm and calculate $e^{-\frac{\epsilon}{2}d_s}$;
3 Normalize to make a distribution ;
4 Draw random vertex w according to the distribution;
5 return w.

References

1. Akiba, T., Iwata, Y., Kawarabayashi, K., Kawata, Y.: Fast shortest-path distance queries on road networks by pruned highway labeling. In: Proceedings of the Sixteenth Workshop on Algorithm Engineering and Experiments (ALENEX), pp. 147–154 (2014)
2. Cao, Y., Xiao, Y., Xiong, L., Bai, L.: PriSTE: from location privacy to spatiotemporal event privacy. arXiv preprint arXiv:1810.09152 (2018)
3. Chatzikokolakis, K., Palamidessi, C., Stronati, M.: Constructing elastic distinguishability metrics for location privacy. Proc. Priv. Enhancing Technol. **2**, 156–170 (2015)
4. Cho, H.J., Chung, C.W.: An efficient and scalable approach to CNN queries in a road network. In: Proceedings of the 31st International Conference on Very Large Data Bases, pp. 865–876 (2005)
5. Dwork, C.: Differential privacy. In: van Tilborg, H.C.A., Jajodia, S. (eds.) Encyclopedia of Cryptography and Security, pp. 338–340. Springer, Boston (2011). https://doi.org/10.1007/978-1-4419-5906-5
6. Chatzikokolakis, K., Palamidessi, C., Stronati, M.: A predictive differentially-private mechanism for mobility traces. In: De Cristofaro, E., Murdoch, S.J. (eds.) PETS 2014. LNCS, vol. 8555, pp. 21–41. Springer, Cham (2014). https://doi.org/10.1007/978-3-319-08506-7_2
7. Chatzikokolakis, K., Andrés, M.E., Bordenabe, N.E., Palamidessi, C.: Broadening the scope of differential privacy using metrics. In: De Cristofaro, E., Wright, M. (eds.) PETS 2013. LNCS, vol. 7981, pp. 82–102. Springer, Heidelberg (2013). https://doi.org/10.1007/978-3-642-39077-7_5
8. Kairouz, P., Oh, S., Viswanath, P.: The composition theorem for differential privacy. IEEE Trans. Inf. Theor. **63**(6), 4037–4049 (2017)
9. Kolahdouzan, M., Shahabi, C.: Voronoi-based k nearest neighbor search for spatial network databases. In: Proceedings of the Thirtieth International Conference on Very Large Data Bases, vol. 30, pp. 840–851 (2004)
10. Andrés, M.E., Bordenabe, N.E., Chatzikokolakis, K., Palamidessi, C.: Geo-indistinguishability: differential privacy for location-based systems. In: Proceedings of the 2013 ACM SIGSAC Conference on Computer and Communications Security, pp. 901–9134 (2013)

11. Manasse, M., McSherry, F., Talwar, K.: Consistent weighted sampling. Technical report, MSR-TR-2010-73, June 2010
12. McSherry, F., Talwar, K.: Mechanism design via differential privacy. In: 48th Annual IEEE Symposium on Foundations of Computer Science (FOCS), pp. 94–103, October 2007
13. Bordenabe, N.E., Chatzikokolakis, K., Palamidessi, C.: Optimal geo-indistinguishable mechanisms for location privacy. In: Proceedings of the 2014 ACM SIGSAC Conference on Computer and Communications Security, New York, NY, USA, pp. 251–262 (2014)
14. Oya, S., Troncoso, C., Pérez-González, F.: Is geo-indistinguishability what you are looking for? In: Proceedings of the 2017 on Workshop on Privacy in the Electronic Society, pp. 137–140 (2017)
15. Papadias, D., Zhang, J., Mamoulis, N., Tao, Y.: Query processing in spatial network databases. In: Proceedings of the 29th International Conference on Very Large Data Bases, pp. 802–813 (2003)
16. Shokri, R., Theodorakopoulos, G., Boudec, J.Y.L., Hubaux, J.P.: Quantifying location privacy. In: Proceedings of the IEEE Symposium on Security and Privacy, pp. 247–262 (2011)
17. Shokri, R., Theodorakopoulos, G., Troncoso, C., Hubaux, J.P., Boudec, J.Y.L.: Protecting location privacy: optimal strategy against localization attacks. In: Proceedings of the 2012 ACM Conference on Computer and Communications Security, pp. 617–627 (2012)
18. To, H., Ghinita, G., Shahabi, C.: A framework for protecting worker location privacy in spatial crowdsourcing. Proc. VLDB Endowment 7(10), 919–930 (2014)
19. Tong, W., Hua, J., Zhong, S.: A jointly differentially private scheduling protocol for ridesharing services. IEEE Trans. Inf. Forensics Secur. 12(10), 2444–2456 (2017)
20. Tyagi, A.K., Sreenath, N.: Location privacy preserving techniques for location based services over road networks. In: Proceedings of International Conference on Communications and Signal Processing (ICCSP), pp. 1319–1326, April 2015
21. Wang, T., Liu, L.: Privacy-aware mobile services over road networks. Proc. VLDB Endowment 2(1), 1042–1053 (2009)
22. Wen, J., Li, Z.: A method of location privacy protection in road network environment. In: 2018 International Conference on Smart Materials, Intelligent Manufacturing and Automation (SMIMA), vol. 173, p. 03048 (2018)
23. Wu, W., Li, B., Chen, L., Zhang, C., Yu, P.S.: Improved Consistent Weighted Sampling Revisited. arXiv:1706.01172 [cs], June 2017
24. Xiao, Y., Xiong, L.: Protecting locations with differential privacy under temporal correlations. In: Proceedings of the 22nd ACM SIGSAC Conference on Computer and Communications Security - CCS 2015, pp. 1298–1309 (2015)
25. Yu, L., Liu, L., Pu, C.: Dynamic differential location privacy with personalized error bounds. In: Proceedings of the Symposium on Network and Distributed System Security (NDSS) (2017)

"When and Where Do You Want to Hide?" – Recommendation of Location Privacy Preferences with Local Differential Privacy

Maho Asada, Masatoshi Yoshikawa, and Yang Cao[✉]

Kyoto University, Kyoto, Japan
asada@db.soc.i.kyoto-u.ac.jp
{yoshikawa,yang}@i.kyoto-u.ac.jp

Abstract. In recent years, it has become easy to obtain location information quite precisely. However, the acquisition of such information has risks such as individual identification and leakage of sensitive information, so it is necessary to protect the privacy of location information. For this purpose, people should know their location privacy preferences, that is, whether or not he/she can release location information at each place and time. However, it is not easy for each user to make such decisions and it is troublesome to set the privacy preference at each time. Therefore, we propose a method to recommend location privacy preferences for decision making. Comparing to existing method, our method can improve the accuracy of recommendation by using matrix factorization and preserve privacy strictly by local differential privacy, whereas the existing method does not achieve formal privacy guarantee. In addition, we found the best granularity of a location privacy preference, that is, how to express the information in location privacy protection. To evaluate and verify the utility of our method, we have integrated two existing datasets to create a rich information in term of user number. From the results of the evaluation using this dataset, we confirmed that our method can predict location privacy preferences accurately and that it provides a suitable method to define the location privacy preference.

Keywords: Privacy preference · Location data · Matrix factorization · Local differential privacy

1 Introduction

In recent years, due to the popularization of smartphones and the development of GPS positioning equipment, location information for people has been able to be obtained quite precisely and easily. Such data can be utilized in various fields such as marketing and urban planning. In addition, there are many applications

S. N. Foley (Ed.): DBSec 2019, LNCS 11559, pp. 164–176, 2019.
https://doi.org/10.1007/978-3-030-22479-0_9

that do not function effectively without location information [9]. Because of such value, market maintenance to buy and sell it has started [9].

However, on the other hand, by publishing accurate location information, there are privacy risks associated with such as individuals being identified [3]. Due to such risks, privacy awareness regarding location information among people is very high. One of the most risky situations is when smartphones are used. This is because we are sending location information to them when using many applications [5].

In order to prevent privacy risks under such circumstances, it is necessary to anonymize or obfuscate location information. One of the countermeasures is a primitive one: turn off location information transmission manually when using a smartphone. There is also a method of applying a location privacy protection technique. Various techniques are used for protection, including k-anonymity [7, 14], differential privacy [4] [8], and encryption [16]. Fawaz [5] proposed a system that applies these privacy protection technologies to smartphones. This system controls the accuracy of the location information sent to each application. The user needs to input how accurate he/she wants to send the respective location information to each application.

In these countermeasures, to avoid a privacy risk by the disclosure of location information, the user has to decide location privacy preference for each location, that is, whether or not he/she publishes the location data at a certain place and time. However, we think that there is a problem in such a situation. What is the best privacy preference is unclear, and it may be different for each user. Therefore, individual users need to determine their location privacy preferences. However, most users find it difficult to determine these preferences themselves [12], and it is troublesome to set the privacy preference at each time.

Therefore, we need a system to recommend location privacy preferences for decision support when choosing a user's location information privacy preference and for the promotion of safe location information release. Recently, one such system was developed using the concept of item recommendation, which is used for online shopping. Item recommendation regards the combination of location and time as an item and whether or not to release location information as a rating of the item, and it predicts the rating of an unknown item using other users' data. Zhang [17] proposed a method of recommending by collaborative filtering.

We focus on the problems of existing location privacy preference recommendation methods and propose a recommendation method to solve these problems.

The contributions of this research are as follows.

1. **Clarifying the definition of location privacy preference:** In location privacy preservation, it is important to define where and when we want to preserve location privacy, which has various granularities. Although many location privacy protection methods have been proposed in the literature, none of them addresses the problem of how to set location privacy preference. Therefore, we generate recommendation models using various granularities for time information and compare their usefulness. From these results, we find

Table 1. A comparison of our method with related work [17].

	Method	How to preserve privacy
Related work [17]	Collaborative filtering (inaccurate for a large amount of data)	Add noise that is not strict mathematically
Our method	Matrix Factorization (accurate for a large amount of data)	Add noise based on local differential privacy

the best granularity that will produce trade-offs between the density of spatial data for the recommendation and the consideration of time.

2. **Applying matrix facrorization to location privacy preference recommendation:** Because location privacy preferences are very sensitive, the system must recommend them accurately. Collaborative filtering, which was used in the method by Zhang [17], experiences problems that when the number of users and products increase; accurate prediction cannot be achieved, and only the nature of either the user or product can be considered well. Therefore, we propose a method to improve by utilizing matrix factorization. As a result of experiments, we confirm that we can predict accurate evaluation values with a probability of 90% for large amount of data.

3. **Recommendation with local differential privacy:** Matrix factorization is involved in privacy risk, because each user needs to send their data to the recommendation system [2,6]. Location privacy preferences encompass location information of the users at a certain times and whether the information is sensitive for him/her. However, a location privacy preference recommendation that achieves highly accurate privacy protection has not been proposed so far. Actually, the method by Zhang [17] did not preserve privacy in a strict mathematical sense. Therefore, we propose a recommendation method that realizes it with local differential privacy, which refers to the method by Shin [13]. A comparison of our method with related work is shown in Table 1. We confirm that our method maintain precision that is the same as that achieved a method without privacy protection.

4. **Generating a location information privacy preference dataset:** A challenge in experiments for testing the performance of our methods is that no appropriate location privacy preference dataset is available in literature. The only available real-world dataset of location privacy preference [11] has few users. Such data is not suitable for the evaluation of the method using matrix factorization [10]. In addition, we need bulk data in the evaluation because there are many users of recommendation in the real world. Therefore, we created an artificial dataset that combines such the location privacy preference dataset and a trajectory dataset with a large number of users.

This paper is organized as follows: We describe the knowledge necessary for realizing our goal in Sect. 2. Then, we describe our method in Sect. 3 and evaluate and discuss about our method in Sect. 4.

2 Preliminaries

2.1 Matrix Factorization

Matrix factorization is one of the most popular methods used for item recommendation, which predicts the ratings of unknown items. This is an extension of collaborative filtering to improve the accuracy for the large amount of data by dimentionality reduction.

We consider the situation in which m users rate any item in n items. We express each user's rating of each item by $\mathcal{M} \subset \{1, \cdots, m\} \times \{1, \cdots, n\}$, and the number of ratings as $M = |\mathcal{M}|$, for the user i's rating of item j. Matrix factorization predicts the ratings of unknown items given $\{r_{ij} : (i,j) \in \mathcal{M}\}$. To make a prediction, we consider a ratings matrix $R = m \times n$, a user matrix $U = d \times m$, and an item matrix $V = d \times n$. The matrices satisfy the formula: $R \approx U^T V$.

In matrix factorization, the user i's element, i.e. the i-th column of U, is expressed by $u_i \in \mathbb{R}^d$, $1 \le i \le m$, and the item j's element, i.e. the j-th column of V, is expressed by $v_j \in \mathbb{R}^d$, $1 \le j \le n$, which are learned from known ratings. The user i's rating of item j is obtained by the inner product of u_i^T and v_j.

In learning, we obtain the matrices U and V, which minimize the following:

$$\frac{1}{M} \sum_{(i,j) \in \mathcal{M}} (r_{ij} - u_i^T v_j)^2 + \lambda_u \sum_{i=1}^{m} ||u_i||^2 + \lambda_v \sum_{j=1}^{n} ||v_j||^2 \tag{1}$$

λ_u and λ_v are positive variables for regularization.

U and V are obtained by updating using the following formulae.

$$u_i^t = u_i^{t-1} - \gamma_t \cdot \{\nabla_{u_i} \phi(U^{t-1}, V^{t-1}) + 2\lambda_u U_i^{t-1}\} \tag{2}$$

$$v_j^t = v_j^{t-1} - \gamma_t \cdot \{\nabla_{v_j} \phi(U^{t-1}, V^{t-1}) + 2\lambda_v V_j^{t-1}\} \tag{3}$$

γ_t is the learning rate at the tth iteration, and $\nabla_{u_i} \phi(U, V)$ and $\nabla_{v_j} \phi(U, V)$ are the gradients of u_i and v_j. They are obtained from derivative of (1) and expressed by the followings:

$$\nabla_{u_i} \phi(U, V) = -\frac{2}{M} \sum_{j:(i,j) \in \mathcal{M}} v_j(r_{ij} - u_i^T v_j) \tag{4}$$

$$\nabla_{v_j} \phi(U, V) = -\frac{2}{M} \sum_{i:(i,j) \in \mathcal{M}} u_i(r_{ij} - u_i^T v_j) \tag{5}$$

We predict the ratings of the unknown items by calculating U and V by these formulae.

2.2   Local Differential Privacy

We use local differential privacy to expand the matrix factorization into a form that satisfies privacy preservation. This approach is an extension of differential privacy [4], in which a trusted server adds noise to the data collected from the users. However, we assume that the server can not be trusted and use local differential privacy, in which users add noise to the data before sending the data to the server.

The idea behind local differential privacy is that for a certain user, regardless of whether or not the user has certain data, the statistical result should not change. The definition is given below:

Definition 1 (Local differential privacy). *We take $x \in Nx' \in N$. A mechanism M satisfies ϵ-local differential privacy if M satisfies the following:*

$$Pr[M(x) \in S] \leq \exp(\epsilon)Pr[M(x') \in S]$$

$\forall S \subseteq Range(M)$ is any output that M may generate. A randomized response [15] is used to realize local differential privacy, which decides the value to output based on the specified probability when inputting a certain value. Each user can add noise to the data according to their own privacy awareness, since he/she can decide the probability.

2.3   Definition of the Location Privacy Preference

The location privacy preference is defined by the following:

Definition 2 (Location privacy preference). *The location privacy preference $p_u(t, l)$, in which the user u wants to hide location information at time t in location l, is expressed by the following:*

$$p_u(t, l) = \begin{cases} 1 \, (Positive) \\ 0 \, (Negative) \end{cases}$$

$1 \, (Positive)$ means that he/she can publish location information, and $0 \, (Negative)$ means that he/she does not want to publish location information.

The time t is expressed as a slot of time divided by a certain standard, and the division method varies depending on the reference time. Additionally, the location l is represented by a combination of geographic information and a category of place or either of these. Geographical information represents the latitude and longitude or certain fixed areas, and the category represents the property of a building located in that place such as a restaurant or a school.

There are various granularity regarding how to represent this information as mentioned in Sect. 1. As the granularity of information changes, the number of items in the recommendation and the degree of consideration of the nature of the time/location change, which influence the utility of the recommendation. However, it is not clear what kind of granularity is the best. Therefore, we confirm the best granularity, that is, the definition of best location information privacy preferences.

3   Recommendation Method

3.1   Framework

We propose a location privacy preference recommendation method that preserves privacy. When a user enters location privacy preferences for a certain number of places and time combinations, the method outputs location privacy preferences for a combination of unknown places and times. We assume the recommendation system exists on an untrusted server. We also assume an attacker who tries to extract the location and time of users' visit and their rating based on the output of the system. Our method aims for compatibility between high availability as a recommendation scheme and privacy protection. We realize the former by matrix factorization and the latter by local differential privacy.

First, we show a rough flow for the recommendation of the location privacy preference using the normal matrix factorization below:

1. The recommendation system sends the user matrix U and item matrix V to the user.
2. The gradients are calculated by using the user's data and step 1 and sent to the system.
3. The system updates U and V by the calculated gradients.

This operation is performed a number of times, and there is a risk that the user's data is leaked to the attacker. Therefore, we add noise to the data in step 2 above to avoid such risks. When the user calculate the gradients to update U and V, noise that satisfies local differential privacy is added to the information regarding "when and where the user visits" and "whether he/she wants to publish the location information." The gradients calculated based on the noise-added data are sent to the recommendation system We show the overview of our method in Fig. 1.

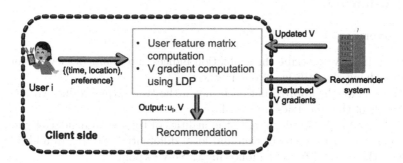

Fig. 1. Overview of our method.

3.2   Addition of Noise

We preserve privacy for the information, that is, when and where the user visits and whether he/she wants to publish their location information. In privacy protection process, we refer the method by Shin [13].

First, we will describe how to add noise to the information regarding the time and location of the user's visits. Let y_{ij} be a value indicating whether or not user i has visited place j, which is 1 if he/she has visited the place and 0 otherwise. The following equation holds: $\sum_{(i,j)\in\mathcal{M}}(r_{il} - u_i^T v_j)^2 = \sum_{i=1}^{n}\sum_{j=1}^{m} y_{ij}(r_{ij} - u_i^T v_j)^2$. Therefore, Eq. (5) can be transformed as follows: $\nabla_{v_j}\phi(U,V) = -\frac{2}{n}\sum_{i:(i,j)\in\mathcal{M}} y_{ij}u_i(r_{ij} - u_i^T v_j)$. To protect information regarding the time and location of the user's visits from privacy attacks, we should add noise to a vector $Y_i = (y_{ij})_{1\leq j\leq m}$. We use a randomized response, and the value y_{ij}^* is obtained by adding noise to y_{ij} as follows.

$$y_{ij}^* = \begin{cases} 0, \ with\, probability\, p/2 \\ 1, \ with\, probability\, p/2 \\ y_{ij}, \ with\, probability\, 1-p \end{cases}$$

Next, we describe a method of privacy protection for information on whether to disclose location information for a certain place and time combination. We add noise η_{ijl}, which is based on a Laplace distribution, to the value $g_{ij} = (g_{ijl})_{1\leq l\leq d} = -2u_i(r_{ij} - u_i^T v_j)$. The noise-added value g_{ijl}^* is expressed as follows: $g_{ijl}^* = g_{ijl} + \eta_{ijl}$.

Each user adds noise to his/her own data in the above way, and the noise-added gradients, $\{(y_{ij}^* g_{ij1}^*, \ldots g_{ijd}^*) : j = 1, \ldots, m\}$, are sent to the server. By repeating the operation k times, updating the value of the matrix using the slope calculated using the data with noise added, we find the matrix for predicting the evaluation value.

4   Evaluation

4.1   Overview

In this section, we describe the evaluation indices and points of view to consider when verifying the utilities of our method.

We evaluate the approximation between the true ratings value. We describe the details of these metrics in Sect. 4.3.

In the evaluations, we compare the utilities of the recommendation methods using normal matrix factorization and local differential privacy. In addition, we evaluate the method from the following three viewpoints:

1. What is the best location privacy preference definition?
2. How much impact does changes in the privacy preservation level make?
3. What impact does changes in the number of unknown evaluation values make?

More detailed results of the evaluation can be found in [1].

Table 2. A measure representing true or false values of the result.

		True rating	
		Positive	Negative
Predicted rating	Positive	TP (True Positive)	FP (False Positive)
	Negative	FN (False Negative)	TN (True Negative)

4.2 Dataset

In the evaluation, we use artificial data combining the location privacy preference dataset and the position information dataset.

For the location privacy preference dataset, we use LocShare acquired from the data archive CRAWDAD [11]. This dataset was obtained from 20 users in London and St. Andrews over one week from April 23 to 29 in 2011, with privacy preference data for 413 places. This dataset has few users, so it is not suitable for the evaluation of our method using matrix factorization [10]. In addition, we need bulk data in the evaluation because there are many users of recommendation in the real world.

Therefore, we generated an artificial dataset by combining the location privacy preference dataset with the trajectory dataset Gowalla, which was acquired from the location information SNS in the U.S. This dataset includes check-in histories of various places from 319,063 users collected from November 2010 to June 2011. The total number of check-ins is 36,001,959, and the number of checked-in places is 2,844,145.

4.3 Metrics

We describe the metrics for verifying the utility of our method.

We measure how accurately the recommendation can predict the ratings. The predicted value in the recommendation can be classified based on the true evaluation value as in Table 2. For example, TP (True Positive) is the number of data that are truly Positive (can be released) and whose predicted values are also Positive. We calculate the number of TP, FP, TN, and FN results from the prediction result and measure the following two indices.

– False Positive Rate: The false positive rate is the percentage of false positives predicted relative to the number of negatives.

$$FPR = \frac{FP}{TN + FP}$$

This metric is an index for verifying whether the location information that the user wants to disclose is not erroneously disclosed. The lower the false positive rate, the higher the accuracy of the privacy protection.

- Recall: Recall is the proportion of data predicted to be positive out of the data that are actually positive.

$$Recall = \frac{TP}{TP + FN}$$

Recall is an index for verifying whether the location information that the user can publish is predicted to be positive, since if the released location information decreases, the benefit decreases. The higher the recall value is, the higher the utility of the recommendation.

4.4   Evaluation Process

We evaluate our method from the three viewpoints mentioned in Sect. 4.1, and we adopt each of the following methods.

1. We used 10-fold cross validation, that divides the users into training data and test data, in which 90% of the users are regarded as training data and 10% of the users are regarded as test data.
2. Among the user's data included in the test data, we regard the known evaluation value as unknown according to the values of the $UnknownRate$ mentioned later.
3. We predict the evaluation value by using the test data and the training data which have undergone conversion processing and calculate the metrics.
4. We repeat the above process 100 times and verify the average of the evaluation indices.

In the evaluation, we change the value of one of the following parameters: $time$, ϵ, $UnknownRate$. $time$ is the length of the standard when dividing time into multiple slots. ϵ is privacy protection level when using local differential privacy. $UnknownRate$ is the ratio of what is regarded as unknown.

4.5   Results

We describe the results of experiments to confirm the utility of the recommendation method.

The Best Location Privacy Preference Definition: When defining the location privacy preference, the best definition regarding the granularity of the information is not yet clear. In this experiment, we examine the influence of changing the granularity of time on the utility. We change the criterion for dividing time into multiple slots as follows: $time = 2, 3, 4, 6, 8, 12$, the other parameters are set as follows: $\epsilon = 0.01$ and the $UnknownRate = 0.1$. The results are shown in Fig. 2.

From these results, we confirm that the precision drops when the granularity is small, that is, when the criterion time is short. This is because the coarser the definition of the granularity is, the smaller the number of goods in the recommendation, and the matrix used for prediction becomes dense. On the other

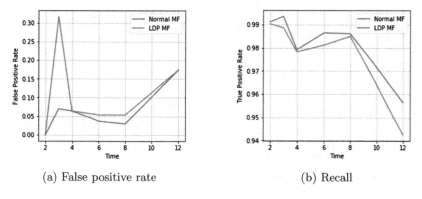

(a) False positive rate   (b) Recall

Fig. 2. Results when the time granularity is changed.

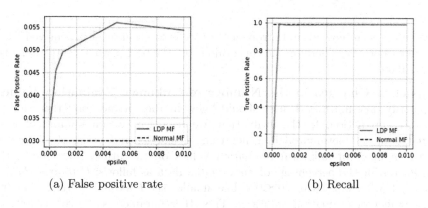

(a) False positive rate   (b) Recall

Fig. 3. Results when the ϵ is changed.

hand, however, we confirm that the accuracy drops even if the granularity is too large. Therefore, in defining the location privacy preference, we should choose the criterion with the highest utility. In this evaluation, the best criterion is 8 h.

Impact of Changes in the Privacy Preservation Level: The strength of the privacy protection can be adjusted with the value ϵ in local differential privacy. The smaller the value of ϵ, the more privacy is protected. On the other hand, there is a risk that the utility of the recommendation decreases as the added noise become large. Therefore, we examine the influence of changing the privacy protection level on the utility.

In this evaluation, we change the privacy protection level as follows: $\epsilon = 0.0001, 0.0003, 0.001, 0.005, 0.01$, the other parameters are set as follows: $time = 6$ and the $UnknownRate = 0.1$. The results are shown in Fig. 3.

From the results, we confirm that a normal recommendation is more useful in general, and as the value of ϵ increases, the usefulness increases. On the other hand, however, the change in the usefulness due to the change in the value of ϵ is small for $\epsilon = 0.003$. Larger values do not have a significant effect on the usefulness. We should select the maximum parameter that can maintain

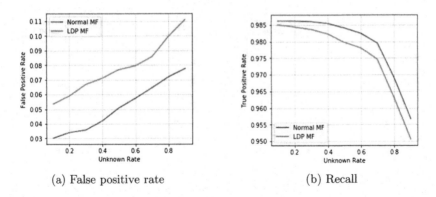

(a) False positive rate   (b) Recall

Fig. 4. Results when the *Unknown Rate* is changed.

prediction accuracy, since a stronger privacy protection level is achieved for a smaller value of ϵ. Therefore, in this evaluation, the best privacy protection level is $\epsilon = 0.001$.

Impact of Changes in the Number of Unknown Evaluation Values: We verify how much each user should know his/her privacy preference for an accurate prediction. In the evaluation, we regard a certain number of evaluated data as unevaluated in generating the model and verify the influence of the number of unknown evaluation values on the utility. We change the parameter for the percentage of unevaluated data as follows: $UnknownRate =$ $0.1, 0.2, 0.3, 0.4, 0.5, 0.6, 0.7, 0.8, 0.9$ The smaller the $UnknownRate$, the higher the number of evaluation values is. The other parameters are set as follows: $time = 6$ and $\epsilon = 0.01$. The results are shown in Figs. 4.

From these results, we confirm that the utility tends to decrease as the number of unevaluated data values increases. This is because the accuracy of the recommendation will be reduced if the training dataset is small. From these results, ideally, the user should know the location privacy preference nearly as much as possible, for about 70% of all products.

In all the evaluation, we compare the utility of the models using normal matrix factorization and local differential privacy. Since appreciable differences were not observed in devising the parameters, we confirm that the recommendation method can maintain an accuracy comparable to that of normal matrix factorization.

5   Conclusion

We propose a location privacy preference recommendation system that uses matrix factorization and achieves privacy protection by local differential privacy. We also confirm how to determine the best location privacy preference definition.

We evaluate our method using an artificial dataset from a location privacy preference dataset and a trajectory dataset. From its results, we confirm that our method can maintain the utility at a level that is the same as a method without privacy preservation. In addition, we confirm the best parameters, the granularity of the location privacy preference definition, the privacy protection level, and the number of rated items.

References

1. Asada, M., Yoshikawa, M., Cao, Y.: When and where do you want to hide? Recommendation of location privacy preferences with local differential privacy. arXiv preprint arXiv:1904.10578 (2019)
2. Calandrino, J.A., Kilzer, A., Narayanan, A., Felten, E.W., Shmatikov, V.: "You might also like:" privacy risks of collaborative filtering. In: 2011 IEEE Symposium on Security and Privacy (SP), pp. 231–246. IEEE (2011)
3. De Montjoye, Y.A., Hidalgo, C.A., Verleysen, M., Blondel, V.D.: Unique in the crowd: the privacy bounds of human mobility. Sci. Rep. **3**, 1376 (2013)
4. Dwork, C.: Differential privacy: a survey of results. In: Agrawal, M., Du, D., Duan, Z., Li, A. (eds.) TAMC 2008. LNCS, vol. 4978, pp. 1–19. Springer, Heidelberg (2008). https://doi.org/10.1007/978-3-540-79228-4_1
5. Fawaz, K., Shin, K.G.: Location privacy protection for smartphone users. In: Proceedings of the 2014 ACM SIGSAC Conference on Computer and Communications Security, pp. 239–250. ACM (2014)
6. Frey, D., Guerraoui, R., Kermarrec, A.M., Rault, A.: Collaborative filtering under a sybil attack: analysis of a privacy threat. In: Proceedings of the Eighth European Workshop on System Security, p. 5. ACM (2015)
7. Huo, Z., Meng, X., Hu, H., Huang, Y.: You can walk alone: trajectory privacy-preserving through significant stays protection. In: Lee, S., Peng, Z., Zhou, X., Moon, Y.-S., Unland, R., Yoo, J. (eds.) DASFAA 2012. LNCS, vol. 7238, pp. 351–366. Springer, Heidelberg (2012). https://doi.org/10.1007/978-3-642-29038-1_26
8. Jiang, K., Shao, D., Bressan, S., Kister, T., Tan, K.L.: Publishing trajectories with differential privacy guarantees. In: SSDBM (2013)
9. Kanza, Y., Samet, H.: An online marketplace for geosocial data. In: Proceedings of the 23rd SIGSPATIAL International Conference on Advances in Geographic Information Systems, p. 10. ACM (2015)
10. Koren, Y., Bell, R., Volinsky, C.: Matrix factorization techniques for recommender systems. Computer **8**, 30–37 (2009)
11. Parris, I., Abdesslem, F.B.: Crawdad st_andrews/locshare dataset (2011). https://crawdad.org/st_andrews/locshare/20111012/
12. Sadeh, N., et al.: Understanding and capturing people's privacy policies in a mobile social networking application. Pers. Ubiquitous Comput. **13**(6), 401–412 (2009)
13. Shin, H., Kim, S., Shin, J., Xiao, X.: Privacy enhanced matrix factorization for recommendation with local differential privacy. IEEE Trans. Knowl. Data Eng. **30**, 1770–1782 (2018)
14. Sweeney, L.: k-anonymity: a model for protecting privacy. Int. J. Uncertainty Fuzziness Knowl. Based Syst. **10**(05), 557–570 (2002)
15. Warner, S.L.: Randomized response: a survey technique for eliminating evasive answer bias. J. Am. Stat. Assoc. **60**(309), 63–69 (1965)

16. Wasef, A., Shen, X.S.: REP: location privacy for vanets using random encryption periods. Mob. Netw. Appl. **15**(1), 172–185 (2010)
17. Zhao, Y., Ye, J., Henderson, T.: Privacy-aware location privacy preference recommendations. In: Proceedings of the 11th International Conference on Mobile and Ubiquitous Systems: Computing, Networking and Services, MOBIQUITOUS 2014, ICST (Institute for Computer Sciences, Social-Informatics and Telecommunications Engineering), ICST, Brussels, Belgium, pp. 120–129 (2014). https://doi.org/10.4108/icst.mobiquitous.2014.258017

Analysis of Privacy Policies to Enhance Informed Consent

Raúl Pardo and Daniel Le Métayer[(⊠)]

Univ Lyon, Inria, INSA Lyon, CITI, 69621 Villeurbanne, France
{raul.pardo-jimenez,daniel.le-metayer}@inria.fr

Abstract. In this paper, we present an approach to enhance informed consent for the processing of personal data. The approach relies on a privacy policy language used to express, compare and analyze privacy policies. We describe a tool that automatically reports the privacy risks associated with a given privacy policy in order to enhance data subjects' awareness and to allow them to make more informed choices. The risk analysis of privacy policies is illustrated with an IoT example.

1 Introduction

One of the most common argument to legitimize the collection of personal data is the fact that the persons concerned have provided their consent or have the possibility to object to the collection. Whether opt-out is considered as an acceptable form of consent (as in the recent California Consumer Privacy Act[1]) or opt-in is required (as in the European General Data Protection Regulation - GDPR[2]), a number of conditions have to be met to ensure that the collection respects the true will of the data subject. In fact, one may argue that this is seldom the case. In practice, internet users generally have to consent on the fly, when they want to use a service, which leads them to accept mechanically the conditions of the provider. Therefore, their consent is not really informed because they do not read the privacy policies of the service providers. In addition, these policies are often vague and ambiguous. This situation, which is already critical, will become even worse with the advent of the internet of things ("IoT") which has the potential to extend to the "real world" the tracking already in place on the internet.

A way forward to address this issue is to allow users to define their own privacy policies, with the time needed to reflect on them, possibly even with the help of experts or pairs. These policies could then be applied automatically to decide upon the disclosure of their personal data and the precise conditions of such disclosures. The main benefit of this approach is to reduce the imbalance of

---

[1] https://leginfo.legislature.ca.gov/faces/billTextClient.xhtml?bill_id=201720180AB 375.

[2] https://eur-lex.europa.eu/legal-content/EN/TXT/?uri=uriserv:OJ.L_.2016.119.01. 0001.01.ENG&toc=OJ:L:2016:119:TOC.

© IFIP International Federation for Information Processing 2019
Published by Springer Nature Switzerland AG 2019
S. N. Foley (Ed.): DBSec 2019, LNCS 11559, pp. 177–198, 2019.
https://doi.org/10.1007/978-3-030-22479-0_10

powers between individuals and the organizations collecting their personal data (hereafter, respectively data subjects, or DSs, and data controllers, or DCs, following the GDPR terminology): each party can define her own policy and these policies can then be compared to decide whether a given DC is authorized to collect the personal data of a DS. In practice, DSs can obviously not foresee all possibilities when they define their initial policies and they should have the opportunity to update them when they face new types of DCs or new types of purposes for example. Nevertheless, their privacy policies should be able to cope with most situations and, as time passes, their coverage would become ever larger.

However, a language to define privacy policies must meet a number of requirements to be able to express the consent of the DSs. For example, under the GDPR, valid consent must be freely given, specific, informed and unambiguous. Therefore, the language must be endowed with a formal semantics in order to avoid any ambiguity about the meaning of a privacy policy. However, the mere existence of a semantics does not imply that DSs properly understand the meaning of a policy and its potential consequences. One way to enhance the understanding of the DSs is to provide them information about the potential risks related to a privacy policy. This is in line with Recital 39 of the GDPR which stipulates that data subjects should be "made aware of the risks, rules, safeguards and rights in relation to the processing of personal data and how to exercise their rights in relation to such processing". This approach can enhance the awareness of the DSs and allow them to adjust their privacy policies in a better informed way.

A number of languages and frameworks have been proposed in the literature to express privacy policies. However, as discussed in Sect. 6, none of them meets all the above requirements, especially the strong conditions for valid consent laid down by the GDPR. In this paper, we define a language, called PILOT, meeting these requirements and show its benefits to define precise privacy policies and to highlight the associated privacy risks. Even though PILOT is not restricted to the IoT, the design of the language takes into account the results of previous studies about the expectations and privacy preferences of IoT users [12].

We introduce the language in Sect. 2 and its abstract execution model in Sect. 3. Then we show in Sect. 4 how it can be used to help DSs defining their own privacy policies and understanding the associated privacy risks. Because the language relies on a well-defined execution model, it is possible to reason about privacy risks and to produce (and prove) automatically answers to questions raised by the DSs. In Sect. 5, we compare PILOT with existing privacy policy languages, and we conclude the paper with avenues for further research in Sect. 6.

2   The Privacy Policy Language PILOT

In this section we introduce, PILOT, a privacy policy language meeting the objectives set forth in Sect. 1. The language is designed so that it can be used both by DCs (to define certain aspects of their privacy rules or general terms regarding data protection) and DSs (to express their consent). DCs can also store the DSs policies that they have received for *accountability* purposes—i.e., to be able to demonstrate that data has been treated in accordance with the choices of DSs.

DCs devices must declare their privacy policies before they collect personal data. We refer to these policies as *DC policies*. Likewise, when a DS device sends data to a DC device, the DS device must always include a policy defining the restrictions imposed by the DS on the use of her data by the DC. We refer to these policies as *DS policies*.

In what follows, we formally define the language PILOT. We start with definitions of the most basic elements of PILOT (Sect. 2.1), which are later used to define the abstract syntax of the language (Sect. 2.2). This syntax is then illustrated with a working example (Sect. 2.3).

2.1   Basic Definitions

Devices and Entities. We start with a set $\mathcal{D}$ of *devices*. Concretely, we consider devices such as smartphones, laptops or access points, that are able to store, process and communicate data.

Let $\mathcal{E}$ denote the set of *entities* such as Google or Alphabet and $\leq_{\mathcal{E}}$ the associated partial order—e.g., since Google belongs to Alphabet we have *Google* $\leq_{\mathcal{E}}$ *Alphabet*. Entities include DCs and DSs. Every device is associated with an entity. However, entities may have many devices associated with them. The function $\texttt{entity} : \mathcal{D} \rightarrow \mathcal{E}$ defines the entity associated with a given device.

Data Items, Datatypes and Values. Let $\mathcal{I}$ be a set of *data items*. Data items correspond to the pieces of information that devices communicate. Each data item has a *datatype* associated with it. Let $\mathcal{T}$ be a set of datatypes and $\leq_{\mathcal{T}}$ the associated partial order. We use function $\texttt{type} : \mathcal{I} \rightarrow \mathcal{T}$ to define the datatype of each data item. Examples of datatypes[3] are: age, address, city and clinical records. Since *city* is one of the elements that the datatype *address* may be composed of, we have *city* $\leq_{\mathcal{T}}$ *address*. We use $\mathcal{V}$ to the denote the set of all values of data items, $\mathcal{V} = (\bigcup_{t \in \mathcal{T}} \mathcal{V}_t)$ where $\mathcal{V}_t$ is the set of values for data items of type t. We use a special element $\bot \in \mathcal{V}$ to denote the undefined value. A data item may be undefined, for instance, if it has been deleted or it has not been collected. The device where a data item is created (its source) is called the *owner* device of the data item. We use a function $\texttt{owner} : \mathcal{I} \rightarrow \mathcal{D}$ to denote the owner device of a given data item.

Purposes. We denote by $\mathcal{P}$ the set of *purposes* and $\leq_{\mathcal{P}}$ the associated partial order. For instance, if newsletter is considered as a specific type of advertisement, then we have *newsletter* $\leq_{\mathcal{P}}$ *advertisement*.

---

[3] Note that here we do not use the term "datatype" as traditionally in programming languages. We use datatype to refer to the semantic meaning of data items.

Conditions. Privacy policies are contextual: they may depend on *conditions* on the information stored on the devices on which they are evaluated. For example, (1) *"Only data from adults may be collected"* or (2) *"Only locations within the city of Lyon may be collected from my smartwatch"* are examples of policy conditions. In order to express conditions we use a simple logical language. Let $\mathcal{F}$ denote a set of functions and *terms t* be defined as follows: $t:: = i \mid c \mid f(\vec{t})$ where $i \in \mathcal{I}$ is data item, $c \in \mathcal{V}$ is a constant value, $f \in \mathcal{F}$ is a function, and $\vec{t}$ is a list of terms matching the arity of f. The syntax of the logical language is as follows: $\varphi:: = t_1 * t_2 \mid \neg\varphi \mid \varphi_1 \wedge \varphi_2 \mid tt \mid ff$ where $*$ is an arbitrary binary predicate, t_1, t_2 are terms; tt and ff represent respectively true and false. For instance, $age \geq 18$ and $smartwatch\_location = Lyon$ model conditions (1) and (2), respectively. We denote the set of well-formed conditions as $\mathcal{C}$. In order to compare conditions, we use a relation, $\vdash: \mathcal{C} \times \mathcal{C}$. We write $\varphi_1 \vdash \varphi_2$ to denote that φ_2 is stronger than φ_1.

2.2 Abstract Syntax of PILOT Privacy Policies

In this section we introduce the abstract syntax of PILOT *privacy policies*, or, simply, PILOT *policies*. We emphasize the fact that this abstract syntax is not the syntax used to communicate with DSs or DCs. This abstract syntax can be associated with a concrete syntax in a restricted form of natural language. We do not describe this mapping here due to space constraints, but we provide some illustrative examples in Sect. 2.3 and describe a user-friendly interface to define PILOT policies in Sect. 4.3. The goal of PILOT policies is to express the conditions under which data can be communicated. We consider two different types of data communications: *data collection* and *transfers*. Data collection corresponds to the collection by a DC of information directly from a DS. A transfer is the event of sending previously collected data to third parties.

Definition 1 (PILOT Privacy Policies Syntax). *Given Purposes $\in 2^{\mathcal{P}}$, retention_time $\in \mathbb{N}$, condition $\in \mathcal{C}$, entity $\in \mathcal{E}$ and datatype $\in \mathcal{T}$, the syntax of PILOT policies is defined as follows:*

$$Pilot\ Privacy\ Policy :: = (datatype, dcr, TR)$$
$$Data\ Communication\ Rule\ (dcr) :: = \langle condition, entity, dur \rangle$$
$$Data\ Usage\ Rule\ (dur) :: = \langle Purposes, retention\_time \rangle$$
$$Transfer\ Rules(TR) :: = \{dcr_1, dcr_2, \ldots\}$$

We use $\mathcal{DUR}$, $\mathcal{DCR}$, $\mathcal{PP}$ to denote the sets of data usage rules, data communication rules and PILOT privacy policies, respectively. The set of transfer rules is defined as the set of sets of data communication rules, $TR \in 2^{\mathcal{DCR}}$. In what follows, we provide some intuition about this syntax and an example of application.

Data Usage Rules. The purpose of these rules is to define the operations that may be performed on the data. *Purposes* is the set of allowed purposes and *retention_time* the deadline for erasing the data. As an example, consider the following data usage rule,

$$dur_1 = \langle \{research\}, 26/04/2019 \rangle.$$

This rule states that the data may be used only for the purpose of research and may be used until *26/04/2019*.

Data Communication Rules. A data communication rule defines the conditions that must be met for the data to be collected by or communicated to an entity. The outer layer of data communication rules—i.e., the condition and entity—should be checked by the sender whereas the data usage rule is to be enforced by the receiver. The first element, *condition*, imposes constraints on the data item and the context (state of the DS device); *entity* indicates the entity allowed to receive the data; *dur* is a data usage rule stating how *entity* may use the data. For example, $\langle age > 18, AdsCom, dur_1 \rangle$. states that data may be communicated to the entity AdsCom which may use it according to dur_1 (defined above). It also requires that the data item *age* is greater than 18. This data item may be the data item to be sent or part of the contextual information of the sender device.

Transfer Rules. These rules form a set of data communication rules specifying the entities to whom the data may be transferred.

PILOT Privacy Policies. DSs and DCs use PILOT policies to describe how data may be used, collected and transferred. The first element, *datatype*, indicates the type of data the policy applies to; *dcr* defines the collection conditions and *TR* the transfer rules. In some cases, several PILOT policies are necessary to fully capture the privacy choices for a given datatype. For instance, a DS may allow only her employer to collect her data when she is at work but, when being in a museum, she may allow only the museum. In this example, the DS must define two policies, one for each location.

2.3   Example: Vehicle Tracking

In this section, we illustrate the syntax of PILOT with a concrete example that will be continued with the risk analysis in Sect. 4.

The use of Automatic Number Plate Recognition (ANPR) [10] is becoming very popular for applications such as parking billing or pay-per-use roads. These systems consist of a set of cameras that automatically recognize plate numbers when vehicles cross the range covered by the cameras. Using this information, it is possible to determine how long a car has been in a parking place or how many times it has traveled on a highway, for example.

ANPR systems may collect large amounts of mobility data, which raises privacy concerns [11]. When data is collected for the purpose of billing, the consent of the customer is not needed since the legal ground for the data processing can be the performance of a contract. However, certain privacy regulations, such as the GDPR, require prior consent for the use of the data for other purposes, such as commercial offers or advertisement.

Consider a DC, Parket, which owns parking areas equipped with ANPR in France. Parket is interested in offering discounts to frequent customers. To this end, Parket uses the number plates recorded by the ANPR system to send commercial offers to a selection of customers. Additionally, Parket transfers some data to its sister company, ParketWW, that operates worldwide with the goal of providing better offers to their customers. Using data for these purposes requires explicit consent from DSs. The PILOT policy below precisely captures the way in which Parket wants to collect and use number plates for these purposes.

$$
(number\_plate, \langle tt, Parket, \langle \{commercial\_offers\}, 21/03/2019 \rangle \rangle, \\
\{\langle tt, ParketWW, \langle \{commercial\_offers\}, 26/04/2019 \rangle \rangle\}) \quad (1)
$$

The condition (tt) in (1) means that Parket does not impose any condition on the number plates it collects or transfers to ParketWW. This policy can be mapped into the following natural language sentence:

Parket may collect data of type *number_plate* and use it for *commercial_offers* purposes until *21/03/2019*.
This data may be transferred to *ParketWW* which may use it for *commercial_offers* purposes until *26/04/2019*.

The parts of the policy in *italic font* correspond to the elements of PILOT's abstract syntax. These elements change based on the content of the policy. The remaining parts of the policy are common to all PILOT policies.

To obtain DSs consent, Parket uses a system which broadcasts the above PILOT policy to vehicles before they enter the ANPR area. The implementation of this broadcast process is outside the scope of this paper; several solutions are presented in [8]. DSs are therefore informed about Parket's policy before data is collected. However, DSs may disagree about the processing of their data for these purposes. They can express their own privacy policy in PILOT to define the conditions of their consent (or denial of consent).

Consider a DS, Alice, who often visits Parket parkings. Alice wants to benefit from the offers that Parket provides in her city (Lyon) but does not want her information to be transferred to third-parties. To this end, she uses the following PILOT policy:

$$
(number\_plate, \langle car\_location = Lyon, Parket, \\
\langle \{commercial\_offers\}, 21/03/2019 \rangle \rangle, \emptyset) \quad (2)
$$

In practice, she would actually express this policy as follows:

Parket may collect data of type *number_plate* if *car_location is Lyon* and use it for *commercial_offers* purposes until *21/03/2019*.

which is a natural language version of the above abstract syntax policy.

In contrast with Parket's policy, Alice's policy includes a condition using *car_location*, which is a data item containing the current location of Alice's car.

In addition, the absence of transfer statement means that Alice does not allow Parket to transfer her data. It is easy to see that Alice's policy is more restrictive than Parket's policy. Thus, after Alice's device[4] receives Parket's policy, it can automatically send an answer to Parket indicating that Alice does not give her consent to the collection of her data in the conditions stated in *Parket*'s policy. In practice, Alice's policy can also be sent back so that Parket can possibly adjust her own policy to match Alice's requirements. Parket would then have the option to send a new DC policy consistent with Alice's policy and Alice would send her consent in return. The new policy sent by Parket can be computed as a join of Parket's original policy and Alice's policy (see Appendix C for an example of policy join which is proven to preserve the privacy preferences of the DS).

This example is continued in Sect. 4 which illustrates the use of PILOT to enhance Alice's awareness by providing her information about the risks related to her choices of privacy policy.

3   Abstract Execution Model

In this section, we describe the abstract execution model of PILOT. The purpose of this abstract model is twofold: it is useful to define a precise semantics of the language and therefore to avoid any ambiguity about the meaning of privacy policies; also, it is used by the verification tool described in the next section to highlight privacy risks. The definition of the full semantics of the language, which is presented in a companion paper [19], is beyond the scope of this paper. In the following, we focus on the two main components of the abstract model: the system state (Sect. 3.1) and the events (Sect. 3.2).

3.1   System State

We first present an abstract model of a system composed of devices that communicate information and use PILOT policies to express the privacy requirements of DSs and DCs. Every device has a set of associated policies. A policy is associated with a device if it was defined in the device or the device received it. Additionally, DS devices have a set of data associated with them. These data may represent, for instance, the MAC address of the device or workouts recorded by the device. Finally, we keep track of the data collected by DC devices together with their corresponding PILOT policies. The system state is formally defined as follows.

Definition 2 (System state). *The system state is a triple $\langle \nu, \pi, \rho \rangle$ where:*

- $\nu : \mathcal{D} \times \mathcal{I} \rightharpoonup \mathcal{V}$ *is a mapping from the data items of a device to their corresponding value in that device.*

---

[4] This device can be Alice's car on-board computer, which can itself be connected to the mobile phone used by Alice to manage her privacy policies [8].

- $\pi : \mathcal{D} \to 2^{\mathcal{D} \times \mathcal{PP}}$ is a function denoting the policy base of a device. The policy base contains the policies created by the owner of the device and the policies sent by other devices in order to state their collection requirements. A pair (d, p) means that PILOT policy p belongs to device d. We write π_d to denote $\pi(d)$.

- $\rho : \mathcal{D} \to 2^{\mathcal{D} \times \mathcal{I} \times \mathcal{PP}}$ returns a set of triples (s, i, p) indicating the data items and PILOT policies that a controller has received. If $(d', i, p) \in \rho(d)$, we say that device d has received or collected data item i from device d' and policy p describes how the data item must be used. We write ρ_d to denote $\rho(d)$.

In Definition 2, ν returns the local value of a data item in the specified device. However, not all devices have values for all data items. When the value of a data item in a device is undefined, ν returns $\bot$. The policy base of a device d, π_d, contains the PILOT policies that the device has received or that have been defined locally. If $(d, p) \in \pi(d)$, the policy p corresponds to a policy that d has defined in the device itself. On the other hand, if $(d, p) \in \pi(e)$ where $d \neq e$, p is a policy sent from device e. Policies stored in the policy base are used to compare the privacy policies of two devices before the data is communicated. The information that a device has received is recorded in ρ. Also, ρ contains the PILOT policy describing how data must be used. The difference between policies in π and ρ is that policies in π are used to determine whether data can be communicated, and policies in ρ are used to describe how a data item must be used by the receiver.

Example 1. Figure 1 shows a state composed of two devices: Alice's car, and Parket's ANPR system. The figure depicts the situation after Alice's car has entered the range covered by the ANPR camera and the collection of her data has already occurred.

The database in Alice's state (ν_{Alice}) contains a data item of type number plate $plate_{Alice}$ whose value is GD-042-PR. The policy base in Alice's device (π_{Alice}) contains two policies: $(Alice, p_{Alice})$ representing a policy that Alice defined, and $(Parket, p_{Parket})$ which represents a policy p_{Parket} sent by Parket. We assume that p_{Alice} and p_{Parket} are the policies applying to data items of type number plate.

Parket's state contains the same components as Alice's state with, in addition, a set of received data (ρ_{Parket}). The latter contains the data item $plate_{Alice}$ collected from Alice and the PILOT policy p_{Parket} that must be applied in order to handle the data. Note that p_{Parket} was the PILOT policy originally defined by Parket. In order for Alice's privacy to be preserved, it must hold that p_{Parket} is more restrictive than Alice's PILOT policy p_{Parket}, which is denoted by $p_{Parket} \sqsubseteq p_{Alice}$.[5] This condition can easily be enforced by comparing the policies before data is collected. The first element in $(Alice, plate_{Alice}, p_{Parket})$ indicates that the data comes from Alice's device. Finally, Parket's policy base has one policy: its own policy p_{Parket}, which was communicated to Alice for data collection. □

---

[5] See Appendix A for the formal definition of $\sqsubseteq$.

Fig. 1. Example system state

3.2   System Events

In this section we describe the set of events E in our abstract execution model. We focus on events that ensure that the exchange of data items is done according to the PILOT policies of DSs and DCs.

Events. The set of events E is composed by the following the events: *request*, *send*, *transfer* and *use*. The events *request*, *send* and *transfer* model valid exchanges of policies and data among DCs and DSs. The event *use* models correct usage of the collected data by DCs. In what follow we explain each event in detail.

request$(sndr, rcv, t, p)$ models request of data from DCs to DSs or other DCs. Thus, $sndr$ is always a DC device, and rcv may be a DC or DS device. A request includes the type of the data that is being requested t and a PILOT policy p. As expected, the PILOT policy is required to refer to the datatype that is requested, i.e., $p = (t, \_, \_)$. As a result of executing *request*, the pair (rcv, p) is added to π_{rcv}. Thus, rcv is informed of the conditions under which $sndr$ will use the requested data.

send$(sndr, rcv, i)$ represents the collection by the DC rcv of a data item i from the DS $sndr$. In order for *send* to be executed, the device $sndr$ must check that π_{sndr} contains: (i) an active policy defined by $sndr$, p_{sndr}, indicating how $sndr$ allows DCs to use her data, and (ii) an active policy sent by rcv, p_{rcv}, indicating how she plans to use the data. A policy is active if it applies to the data item to be sent, to rcv's entity, the retention time has not yet been reached, and its condition holds.[6] Data can only be sent if p_{rcv} is more restrictive than p_{sndr} (i.e., $p_{rcv} \sqsubseteq p_{sndr}$), which must be checked by $sndr$. We record the data exchange in ρ_{rcv} indicating: the sender, the data item and rcv's PILOT policy, $(sndr, i, p_{rcv})$. We also update rcv's database with the value of i in $sndr$'s state, $\nu(rcv, i) = \nu(sndr, i)$.

transfer$(sndr, rcv, i)$ is executed when a DC $(sndr)$ transfers a data item i to another DC (rcv). First, $sndr$ checks whether π_{sndr} contains an active policy, from rcv, p_{rcv}. Here we do not use a PILOT policy from $sndr$, instead we use the PILOT policy p sent along with the data—defined by the owner of i. Thus, $sndr$ must check whether there exists an active transfer rule (tr) in the set

---

[6] See Appendix B for the formal definition of active policy and active transfer.

of transfers rules of the PILOT policy p. As before, $sndr$ must check that the policy sent by rcv is more restrictive than those originally sent by the owner of the data, i.e., $p_{rcv} \sqsubseteq p_{tr}$ where p_{tr} is a policy with the active transfer tr in the place of the data communication rule and with the same set of transfers as p. Note that data items can be transferred more than once to the entities in the set of transfers as long as the retention time has not been reached. This is not an issue in terms of privacy as data items are constant values. In the resulting state, we update ρ_{rcv} with the sender, the data item and rcv's PILOT policy, $(sndr, i, p_{rcv})$. Note that, in this case, the owner of the data item is not $sndr$ since transfers always correspond to exchanges of previously collected data, $\mathtt{owner}(i) \neq sndr$. The database of rcv is updated with the current value of i in ν_{sndr}.

$use(dev, i, pur)$ models the use of a data item i by a DC device dev for purpose pur. Usage conditions are specified in the data usage rule of the policy attached to the data item, denoted as p_i, in the set of received data of dev, ρ_{dev}. Thus, in order to execute use we require that: (i) the purpose pur is allowed by p_i, and (ii) the retention time in p_i has not elapsed.

4   Risk Analysis

As described in the introduction, an effective way to enhance informed consent is to raise user awareness about the risks related to personal data collection. Privacy risks may result from different sorts of misbehavior such as the use of data beyond the allowed purpose or the transfer of data to unauthorized third parties [15].

In order to assess the risks related to a given privacy policy, we need to rely on assumptions about potential risk sources, such as:

- Entities e_i that may have a strong interest to use data of type t for a given purpose pur.
- Entities e_i that may have facilities and interest to transfer data of type t to other entities e_j.

In practice, some of these assumptions may be generic and could be obtained from databases populated by pairs or NGOs based on history of misconducts by companies or business sectors. Others risk assumptions can be specific to the DS (e.g., if she fears that a friend may be tempted to transfer certain information to another person). Based on these assumptions, a DS who is wondering whether she should add a policy p to her current set of policies can ask questions such as: "if I add this policy p:

- Is there a risk that my data of type t is used for purpose pur?
- Is there a risk that, at some stage, entity e gets my data of type t? "

In what follows, we first introduce our approach to answer the above questions (Sect. 4.1); then we illustrate it with the example introduced in Sect. 2.3. (Section 4.2) and we present a user-friendly interface to define and analyze privacy policies (Sect. 4.3).

4.1 Automatic Risk Analysis with SPIN

In order to automatically answer questions of the type described above, we use the verification tool SPIN [14]. SPIN belongs to the family of verification tools known as *model-checkers*. A model-checker takes as input a model of the system (i.e., an abstract description of the behavior of the system) and a set of properties (typically expressed in formal logic), and checks whether the model of the system satisfies the properties. In SPIN, the model is written in the modeling language PROMELA [14] and properties are encoded in *Linear Temporal Logic* (LTL) (e.g., [4]). We chose SPIN as it has successfully been used in a variety of contexts [18]. However, our methodology is not limited to SPIN and any other formal verification tool such as SMT solvers [5] or automated theorem provers [24] could be used instead.

Our approach consists in defining a PROMELA model for the PILOT events and privacy policies, and translating the risk analysis questions into LTL properties that can be automatically checked by SPIN. For example, the question *"Is there a risk that Alice's data is used for the purpose of profiling by ParketWW?"* is translated into the LTL property *"ParketWW never uses Alice's data for profiling"*. Devices are modeled as processes that randomly try to execute events defined as set forth in Sect. 3.2.

In order to encode the misbehavior expressed in the assumptions, we add "illegal" events to the set of events that devices can execute. For instance, consider the assumption *"use of data beyond the allowed purpose"*. To model this assumption, we introduce the event *illegal_use*, which behaves as *use*, but disregards the purpose of the DS policy for the data.

SPIN explores all possible sequences of executions of events (including misbehavior events) trying to find a sequence that violates the LTL property. If no sequence is found, the property cannot be violated, which means that the risk corresponding to the property cannot occur. If a sequence is found, the risk corresponding to the property can occur, and SPIN returns the sequence of events that leads to the violation. This sequence of events can be used to further clarify the cause of the violation.

4.2 Case Study: Vehicle Tracking

We illustrate our risk analysis technique with the vehicle tracking example introduced in Sect. 2.3. We first define the PROMELA model and the assumptions on the entities involved in this example. The code of the complete model is available in [23].

Promela Model. We define a model involving the three entities identified in Sect. 2.3 with, in addition, the car insurance company *CarInsure* which is identified as a potential source of risk related to *ParketWW*, i.e., $\mathcal{E} = \{$*Alice, Parket, ParketWW, CarInsure*$\}$. Each entity is associated with a single device: $\mathcal{D} = \mathcal{E}$ and $\texttt{entity}(x) = x$ for $x \in \{$*Alice, Parket, ParketWW, CarInsure*$\}$. We focus on one datatype $\mathcal{T} = \{$*number_plate*$\}$ with its set of values defined as $\mathcal{V}_{number\_plate} =$

{GD-042-PR}. We consider a data item $plate_{Alice}$ of type $number\_plate$ for which $Alice$ is the owner. Finally, we consider a set of purposes $\mathcal{P} = \{\,commercial\_offers,\ profiling\,\}$.

Risk Assumptions on Entities. In this case study, we consider two risk assumptions:

1. ParketWW may transfer personal data to CarInsure disregarding the associated DS privacy policies.
2. CarInsure has strong interest in using personal data for profiling.

In practice, these assumptions, which are not specific to Alice, may be obtained automatically from databases populated by pairs or NGOs for example.

Set of Events. The set of events that we consider is derived from the risk assumptions on entities. On the one hand, we model events that behave correctly, i.e., as described in Sect. 3.2. In order to model the worst case scenario in terms of risk analysis, we consider that: the DCs in this case study (i.e., Parket, ParketWW and CarInsure) can request data to any entity (including Alice), the DCs can collect Alice's data, and the DCs can transfer data among them. On the other hand, the risk assumptions above are modeled as two events: ParketWW may transfer data to CarInsure disregarding Alice's policy, and CarInsure may use Alice's data for profiling even if it is not allowed by Alice's policy. Let $DC, DC' \in \{Parket, ParketWW, CarInsure\}$, the following events may occur:

- $request(DC, Alice, number\_plate, p)$ - A DC requests a number plate from Alice and p is the PILOT policy of the DC.
- $request(DC, DC', number\_plate, p)$ - A DC requests data items of type number plate from another DC and p is the PILOT policy of the requester DC.
- $send(Alice, DC, i)$ - Alice sends her item i to a DC.
- $transfer(DC, DC', i)$ - A DC transfers a previously received item i to another DC.
- $illegal\_transfer(ParketWW, CarInsure, i)$ - ParketWW transfers a previously received item i to CarInsure disregarding the associated PILOT policy defined by the owner of i.
- $illegal\_use(CarInsure, i, profiling)$ - CarInsure uses data item i for profiling disregarding the associated privacy policy defined by the owner of i.

Alice's Policies. In order to illustrate the benefits of our risk analysis approach, we focus on the following two policies that Alice may consider.

$$p\_trans_{Alice} = (number\_plate, \langle tt, Parket, \langle\{commercial\_offers\}, 21/03/2019\rangle\rangle, \\ \{\langle tt, ParketWW, \langle\{commercial\_offers\}, 26/04/2019\rangle\rangle\}).$$

$$p\_no\_trans_{Alice} = (number\_plate, \langle tt, Parket, \langle\{commercial\_offers\}, 21/03/2019\rangle\rangle, \emptyset).$$

The policy p_trans_{Alice} states that Parket can collect data of type number plate from Alice, use it for commercial offers and keep it until 21/03/2019. It also allows Parket to transfer the data to ParketWW. ParketWW may use the data for commercial offers and keep it until 26/04/2019. The policy $p\_no\_trans_{Alice}$ is similar to p_trans_{Alice} except that it does not allow Parket to transfer the data. We assume that Alice has not yet defined any other privacy policy concerning Parket and ParketWW.

Parket's Policy. We set Parket's privacy policy equal to Alice's. By doing so, we consider the worst case scenario in terms of privacy risks because it allows Parket to collect Alice's data and use it in all conditions and for all purposes allowed by Alice.

ParketWW's Policy. Similarly, ParketWW's policy is aligned with the transfer rule in p_trans_{Alice}:

$$p_{ParketWW} = (number\_plate, \langle tt, ParketWW, \langle \{commercial\_offers\}, 26/04/2019 \rangle \rangle, \emptyset).$$

The above policy states that ParketWW may use data of type number plate for commercial offers and keep it until *26/04/2019*. It also represents the worst case scenario for risk analysis, as it matches the preferences in Alice's first policy.

Results of the Risk Analysis

Table 1 summarizes some of the results of the application of our SPIN risk analyzer on this example. The questions in the first column have been translated into LTL properties used by SPIN (see [23]). The output of SPIN appears in columns 2 to 5. The green boxes indicate that the output is in accordance with Alice's policy while red boxes correspond to violations of her policy.

Columns 2 and 3 correspond to executions of the system involving correct events, considering respectively p_trans_{Alice} and $p\_no\_trans_{Alice}$ as Alice's policy. As expected, all these executions respect Alice's policies.

Columns 3 and 4 consider executions involving *illegal_transfer* and *illegal_use*. These columns show the privacy risks taken by Alice based on the above risk assumptions. Rows 3 and 6 show respectively that CarInsure may get Alice's data and use it for profiling. In addition, the counterexamples generated by SPIN, which are not pictured in the table, show that this can happen only after ParketWW executes *illegal_transfer*.

From the results of this privacy risk analysis Alice may take a better informed decision about the policy to choose. In a nutshell, she has three options:

1. Disallow Parket to use her data for commercial offers, i.e., choose to add neither p_trans_{Alice} nor $p\_no\_trans_{Alice}$ to her set of policies (Parket will use the data only for billing purposes, based on contract).
2. Allow Parket to use her data for commercial offers without transfers to ParketWW, i.e., choose $p\_no\_trans_{Alice}$.

3. Allow Parket to use her data for commercial offers and to transfer to ParketWW, i.e., choose p_trans_{Alice}.

Therefore, if Alice wants to receive commercial offers but does not want to take the risk of being profiled by an insurance company, she should take option two.

Table 1. Risk Analysis of Alice's policies p_trans_{Alice} and $p\_no\_trans_{Alice}$. Red boxes denote that Alice's policy is violated. Green boxes denote that Alice's policy is respected.

Question	Normal behavior		Misbehavior Assumptions	
	p_trans_{Alice}	$p\_no\_trans_{Alice}$	p_trans_{Alice}	$p\_no\_trans_{Alice}$
Can Parket receive Alice's data?	Yes	Yes	Yes	Yes
Can ParketWW receive Alice's data?	Yes	No	Yes	No
Can CarInsure receive Alice's data?	No	No	Yes	No
Can Parket use Alice's data for other purpose than commercial offers?	No	No	No	No
Can ParketWW use Alice's data for other purpose than commercial offers?	No	No	No	No
Can CarInsure use Alice's data for profiling?	No	No	Yes	No

4.3   Usability

In order to show the usability of the approach, we have developed a web application to make it possible for users with no technical background to perform risk analysis as outlined in Sect. 4.2 for the ANPR system.

Figure 2 shows the input forms of the web application. First, DSs have access to a user-friendly form to input PILOT policies. In the figure we show an example for the policy p_trans_{Alice}. Then DSs can choose the appropriate risk assumptions from the list generated by the system. Finally, they can ask questions about the potential risks based on these assumptions. When clicking on "Verify!", the web application runs SPIN to verify the LTL property corresponding to the question. The text "Not Analyzed" in grey is updated with "Yes" or "No" depending on the result. The figure shows the results of the three first questions with p_trans_{Alice} and no risk assumption chosen (first column in Table 1).

The web application is tailored to the ANPR case study we use throughout the paper. The PROMELA model and the policies defined in Sect. 4.2 are implemented in the application. This prototype can be generalized in different

PILOT Privacy Policy

Enter the PILOT privacy policy you would like to analyse:

| Parket ⇕ | may collect data of type | number plate ⇕ | and use it for | commercial offers ⇕ | until |

| 21 / 03 / 2019   ⊗ |

This data may be transferred to:

| ParketWW ⇕ | which may use it for | commercial offers ⇕ | until | 26 / 04 / 2019   ⊗ | Remove transfer |

Add transfer

Risk Analysis

Risk Assumptions

Choose the assumptions for the model:

☐ ParketWW ⇕ may transfer personal data to Carinsure ⇕ disregarding the associated DS policies.

☐ Carinsure ⇕ has strong interest in using data for profiling ⇕ .

Risk Questions

Click on *Verify!* to get answers to the questions below. The answer depends on the PILOT policy and the assumptions you have chosen.

- Can *Parket* receive my data?
  Yes Verify!

- Can *ParketWW* receive my data?
  Yes Verify!

- Can *CarInsurance* receive my data?
  No Verify!

- Can *Parket* use my data for other purpose than *commercial offers*?
  Not analyzed Verify!

- Can *ParketWW* use my data for other purpose than *commercial offers*?
  Not analyzed Verify!

- Can *CarInsure* use my data for *profiling*?
  Not analyzed Verify!

Fig. 2. Input forms of risk analysis web application.

directions, for example by allowing users to enter specific risk assumptions on third parties. The range of questions could also be extended to include questions such as "Can X use Y's data for other purpose than pur?" The code of the web application is available at [23].

5 Related Work

Several languages or frameworks dedicated to privacy policies have been proposed. A pioneer project in this area was the "Platform for Privacy Preferences" (P3P) [22]. P3P makes it possible to express notions such as purpose, retention time and conditions. However, P3P is not really well suited to the IoT as it was conceived as a policy language for websites. Also, P3P does not offer support for defining data transfers. Other languages close to P3P have been proposed, such as the "Enterprise Policy Authorization Language" (EPAL) [2] and "An Accountability Policy Language" (A-PPL) [3]. The lack of a precise execution model for these languages may also give rise to ambiguities and variations in their implementations.

Even if its first target was the interactions with service providers rather than IoT environments, the language that is the closest to the spirit of PILOT is the Data Handling Policy (DHP) language [1]. DHP also allows users to express the actions and purposes that are authorized for (specific or generic) recipients. DHP does not include explicitly transfer rules and retention time but, in contrast to PILOT, it makes it possible to specify obligations. Obligations can be used, for example, to require the deletion of data after a given period of time.

None of the above works include tools to help users understand the privacy risks associated with a given a policy, which is a major benefit of PILOT as discussed in Sect. 4. In the same spirit, De et al. [16] have proposed a methodology where DSs can visualize the privacy risks associated to their privacy settings. Here the authors use harm trees to determine the risks associated with privacy settings. The main difference with PILOT is that harm trees must be manually defined for a given application whereas we our analysis is fully automatic.[7]

Another line of work is that of formal privacy languages. Languages such as S4P [7] and SIMPL [17] define unambiguously the behavior of the system—and, consequently, the meaning of the policies—by means of trace semantics. The goal of this formal semantics is to be able to prove global correctness properties such as "DCs always use DS data according to their policies". While this semantics is well-suited for its intended purpose, it cannot be directly used to develop policy enforcement mechanisms. In contrast, we provide a PROMELA model in Sect. 4—capturing the execution model of PILOT (cf. Sect. 3)—that can be used as a reference to implement a system for the enforcement for PILOT policies. In addition, these languages, which were proposed before the adoption of the GDPR, were not conceived with its requirements in mind.

Other languages have been proposed to specify privacy regulations such as HIPAA, COPAA and GLBA. For instance, CI [6] is a dedicated linear temporal logic based on the notion of contextual integrity. CI has been used to model certain aspects of regulations such as HIPAA, COPPA and GLBA. Similarly, PrivacyAPI [18] is an extension of the access control matrix with operations such as notification and logging. The authors also use a PROMELA model of HIPAA to be able to verify the "correctness" and better understand the regulation. PrivacyLFP [9] uses first-order fixed point logic to increase the expressiveness of previous approaches. Using PrivacyLFP, the authors formalize HIPAA and GLBA with a higher degree of coverage than previous approaches. The main difference between PILOT and these languages is their focus. PILOT is focused on modeling DSs and DCs privacy policies and enhancing DSs awareness whereas these languages focus on modeling regulations.

Usage control (UCON) [20,21] appeared as an extension of access control to express how the data may be used after being accessed. To this end, it introduces *obligations*, which are actions such as "do not transfer data item i".

---

[7] Only risk assumptions must be defined, which is useful to answer different "what-if" questions.

The Obligation Specification Language (OSL) [13] is an example of enforcement mechanism through digital right management systems. However, UCON does not offer any support to compare policies and does not differentiate between DSs and DCs policies, which is a critical feature in the context of privacy policies. For DSs to provide an informed consent, they should know whether DCs policies comply with their own policies.

Some work has also been done on privacy risk analysis [15], in particular to address the needs of the GDPR regarding Privacy Impact Assessments. We should emphasize that the notion of risk analysis used in this paper is different in the sense that it applies to potential risks related to privacy policies rather than systems or products. Hence, the risk assumptions considered here concern only the motivation, reputation and potential history of misbehavior of the parties (but not the vulnerabilities of the systems, which are out of reach and expertise of the data subjects).

6 Conclusion

In this paper, we have presented the privacy policy language PILOT, and a novel approach to analyzing privacy policies which is focused on enhancing informed consent. An advantage of a language like PILOT is the possibility to use it as a basis to implement "personal data managers", to enforce privacy policies automatically, or "personal data auditors", to check a posteriori that a DC has complied with the DS policies associated with all the personal data that it has processed. Another orthogonal challenge in the context of the IoT is to ensure that DSs are always informed about the data collection taking place in their environment and can effectively communicate their consent (or objection) to the surrounding sensors. Different solutions to this problem have been proposed in [8] relying on PILOT as a privacy policy language used by DCs to communicate their policies and DSs to provide their consent. These communications can either take place directly or indirectly (through registers in which privacy policies can be stored).

The work described in this paper can be extended in several directions. First, the risk analysis model used here is simple and could be enriched in different ways, for example by taking into account risks of inferences between different types of data. The evaluation of these risks could be based on past experience and research such as the study conducted by Privacy International.[8] The risk analysis could also involve the history of the DS (personal data already collected by DCs in the past). On the formal side, our objective is to use a formal theorem prover to prove global properties of the model. This formal framework could also be used to implement tools to verify that a given enforcement system complies with the PILOT policies.

---

[8] https://privacyinternational.org/sites/default/files/2018-04/data%20points%20used%20in%20tracking_0.pdf.

Acknowledgments. This work has been partially funded by the ANR project CISC (Certification of IoT Secure Compilation) and by the Inria Project Lab SPAI.

Appendix

A   Policy Subsumption

We formalize the notion of *policy subsumption* as a relation over PILOT policies. We start by defining subsumption of data usage and data communication rules, which is used to define PILOT policy subsumption.

Definition 3 (Data Usage Rule Subsumption). *Given two data usage rules $dur_1 = \langle P_1, rt_1 \rangle$ and $dur_2 = \langle P_2, rt_2 \rangle$, we say that dur_1 subsumes dur_2, denoted as $dur_1 \preceq_{\mathcal{DUR}} dur_2$, iff (i) $\forall p_1 \in P_1 \cdot \exists p_2 \in P_2$ such that $p_1 \leq_{\mathcal{P}} p_2$; and (ii) $rt_1 \leq rt_2$.*

Definition 4 (Data Communication Rule Subsumption). *Given two data communication rules $dcr_1 = \langle c_1, e_1, dur_1 \rangle$ and $dcr_2 = \langle c_2, e_2, dur_2 \rangle$, we say that dcr_1 subsumes dcr_2, denoted as $dcr_1 \preceq_{\mathcal{DCR}} dcr_2$, iff (i) $c_1 \vdash c_2$; (ii) $e_1 \leq_{\mathcal{E}} e_2$; and (iii) $dur_1 \preceq_{\mathcal{DUR}} dur_2$.*

Definition 5 (PILOT Privacy Policy Subsumption). *Given two PILOT privacy policies $\pi_1 = \langle t_1, dcr_1, TR_1 \rangle$ and $\pi_2 = \langle t_2, dcr_2, TR_2 \rangle$, we say that π_1 subsumes π_2, denoted as $\pi_1 \sqsubseteq \pi_2$ iff (i) $t_1 \leq_{\mathcal{T}} t_2$; (ii) $dcr_1 \preceq_{\mathcal{DCR}} dcr_2$; and (iii) $\forall tr_1 \in TR_1 \cdot \exists tr_2 \in TR_2$ such that $tr_1 \preceq_{\mathcal{DCR}} tr_2$.*

B   Active Policies and Transfer Rules

Here we formally define when PILOT policies and transfer rules are active. Let $\mathtt{eval}(\nu, d, \varphi)$ denote an *evaluation function* for conditions. $\mathtt{eval}(\nu, d, \varphi)$ is defined as described in Table 2. We use a function $\mathtt{time}(e) : E \to \mathbb{N}$ to assign a timestamp—represented as a natural number $\mathbb{N}$—to each event of a trace.

Active policy. Formally, $\mathtt{activePolicy}(p, send(sndr, rcv, i), st) = \mathtt{type}(i) \leq_{\mathcal{T}}$
$t \wedge \mathtt{eval}(\nu, sndr, \varphi) \wedge \mathtt{time}(st, send(sndr, rcv, i)) < rt \wedge \mathtt{entity}(rcv) \leq_{\mathcal{E}} e$
where $p = (t, \langle \varphi, e, \langle \_, rt \rangle \rangle, \_)$ and $st = \langle \nu, \_, \_ \rangle$. Intuitively, given $p = (t, \langle \varphi, e, \langle P, rt \rangle \rangle, TR)$, we check that: (i) the type of the data to be sent corresponds to the type of data the policy is defined for ($\mathtt{type}(i) \leq_{\mathcal{T}} t$); (ii) the condition of the policy evaluates to true ($\mathtt{eval}(\nu, sndr, \varphi)$); the retention time for the receiver has not expired ($\mathtt{time}(send(sndr, rcv, i)) < rt$); and (iii) the entity associated with the receiver device is allowed by the policy ($\mathtt{entity}(rcv) \leq_{\mathcal{E}} e$).

Active transfer rule. In order for a transfer rule to be active, the above checks are performed on the transfer rule tr, and, additionally, it is required that the retention time for the sender has not elapsed ($\mathtt{time}(transfer(sndr, rcv, i)) < rt$).

Table 2. Definition of $\mathtt{eval}(\nu, d, \varphi)$. We use $\hat{c}$, $\hat{f}$ and $\hat{*}$ to denote the interpretation of constants, functions and binary predicates, respectively.

$$
\begin{aligned}
\mathtt{eval}(\nu, d, tt) &= \mathtt{true} \\
\mathtt{eval}(\nu, d, ff) &= \mathtt{false} \\
\mathtt{eval}(\nu, d, i) &= \nu(d, i) \\
\mathtt{eval}(\nu, d, c) &= \hat{c} \\
\mathtt{eval}(\nu, d, f(t_1, t_2, \ldots)) &= \hat{f}(\mathtt{eval}(\nu, d, t_1), \mathtt{eval}(\nu, d, t_2), \ldots) \\
\mathtt{eval}(\nu, d, t_1 * t_2) &= \begin{cases} \mathtt{eval}(\nu, d, t_1) \,\hat{*}\, \mathtt{eval}(\nu, d, t_2) \text{ if } \mathtt{eval}(\nu, d, t_i) \neq \bot \\ \bot \text{ otherwise} \end{cases} \\
\mathtt{eval}(\nu, d, \varphi_1 \wedge \varphi_2) &= \begin{cases} \mathtt{eval}(\nu, d, \varphi_1) \text{ and } \mathtt{eval}(\nu, d, \varphi_2) \text{ if } \mathtt{eval}(\nu, d, \varphi_i) \neq \bot \\ \bot \text{ otherwise} \end{cases} \\
\mathtt{eval}(\nu, d, \neg\varphi) &= \begin{cases} \mathtt{not}\ \mathtt{eval}(\nu, d, \varphi) \text{ if } \mathtt{eval}(\nu, d, \varphi) \neq \bot \\ \bot \text{ otherwise} \end{cases}
\end{aligned}
$$

C   Policy Join

We present a *join operator* for PILOT policies and prove that the resulting policy is more restrictive than the policies used to compute the join. We first define join operators for data usage rules and data communication rules, and use them to the join operator for PILOT policies. Let $min(e, e')$ be a function that, given two elements $e, e' \in \mathcal{X}$ returns the minimum in the corresponding partial order $\leq_{\mathcal{X}}$. Let $\pitchfork$ denote the intersection keeping the minimum of comparable elements in the partial order of purposes. Formally, given $P, P' \in \mathcal{P}$, $P \pitchfork P' \triangleq (P \cap P') \cup P''$ where $P'' = \{p \in P \mid \exists p' \in P' \text{ s.t. } p < p'\}$.

Definition 6 (Data Usage Rule Join). *Given two data usage rules* $dur_1 = \langle P_1, rt_1 \rangle$ *and* $dur_2 = \langle P_2, rt_2 \rangle$, *the data usage rule join operator is defined as:* $dur_1 \sqcup_{\mathcal{DUR}} dur_2 = \langle P_1 \pitchfork P_2, min(rt_1, rt_2) \rangle$.

Definition 7 (Data Communication Rule Join). *Given two data communication rules* $dcr_1 = \langle c_1, e_1, dur_1 \rangle$ *and* $dcr_2 = \langle c_2, e_2, dur_2 \rangle$, *the data communication rule join operator is defined as:* $dur_1 \sqcup_{\mathcal{DCR}} dur_2 = \langle c_1 \wedge c_2, min(e_1, e_2), dur_1 \sqcup_{\mathcal{DUR}} dur_2 \rangle$.

Definition 8 (PILOT Policy Join $\sqcup$). *Given two PILOT policies* $p = (t_1, dcr_1, TR_1)$ *and* $q = (t_2, dcr_2, TR_2)$, *the policy join operator is defined as:* $dur_1 \sqcup_{\mathcal{DCR}} dur_2 = (min(t_1, t_2), dcr_1 \sqcup_{\mathcal{DCR}} dcr_2, \{t \sqcup_{\mathcal{DCR}} t' \mid t \in TR_1 \wedge t' \in TR_2 \wedge t \preceq_{\mathcal{DCR}} t'\})$.

We say that an join operation is privacy preserving if the resulting policy is more restrictive than both operands. Formally,

Definition 9 (Privacy Preserving Join). *We say that* $\sqcup$ *is privacy preserving iff* $\forall p, q \in \mathcal{PP}. (p \sqcup q) \sqsubseteq p \wedge (p \sqcup q) \sqsubseteq q$.

In what follows we prove that the operation $\sqcup$ is privacy preserving, Lemma 3.

Lemma 1. *Given two data usage rules* $dur_1, dur_2 \in DUR$ *it holds that* $dur_1 \sqcup_{DUR} dur_2 \preceq_{DUR} dur_1$ *and* $dur_1 \sqcup_{DUR} dur_2 \preceq_{DUR} dur_2$.

Proof. We split the proof into the two conjuncts of Lemma 1.

$\underline{dur_1 \sqcup_{DUR} dur_2 \preceq_{DUR} dur_1}$ - We split the proof into the elements of data usage
rules, i.e., purposes and retention time.

- We show that $\forall p \in dur_1.P \mathbin{\widehat{\cap}} dur_2.P \cdot \exists p' \in dur_1.P$ such that $p \leq_P p'$.
  - $\forall p \in [(dur_1.P \cap dur_2.P) \cup \{p_1 \in dur_1.P \mid \exists p_2 \in dur_2.P \text{ s.t. } p_1 \leq_P p_2\}] \cdot \exists p' \in dur_1.P$ such that $p \leq_P p'$ [By Def. $\widehat{\cap}$]
  - We split the proof for each operand in the union.
    * We show that $\forall p \in (dur_1.P \cap dur_2.P) \cdot \exists p' \in dur_1.P$ such that $p \leq_P p'$. Assume $p \in (dur_1.P \cap dur_2.P)$. Then $\exists p' \in dur_1.P$ s.t. $p = p'$ [By Def. $\cap$]. Therefore, $p \leq_P p'$.
    * We show that $\forall p \in \{p_1 \in dur_1.P \mid \exists p_2 \in dur_2.P \text{ s.t. } p_1 \leq_P p_2\} \cdot \exists p' \in dur_1.P$ such that $p \leq_P p'$. Assume $p \in \{p_1 \in dur_1.P \mid \exists p_2 \in dur_2.P \text{ s.t. } p_1 \leq_P p_2\}$. Then, $p \in dur_1.P$, and, consequently, $\exists p' \in dur_1.P$ s.t. $p \leq_P p'$.
- $min(dur_1.rt, dur_2.rt) \leq dur_1.rt$ [By Def. min]

$\underline{dur_1 \sqcup_{DUR} dur_2 \preceq_{DUR} dur_2}$ - Retention time is symmetric to the previous case,
therefore we only show purposes.

- We show that $\forall p \in dur_1.P \mathbin{\widehat{\cap}} dur_2.P \cdot \exists p' \in dur_2.P$ such that $p \leq_P p'$.
  - $\forall p \in [(dur_1.P \cap dur_2.P) \cup \{p_1 \in dur_1.P \mid \exists p_2 \in dur_2.P \text{ s.t. } p_1 \leq_P p_2\}] \cdot \exists p' \in dur_2.P$ such that $p \leq_P p'$ [By Def. $\widehat{\cap}$]
  - We split the proof for each operand in the union.
    * The case $\forall p \in (dur_1.P \cap dur_2.P) \cdot \exists p' \in dur_2.P$ such that $p \leq_P p'$ is symmetric to the case above.
    * We show that $\forall p \in \{p_1 \in dur_1.P \mid \exists p_2 \in dur_2.P \text{ s.t. } p_1 \leq_P p_2\} \cdot \exists p' \in dur_2.P$ such that $p \leq_P p'$. Assume $p \in \{p_1 \in dur_1.P \mid \exists p_2 \in dur_2.P \text{ s.t. } p_1 \leq_P p_2\}$. Then $\exists p_2 \in dur_2.P$ s.t. $p \leq_P p_2$, and, consequently, $\exists p' \in dur_2.P$ s.t. $p \leq_P p'$. $\qquad\square$

Lemma 2. *Given two data communication rules* $dcr_1, dcr_2 \in DCR$ *it holds that* $dcr_1 \sqcup_{DUR} dcr_2 \preceq_{DUR} dcr_1$ *and* $dcr_1 \sqcup_{DCR} dcr_2 \preceq_{DCR} dcr_2$.

Proof. We split the proof into the two conjuncts of Lemma 2.

$\underline{dcr_1 \sqcup_{DUR} dcr_2 \preceq_{DUR} dcr_1}$ - We split the proof into the elements of data communication rules, i.e., conditions and entities and data usage rules.

- $dcr_1.c \wedge dcr_2.c \vdash dcr_1.c$ [By $\wedge$-elimination]
- $min(dcr_1.e, dcr_2.e) \leq_{\mathcal{E}} dcr_1.e$ [By Def. min]
- $dcr_1.dur \sqcup_{DUR} dcr_2.dur \preceq_{DCR} dcr_1.dur$ [By Lemma 1]

$\underline{dcr_1 \sqcup_{DUR} dcr_2 \preceq_{DUR} dcr_2}$ - The proof is symmetric to the previous case. $\quad\square$

Lemma 3. *The operation* $\sqcup$ *in Definition 8 is privacy preserving.*

Proof. Let p_1 and p_2 be two PILOT privacy policies. We show that $p_1 \sqcup p_2 \sqsubseteq p_1$ and $p_1 \sqcup p_2 \sqsubseteq p_2$. We proof each conjunct separately.

$p_1 \sqcup p_2 \sqsubseteq p_1$ - We split the proof in cases based on the structure of PILOT policies, i.e., datatype $(p_1.t, p_2.t)$, data communication rules $(p_1.dcr, p_2.dcr)$ and transfers $(p_1.TR, p_2.TR)$.

- $min(p_1.t, p_2.t) \leq_{\mathcal{T}} p_1.t$. [By Def. of min]
- $p_1.dcr \sqcup_{\mathcal{DCR}} p_2.dcr \preceq_{\mathcal{DCR}} p_1.dcr$. [By Lemma 2]
- $\forall tr_\sqcup \in \{tr_1 \sqcup_{\mathcal{DCR}} tr_2 \mid tr_1 \in p_1.TR \wedge tr_2 \in p_2.TR \wedge tr_1 \preceq_{\mathcal{DCR}} tr_2\} \cdot \exists tr \in TR_1$ s.t. $tr_\sqcup \preceq_{\mathcal{DCR}} tr$. Assume $tr_\sqcup \in \{tr_1 \sqcup_{\mathcal{DCR}} tr_2 \mid tr_1 \in p_1.TR \wedge tr_2 \in p_2.TR \wedge tr_1 \preceq_{\mathcal{DCR}} tr_2\}$. Then $tr_1 \in p_1.TR$ and $tr_\sqcup \preceq_{\mathcal{DCR}} tr_1$. [By Lemma 2]

$p_1 \sqcup p_2 \sqsubseteq p_2$ - The proof is symmetric to the previous case. $\qquad\square$

References

1. Ardagna, C.A., De Capitani di Vimercati, S., Samarati, P.: Enhancing user privacy through data handling policies. In: Damiani, E., Liu, P. (eds.) DBSec 2006. LNCS, vol. 4127, pp. 224–236. Springer, Heidelberg (2006). https://doi.org/10.1007/11805588_16
2. Ashley, P., Hada, S., Karjoth, G., Powers, C., Schunter, M.: Enterprise privacy authorization language (EPAL). IBM Research (2003)
3. Azraoui, M., Elkhiyaoui, K., Önen, M., Bernsmed, K., De Oliveira, A.S., Sendor, J.: A-PPL: an accountability policy language. In: Garcia-Alfaro, J., et al. (eds.) DPM/QASA/SETOP -2014. LNCS, vol. 8872, pp. 319–326. Springer, Cham (2015). https://doi.org/10.1007/978-3-319-17016-9_21
4. Baier, C., Katoen, J.: Principles of Model Checking. MIT Press, Cambridge (2008)
5. Barrett, C., Tinelli, C.: Satisfiability modulo theories. In: Clarke, E., Henzinger, T., Veith, H., Bloem, R. (eds.) Handbook of Model Checking, pp. 305–343. Springer, Cham (2018). https://doi.org/10.1007/978-3-319-10575-8_11
6. Barth, A., Datta, A., Mitchell, J.C., Nissenbaum, H.: Privacy and contextual integrity: framework and applications. In: Proceedings of the 27th IEEE Symposium on Security and Privacy, S&P 2006, pp. 184–198 (2006)
7. Becker, M., Malkis, A., Bussard, L.: S4P: a generic language for specifying privacy preferences and policies. Research report, Microsoft Research (2010)
8. Cunche, M., Le Métayer, D., Morel, V.: A generic information and consent framework for the IoT. Research report RR-9234, Inria (2018). https://hal.inria.fr/hal-01953052
9. DeYoung, H., Garg, D., Jia, L., Kaynar, D.K., Datta, A.: Experiences in the logical specification of the HIPAA and GLBA privacy laws. In: Proceedings of the 2010 ACM Workshop on Privacy in the Electronic Society, WPES 2010, pp. 73–82 (2010)
10. Du, S., Ibrahim, M., Shehata, M.S., Badawy, W.M.: Automatic license plate recognition (ALPR): a state-of-the-art review. IEEE Trans. Circuits Syst. Video Technol. 23(2), 311–325 (2013)
11. Electronic Fountrier Foundatino (EFF): Automated License Plate Readers (ALPR) (2017). https://www.eff.org/cases/automated-license-plate-readers
12. Emami-Naeini, P., et al.: Privacy expectations and preferences in an IoT world. In: Proceedings of the 13th Symposium on Usable Privacy and Security, SOUPS 2017, pp. 399–412 (2017)
13. Hilty, M., Pretschner, A., Basin, D., Schaefer, C., Walter, T.: A policy language for distributed usage control. In: Biskup, J., López, J. (eds.) ESORICS 2007. LNCS, vol. 4734, pp. 531–546. Springer, Heidelberg (2007). https://doi.org/10.1007/978-3-540-74835-9_35

14. Holzmann, G.J.: The SPIN Model Checker - Primer and Reference Manual. Addison-Wesley, Boston (2004)
15. De, S.J., Le Métayer, D.: Privacy Risk Analysis. Morgan & Claypool Publishers, San Rafael (2016)
16. De, S.J., Le Métayer, D.: Privacy risk analysis to enable informed privacy settings. In: 2018 IEEE European Symposium on Security and Privacy, Workshops, EuroS&P Workshops, pp. 95–102 (2018)
17. Métayer, D.: A formal privacy management framework. In: Degano, P., Guttman, J., Martinelli, F. (eds.) FAST 2008. LNCS, vol. 5491, pp. 162–176. Springer, Heidelberg (2009). https://doi.org/10.1007/978-3-642-01465-9_11
18. May, M.J., Gunter, C.A., Lee, I.: Privacy APIs: access control techniques to analyze and verify legal privacy policies. In: Proceedings of the 19th IEEE Computer Security Foundations Workshop, CSFW 2006, pp. 85–97. IEEE Computer Society (2006)
19. Pardo, R., Le Métayer, D.: Formal verification of legal privacy requirements (Submitted for Publication)
20. Park, J., Sandhu, R.S.: The $UCON_{ABC}$ usage control model. ACM Trans. Inf. Syst. Secur. **7**(1), 128–174 (2004)
21. Pretschner, A., Hilty, M., Basin, D.A.: Distributed usage control. Commun. ACM **49**(9), 39–44 (2006)
22. Reagle, J., Cranor, L.F.: The platform for privacy preferences. Commun. ACM **42**(2), 48–55 (1999)
23. PILOT Risk Analysis Model. https://github.com/raulpardo/pilot-risk-analysis-model
24. Robinson, A.J., Voronkov, A.: Handbook of Automated Reasoning, vols. 1 and 2. Elsevier, Amsterdam (2001)

Security Protocol Practices

Lost in TLS? No More!
Assisted Deployment of Secure
TLS Configurations

Salvatore Manfredi[1,2]([✉]) [iD], Silvio Ranise[1] [iD], and Giada Sciarretta[1] [iD]

[1] Security & Trust, FBK, Trento, Italy
{smanfredi,ranise,giada.sciarretta}@fbk.eu
[2] University of Trento, Trento, Italy

Abstract. Over the last few years, there has been an almost exponential growth of TLS popularity and usage, especially among applications that deal with sensitive data. However, even with this widespread use, TLS remains for many system administrators a complex subject. The main reason is that they do not have the time to understand all the cryptographic algorithms and features used in a TLS suite and their relative weaknesses. For these reasons, many different tools have been developed to verify TLS implementations. However, they usually analyze the TLS configuration and provide a list of possible attacks, without specifying their mitigations. In this paper, we present TLSAssistant, a fully-featured tool that combines state-of-the-art TLS analyzers with a report system that suggests appropriate mitigations and shows the full set of viable attacks.

Keywords: TLS misconfiguration · Vulnerability detection · Assisted mitigations

1 Introduction

Transport Layer Security (TLS) consists of a set of cryptographic protocols designed to provide secure communications over a network. Developed as a successor of the Secure Socket Layer (SSL) protocol, TLS has gained popularity and widespread usage since the release of its first version in 1999 [21]. According to [37], more than 130,000 of the top Alexa websites [43] support one or multiple versions of the TLS protocol.

The popularity of TLS has encouraged attackers to find vulnerabilities and develop exploits as documented by a long line of reported attacks and corresponding fixes [1–5, 16, 17, 28, 31–33, 42, 46, 48] together with the evolution of the standard TLS specification from 1.0 to 1.3 as a result of the strategies put in place by Internet service providers such as Apple, Google, Amazon, and Mozilla to deprecate the use of TLS versions 1.0 and 1.1 [7] and of the SHA-1 hash function [6]. The types of attacks vary widely and include the renegotiation of cipher

© IFIP International Federation for Information Processing 2019
Published by Springer Nature Switzerland AG 2019
S. N. Foley (Ed.): DBSec 2019, LNCS 11559, pp. 201–220, 2019.
https://doi.org/10.1007/978-3-030-22479-0_11

suites to exploit weak encryption algorithms [31], the knowledge of initialization vectors to retrieve symmetric keys [17], and the use of libraries to exploit poor certificate validation in deployments where clients are non-browsers [15]. The variety attacks is the result of *(i)* maintaining backward compatibility and *(ii)* evolving use case scenarios in which TLS is deployed. The main problem with *(i)* is illustrated by the following observation from [37]: more than 108,000 web sites still support TLS 1.0 that is vulnerable to a set of well-known attacks including Man In The Middle (MITM). The problem with *(ii)* has already been pointed out in [15] for SSL where it is shown that certificate validation—as supported by available libraries for developing clients not based on browsers (e.g., native mobile applications)—is flawed and permits to mount MITM attacks.

To help administrators in deploying secure TLS instances, a variety of tools [11,14,29,39,45,47,50,51] have been developed for identifying weaknesses that may lead to one or more known attacks. While such tools are quite effective in automatically finding vulnerabilities and issuing warning about possible attacks, the burden of finding adequate mitigation measures is completely left on the administrator who must first collect information about the identified problem and related fixes. Typically, such information is distributed in several sources ranging from scientific papers to blog posts. Even disregarding the effort to collect enough material to mitigate a security problem—notice that available tools have varying coverage of the known TLS attacks—administrators should have enough skills to understand the (often subtle) details and turn the information in a concrete strategy to fix the problem. In other words, there is a problem in making actionable the reports returned by available tools. To overcome this problem, we make the following four main contributions:

- we build an exhaustive catalogue of known attacks to TLS deployments;
- we perform a comparison of the state-of-the-art tools capable of identifying attacks of TLS deployments and characterize the coverage with respect to the catalogue compiled in the previous point;
- we design and build an open-source tool, called TLSAssistant[1], that reuses some of the tools for identifying attacks considered in the previous point to maximize coverage and enriches reports with possible mitigations and fixes, including code snippets when the TLS entities are among the most widely used (e.g., the TLS server is Apache);
- we experimentally evaluate the effectiveness of TLSAssistant by reporting our experience in using it in the context of the deployment of a large scale infrastructure for identity management and, most importantly, in a user study involving users with little or no security skills. The findings of the user study provide encouraging first evidence of the effectiveness of the tool as even un-experienced users were able to successfully mitigate complex attacks.

While many of the problems reported by TLSAssistant are server-side, particular attention has been devoted to the security issues that result from inadequate

---

[1] Available at sites.google.com/fbk.eu/tlsassistant.

certificate validation in mobile applications. The main motivation for this choice is the increasing role of mobile applications in accessing Internet resources combined with the serious security consequences of managing certificates without the help of browsers (as already observed in [15]).

Plan of the paper. Section 2 provides the necessary background notions on TLS and a brief overview of its vulnerabilities. Section 3 contains a comparison of the state-of-the-art automated tools for identifying TLS attacks both server and (mobile) client-side. Section 4 contains a comprehensive catalogue of attacks and related mitigations in the various version of TLS. Section 5 introduces the tool TLSAssistant with its architecture, usage, and some details about the implementation. Section 6 reports the use of TLSAssistant in the deployment of an eIDAS solution based on the new Italian identity cards and a user study involving bachelor and master degree students that were asked to fix two non-trivial vulnerabilities. Section 7 concludes the paper and highlights future work.

2   Background

We provide some background notions about TLS needed to better understand the security implications of its use. We briefly describe the general structure of the TLS suite and some details in Sect. 2.1 and then give a concise guide to the main vulnerabilities in Sect. 2.2.

2.1   TLS

The Transport Layer Security (TLS) suite is composed of two main protocols:

Handshake: allows the parties to exchange all information required to establish a reliable session. Depending on the configuration, the handshake can provide either mutual or one-way authentication (usually is one-way, thus only server provides a certificate). The protocol supports two special messages: *(i)* `Change Cipher Spec` that signals the transition of the session to a different ciphering strategy and *(ii)* `Alert` which propagates potential alert messages.

Record: encapsulates the messages to be transmitted to ensure their security. The record protocol is composed of the following steps: *(i)* splitting of the data stream into chunks; *(ii)* compression of the chunks; *(iii)* generation of the Message Authentication Code (MAC) with the algorithm agreed during the handshake; *(iv)* encryption of the payload using the cipher chosen during the handshake; and *(v)* addition of the header to enable the packet to be transmitted.

Over the years, TLS has seen the release of four versions: v1.0 released in 1999 [21], v1.1 released in 2006 [22], v1.2 released in 2008 [23], v1.3 released in 2018 (August) [26]. In the following we will detail v1.2, as is the most widely supported, and v1.3, as it has introduced a set of changes to mitigate known vulnerabilities that affected the previous TLS versions.

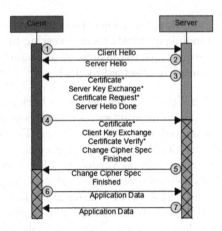

Fig. 1. TLS 1.2 full handshake.

TLS 1.2. According to Qualys' March monthly scan [37], TLS 1.2 is currently the most widely supported protocol with a coverage of 94.8%. Each message of the full handshake is shown in Fig. 1 (the striped sections show the encrypted transmission, asterisks indicate optional messages). For lack of space, we refer to the TLS specification [23] for the description of all the messages. Here, we specify the first two messages: *Client Hello* and *Server Hello*. As we will describe in Sect. 5, our tool analyzer will send different *Client Hello* messages and analyze the *Server Hello* answers to check the presence of vulnerabilities. The remaining messages are used to authenticate the parties, calculate the symmetric key and to apply the ciphering strategies.

Client Hello. The client can start the handshake at any time by sending a *Client Hello* message to the server, it contains: *(i)* the version of the protocol that the client wants to use (it should be the highest available); *(ii)* a random value obtained by chaining the timestamp (32 bit in UNIX time) and a randomly generated nonce (28 bytes); *(iii)* a session identifier: empty field indicating the will to create a new session (in case of session resumption the client behaves differently); *(iv)* list of supported cipher suites, each element has the following structure: TLS_⟨KeyExchange⟩_WITH_⟨Cipher⟩_⟨Mac⟩; *(v)* a list of supported compression methods; and *(vi)* a list of requested extensions (set of additional functionalities the server has to provide).

Server Hello. In response to the *Client Hello*, the server sends its hello message that contains: *(i)* a chosen protocol (the highest version supported by both parties); *(ii)* a random value obtained by chaining the timestamp (32 bit in UNIX time) and a randomly generated nonce (28 bytes); *(iii)* a freshly-generated value that will identify the new session (in case of session resumption, the server behaves differently); *(iv)* a chosen cipher suite; *(v)* a chosen compression method; and *(vi)* a list of required extensions (additional features).

Table 1. Main differences introduced with TLS 1.3

Status	What	Why
Removed	not-AEAD ciphers	avoid attacks on legacy ciphers
	RSA key exchange	always provide forward secrecy
	broken hash algorithms (MD5, SHA-1)	avoid SLOTH and similar attacks
	Change Cipher Spec message	streamline the handshake
	data compression	avoid CRIME attack
	session renegotiation	avoid renegotiation attacks
Added	0-RTT mode for a quick resumption	increase resumption speed
	EncryptedExtensions msg	avoid transmitting preferences in plaintext
Changed	msg encryption starts after the *Server Hello*	allow Client certificate encryption
	Hello content (structure unchanged)	extend Handshake capabilities

TLS 1.3. After years in the making, the final version of the standard has been published this August, 2018. Table 1 summarizes the key differences with TLS 1.2 according to the RFC [26]. Thanks to these changes, TLS 1.3 is not prone to any of the known attacks such as the ones related to legacy ciphers (e.g., Lucky 13) or broken hash algorithms (e.g., SLOTH).

2.2 Vulnerabilities

There exist many TLS-related vulnerabilities: some of them exploit the support of weak cryptographic aspects (e.g., weak ciphers and hash functions), others use an (un)voluntary weakening of security properties to bypass the authentication process (e.g., accepting self-signed certificates [34] and setting a permissive hostname verifier [34]) or the loss of trust in the PKI system due to a improper certificate generation (e.g., CA impairment [19] and Certificate Spoofing [25]). In Table 2 we detail a set of well-known TLS attacks, each line contains: *(i)* the name given by the authors; *(ii)* the feature or weakness exploited; *(iii)* a brief description on how the attack can be mounted, and *(iv)* which version of TLS can be affected by such attack. To better understand the described vulnerability exploitation, we review some cryptographic aspects:

Export ciphers. Weakened ciphers introduced by the U.S. government to limit the security of foreign countries' transmissions [20];

Stream ciphers. Symmetric key ciphers in which each digit is encrypted combining it with a pseudorandom cipher [40];

Block ciphers. Symmetric key ciphers in which a set of bits with a fixed length (called block) is encrypted all at once [44];

Compression mechanism. TLS feature used to reduce the amount of data sent through the network [24];

Hash functions. Function that takes as input data of arbitrary size and produces as output a string with fixed length [49];

Renegotiation. TLS feature used to enhance the security of an already established session without dropping the current connection [27];

Table 2. Known TLS attacks.

Name	Vulnerability	Attack	Affects
3SHAKE [31]	Renegotiation feature	Completing three handshakes with incorrectly placed certificates	ANY
Bar Mitzvah [28]	RC4 steam cipher	Extracting weak keys by targeting the first 100 bytes of the ciphertext	ANY
BEAST [17]	Initialization vector in cipher block chain	Guessing the plaintext to retrieve the symmetric key	TLS 1.0
CRIME [33]	TLS header compression mechanism (DEFLATE)	Continuously requesting data from the server in order to decrypt the session cookies (inferring the encryption)	ANY
DROWN [3,36]	SSLv2 weakness due to the use of export ciphers	Decrypting intercepted TLS connections by connecting to an SSLv2 server that uses the same private key	SSLv2
Logjam [1,18]	Weakness of export cipher suites	Negotiating the use of weak cipher suites (`DHE_EXPORT`)	TLS 1.1
Lucky 13 [2,12]	CBC-mode weakness due to HMAC-SHA1 decryption failure information leakage	Replacing the last bytes with chosen bytes and monitoring the transmission time	ANY
POODLE [32]	SSLv3 weakness due to the missing validation of padding bytes	Downgrading to SSLv3 and guessing the padding in order to slowly recover plaintext	SSLv3
RC4 NOMORE [48]	Bias in the generation of the "random" keys of the RC4 stream cipher	Statistically analyzing the Fluhrer-McGrew biases	ANY
SLOTH [4,13]	Availability of weak hash functions	Requesting a RSA-MD5 certificate signature and looking for collisions	TLS 1.2
Reneg. [42]	Renegotiation feature	Blocking the handshake process of the victim and use it to complete the attacker's transaction	ANY
Sweet32 [5]	64-bit block ciphers	Mounting a birthday attack which creates collisions	TLS 1.2
Truncation [46]	Server incorrect handling of the TLS termination mode through multiple connections	Keeping the victim's session alive (by blocking the logout request sent to the server)	ANY
BREACH [16]	HTTP compression mechanism	Requesting data from the server in order to guess the response body (note: without downgrading the SSL/TLS connection)	ANY

Termination protocol. Exchange of alert messages which signals the end of the message sending [21, §7.2.1].

Among all the attacks, here we detail the two used in our experimentation (see Sect. 6.2): CRIME [33] and BREACH [16]. Both attacks are related to the availability of DEFLATE [24], a compression algorithm that reduces the size of an input by replacing duplicate strings with a reference to their last occurrence. Given that neither TLS nor HTTP hide the size of each message, an attacker can exploit this information leakage to steal sensitive data. Supposing the will to steal session cookies, the attack is performed by injecting (e.g., using a controlled JavaScript loaded by the victim) different characters into the client's messages trying to guess the cookie. Thanks to DEFLATE, if the guess is wrong and the characters are not part of the cookie, the size of the response will be bigger. On the other hand, if the attacker guessed correctly, the size will remain the same. This attack is referred as CRIME if it exploits the compression within TLS, BREACH otherwise.

3   Tools Comparison

There are many TLS analyzers on the market and we wanted to understand which one suited better our purposes. For this reason, we decided to compare them to find the one who had the highest amount of features.

Table 3. Tool comparison - server.

Checks	sslscan	sslenum	TLSSLed	TLS-atk	3Shake_chk	testssl
SSLv3, TLS 1.0, 1.1 and 1.2, RC4	●	●	●	●	○	●
AES ciphers	◐	◐	●	●	○	●
Weak ciphers	◐	◐	◐	◐	○	●
SSLv2, Secure renegotiation	●	○	●	●	○	●
POODLE, CBC-mode cipher, 3DES	●	●	○	●	○	●
MD5/SHA1 signature alert	●	●	●	○	○	●
Sweet32	●	◐	○	◐	○	●
Certificate expiration	●	○	●	○	○	●
Weak DH parameters	●	●	○	○	○	●
Heartbleed, TLS compression	●	○	○	●	○	●
BEAST	◐	○	●	○	○	●
TLS 1.3, DROWN	○	○	○	●	○	●
Qualys scoring	○	●	○	○	○	●
More analysis[a]	○	○	○	○	○	●
3SHAKE	○	○	○	○	●	○

[a] server's default picks, certificate info, HSTS, HPKP, security headers, cookie, reverse proxy, client simulations, SPDY and HTTP2 availability

Table 3 shows the comparison between six tools that perform server-related TLS vulnerabilities[2]. Each detection is identified depending on the type of information resulting. In particular, ●,◗ and ○ mean an explicit, implicit (which can be inferred using other explicit detections) or missing detection, respectively. The evaluated tools are:

sslscan [39]: the analyzer is able to detect the full set of available ciphers on a webserver. The default output shows the full list of accepted/rejected connections, detailing each line with the cipher's name, its key length and the used protocol;

ssl-enum-ciphers [29]: script developed for the **nmap** security scanner [30] that lists all the available cipher suites, compression methods and a small set of possible misconfigurations. The generated report shows the set of ciphers (available per protocol) with the relative Qualys' rating [38], a grade which goes from A+ to F depending on the level of provided security;

TLSSLed.sh [45]: built on top of an older version of **sslscan** [39], this script check if the server supports old protocols, weak ciphers and for the certificate signature. The verbose output highlights the results using different colours;

TLS-Attacker [47]: open source framework for analyzing TLS libraries. It can be fully-customized to perform any kind of connection and contains a set of pre-configured attacks for testing purposes. By running each attack, the user can understand whether or not the server is vulnerable;

3SHAKE checker [51]: is a simple script that checks if the target server supports **extended_master_secret**, an extension specifically designed to mitigate the 3SHAKE [31] attack. The output shows, for each available version of TLS, if the extension request has been accepted;

testssl.sh [50]: is a fully-featured open source command-line tool able to analyze a server's configuration. The tool is mainly focused on detecting weaknesses and various configuration issues while being able to perform a wider set of tests. Among these, **testssl.sh** is able to list the set of ciphers available per protocol, analyze the chain of trust of a provided certificate, simulate handshakes and much more. These features make **testssl.sh** the most powerful tool among the evaluated. The generated report contains the results for all the performed analysis, associated with a colour that signals the severity of the detected result.

All the listed tools work by repeatedly connecting to the target server using specifically crafted *ClientHello* messages. By checking the server's responses (i.e. *ServerHello*), the tools are able to understand the server's configuration. Besides the amount of provided features, the compared tools have a major limitation: all of them offer little or no explanation on how to actually mitigate the detected weaknesses. This somehow defies their purpose given that a system administrator will still have to spend a lot of time and effort researching the most appropriate set of mitigations to apply.

---

[2] Given the need for modularity, we focused on local analyzers rather than their online counterparts.

Table 4. Tool comparison - mobile clients.

Checks	Mallodroid	Tapioca
Detect non-default trust managers	●	○
Check client's certificate validation	◗	●
Enumerate contacted hosts	○	●
Validate HTTPS negotiations	○	●
Read encrypted traffic	○	●

3.1 Mobile Clients

As mentioned in the introduction, while in a browser the handle of TLS and its certificates is built-in, this is not the case for mobile native applications: a developer can either choose to use one of the many available TLS libraries or to implement his own methods. In both cases, an incorrect certificate handling may lead to several authentication-related issues. For this reason, there is the need for specific tools.

Table 4 shows the differences between two Android-related analyzers:

Mallodroid [14]: Python script (built on top of Androguard [10]) that performs static analysis on the code of an Android application. Taking as input the app installer (.apk), Mallodroid uses the capabilities inherited from Androguard to decompile the application. Once the script acquires the source code, it *(i)* extracts the set of URLs the app is instructed to connect and checks the validity of their certificates, and *(ii)* identifies if the app is using an non-standard trust manager and checks the related methods;

Tapioca [11]: testing framework that performs a series of unique checks by simulating a MITM. Using different types of packet capture, the tools is able to: *(i)* validate the negotiation between server and client; *(ii)* enumerate all the URLs the app tries to connect; *(iii)* verify if the client correctly validates the received certificates; and *(iv)* (prior packet decryption) search among the messages to locate known strings.

4 Mitigations Identification

Given the known vulnerabilities described in Sect. 2.2, system administrators should identify and follow a set of mitigations. To assist them we have collected in Table 5 the current best practice to mitigate the known vulnerabilities of TLS 1.2. The vast majority of the mitigations is applied by changing some lines in the server's configuration file while the remaining are related to vulnerable/outdated support libraries. The identification of such mitigations is not trivial because the currently available reports (see Appendix B) lack of clear indications on which is the source of misconfiguration.

Table 5. List of Mitigations for TLS 1.2.

Mitigation	Attack
Disable renegotiation	3SHAKE [31]
	Renegotiation attack [42]
Enable the use of extended_master_secret TLS extension	3SHAKE [31]
Disable RC4	Bar Mitzvah [28]
	RC4 NOMORE [48]
Disable the compression mechanism	CRIME [33]
Disable SSLv2	DROWN [3]
Use AEAD ciphers	Lucky 13 [2]
Disable SSLv3	POODLE [32]
Disable RSA-MD5 certificate signature	SLOTH [4]
Enforce AES usage (and disable 3DES when possible)	Sweet32 [5]
Enforce the termination mode	TLS Truncation [46]
Disable HTTP compression (may slow down the transmission)	BREACH [16]
Ignore self-signed certificates and perform a complete validation (up to the trusted root)	Accept self-signed certs [34]
Check if the hostname (from the certificate) matches the one related to the transmission	Setting a permissive hostname verifier [34]

5   TLSAssistant

During our study of TLS-related vulnerabilities, we noticed that all the currently available TLS analyzers have two major limitations. Putting aside the amount of provided features, all the examined tools gave little or no explanation on how to actually mitigate the detected weaknesses. On the other hand, every tool focuses on a specific party of the communication (either server or client) thus making its usage only part of a complete analysis.

To assist average system administrators and app developers to deploy resilient instances of the TLS protocol we propose TLSAssistant. By bringing together different powerful analyzers, our tool is able to cover a full-range of analysis on all the parties involved in a secure communication and to provide a set of mitigation measures that aim to thwart the impact of the identified vulnerabilities.

5.1 Architecture

TLSAssistant is written in Bash and can thus be invoked via command-line. Among the available parameters, the tool takes as input the target to be evaluated (e.g., the IP address of a server) and outputs a single report file. The content of the report depends on the detected weaknesses and on the level of verbosity the user chose. Being built on top of other works, our TLSAssistant has been designed to be modular and easily upgradable. Figure 2 shows the architecture with its two main components: ANALYZER and EVALUATOR.

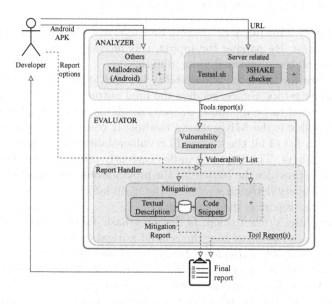

Fig. 2. TLSAssistant architecture

ANALYZER. Takes as input a series of parameters depending on which analysis the user wants to run. By design, our tool has a flexible architecture that allows a continuous integration of newer and more sophisticated tools. Currently, the set of integrated tools consists of command-line scripts written either in Bash or Python. At the time of writing, the ANALYZER integrates the following tools:

Testssl.sh [50] chosen among many others (as shown in Sect. 3) due to the enormous amount of features and for its ongoing development;

3SHAKE checker [51] added to make the ANALYZER able to test whether a server is vulnerable to the Triple Handshake Attack or not. This is an example of the continuous integration that has driven the design of TLSAssistant: being able to integrate different analyzers to become a useful toolbox for a complete TLS-vulnerability detection;

Mallodroid [14] even if less powerful than Tapioca (see Sect. 3), it was more suitable for our modularity requirement. Indeed, the current version of the installer of Tapioca turns the client machine into a dedicated appliance; a design choice incompatible with our tool.

The integrated tools allow the ANALYZER to take as input: *(i)* a hostname/IP address (optionally specifying the port to scan); *(ii)* an apk installer or *(iii)* both of the previous. Once loaded, the module will run each of the tools related to the required scan, collect their reports and transmit them to the EVALUATOR.

EVALUATOR. Core of TLSAssistant and our main contribution, it is responsible for the enumeration of the detected vulnerabilities and the generation of the report that will guide the system administrator towards all the mitigations to be applied. It can be seen as two dependent modules:

Vulnerability enumerator collects and analyzes the reports generated by the ANALYZER. By parsing the inputs, this module is able to compile a list containing all the discovered vulnerabilities.

Report handler takes the vulnerability list and, in accordance with the system administrators' choice, renders the final output. While TLSAssistant has been developed to be modular, the only available source of information currently available is the **Mitigations** module. It consists of a shared database containing a list of all the known TLS vulnerabilities with their descriptions and related fixes. The Report handler currently offers three kinds of report, each version provides the content of the previous one and adds more technical details. For every detected weakness, the main information contained in each version of the report is the following:

v0 mitigations' description. Is the most basic form of report, it only contains a description of how the related mitigation works;

v1 code snippet. Provides a fragment of code that can be copy-pasted into the webserver's configuration to seamlessly fix the weakness. TLSAssistant can detect any webserver but is currently only able to provide snippets for Apache HTTP server. We plan to extend the code coverage to all the most common webservers available on the market;

v2 tools' individual reports. In addition to our detailed contribution, this kind of report also provides the full set of individual reports generated by each tool.

6   Experimental Evaluation

To evaluate TLSAssistant's efficacy, we have analyzed a real use-case scenario involving the Italian eID card (CIE 3.0) [9] (Sect. 6.1) and conducted a user-study experimentation involving university students (Sect. 6.2). These two instances helped us prove that the result of our work is effective both for security experts, who may benefit from an additional support, and for unexperienced users who seamlessly became able to perform complex mitigations without the need for an in-depth knowledge.

6.1   Use-Case: CIE 3.0

In a joint collaboration between FBK and IPZS (acronym for "Istituto Poligrafico e Zecca dello Stato") [35], which is the Italian state printing office and mint, we implemented a mobile authentication mechanisms that uses the Italian electronic identity card (CIE 3.0 - Carta d'Identitá Elettronica) [9] to access public administration online services. Being the use of TLS the basic building block of the solution, any unpatched vulnerability (see Sect. 2.2) may compromise the entire authentication process.

For this reason, we run TLSAssistant targeting a prototype of the infrastructure and found that the deployment (which was entering in the final development stages) was prone to *Lucky 13*, *3SHAKE* and an incorrect certificate handling. These three issues, that can be easily go unnoticed for a variety of reasons, have now been fixed. This example clearly shows how running a tool like TLSAssistant can help even expert system administrators to determine if a new deployment contains some severe misconfigurations.

6.2   User Study

The following paragraphs will detail the settings of the experimentation (designed following the template and guidelines by Wohlin et al. [8]) and a summary of the main results.

Experiment Scoping and Planning. As described in Sect. 5, TLSAssistant is based on the most powerful TLS analyzers available on the market. The main additional feature is the generation of a report that assist the user during the mitigation process: together with the list of vulnerabilities, a textual description of the mitigations and (when is possible) a corresponding code snippet is provided.

The *goal* of this study is to analyze the effect of providing a set of mitigations with the *purpose* of evaluating the support offered by TLSAssistant in patching a TLS configuration.

The *context* of this study consists of:

Subjects: 16 Bachelor and Master students from University of Trento (with background on information security) playing the role of an unexperienced system administrator;
Objects: two VMs with custom-compiled misconfigured versions of Apache HTTP Server v2.4.37 and OpenSSL v1.0.2:
  O_1 a TLS configuration vulnerable to BREACH;
  O_2 a TLS configuration vulnerable to CRIME;

It is important to note that the proposed objects are representative of realistic TLS misconfigurations. To fit the time constraint of our experiment, only one vulnerability is present in each object. The selected objects are comparable in terms of complexity of the operation required to patch the problem.

Research Questions and Hypothesis Formulation. In this study, we want to evaluate whether the report provided by TLSAssistant (with textual descriptions of the mitigations and code snippets) facilitates the patching task in terms of time and correctness. Thus, our research questions are:

RQ_1 *(on time)*: does the time spent by a system administrator in patching an error decrease when the tool provides a text description of the mitigation and the corresponding code snippet?

RQ_2 *(on correctness)*: does the capabilities of a system administrator in patching an error increase when the tool provides a text description of the mitigation and the corresponding code snippet?

Thus, the null hypothesis can be formulated as follows:

H_{01} *(on time)*: providing a text description of the mitigation and the corresponding code snippet does not significantly decrease the time spent by a system administrator to patch the error;

H_{02} *(on correctness)*: providing a text description together with a code snippet of the mitigation does not significantly increase the capability of a system administrator to patch the error.

Variables Selection. To measure the subject's capability to perform a patching task (*vulnerability detected and solved*) and the time spent, we asked subjects to run the provided tool, look at the resulting report, and perform the patching task (perform the required operations to patch the misconfiguration).

The main factor of the experiment — that acts as an independent variable — is the presence of the treatment during the execution of the task. In our experiment, we have considered the following alternative treatments:

Treatment 1 (Tr₁): TLSAssistant provides as report a list of vulnerabilities plus a textual description of the mitigations and a suggested code snippet to perform the mitigation.

Treatment 2 (Tr₂): TLSAssistant provides as report the original reports of the tools that are composing the server-related module of the ANALYZER (Testssl.sh and 3SHAKE checker).

Experiment Design and Procedure. We adopt a counter-balanced experiment design intended to fit two lab sessions. Subjects are classified into four groups (despite they work alone), each one working in two labs on different objects with different treatments. The design allows for considering different combinations of objects and treatments in different order across labs (see Table 6).

Before our experiment, subjects were properly trained with lectures and exercises on TLS. The purpose of training is to make subjects confident about the kind of tasks they are going to perform and the environment they will have available.

The experiment was carried out according to the following procedure. Subjects had to:

Table 6. Labs.

	Group A	Group B	Group C	Group D
Lab 1	O_1 with Tr_1	O_2 with Tr_2	O_2 with Tr_1	O_1 with Tr_2
Lab 2	O_2 with Tr_2	O_1 with Tr_1	O_1 with Tr_2	O_2 with Tr_1

1. complete a pre-experiment survey questionnaire;
2. for each of the two labs to be performed: (i) mark the start time; (ii) perform the patching task; and (iii) mark the stop time;
3. complete a post-experiment survey questionnaire.

Post-experiment survey questionnaire (reported in Appendix A) deals with object clarity of the tasks, cognitive effects of the treatments on the behaviour of the subjects and perceived usefulness of TLSAssistant.

Results. The amount of time required to correctly patch a vulnerability is significantly longer when working with the report provided in Tr_2 than when working with the report with the mitigations (Tr_1): 25 min on average to fix a vulnerability with Tr_2, 7 min on average to fix a vulnerability with Tr_1. Thus, hypothesis H_{01} on time can be rejected. Therefore, we can formulate the following alternative hypothesis:

H_{A1}: providing a text description of the mitigation and the corresponding code snippet decreases the time spent by a system administrator to patch the error.

Regarding the task correctness all students were able to correctly patch the vulnerability with Tr_1; however, just the 68.75% of students was able to perform a proper vulnerability patch with Tr_2, which corresponds to a 31.25% difference on the overall sampled population. For this reason, we can accept the following alternative hypothesis:

H_{A2}: providing a text description together with a code snippet of the mitigation increases the capability of a system administrator to patch the error.

Moreover, from the post-experiment survey we can learned that the 81.25% of the students considers Tr_1 more useful and the 93.25% assessed that Tr_2 is more complex to understand. In addition, all the students positively recommend our tool. Here we report some comments:

"Fast, correct and easy to use. It found the vulnerability and helped me solving it"

"It would be very easy to fix such vulnerabilities following the given instructions. Also, you can search for more info about the vulnerability itself, which can help you to learn more about TLS."

"I won't waste a lot of time looking for all vulnerabilities"

7   Conclusions and Future Work

To assist system administrators with limited security skills to deploy resilient instances of the TLS protocol suite we propose TLSAssistant, a fully-featured tool that combines state-of-the-art tools with a report system that provides appropriate mitigations.

To design this tool, we have: *(i)* compared the state-of-the-art tools for TLS analysis, *(ii)* classified known TLS vulnerabilities and *(iii)* identified their mitigations. Finally, to validate the efficacy of our tool, we performed a user-study experimentation involving university students and analyzed a real use-case scenario involving the Italian eID card (CIE 3.0) [9].

As future work, we plan to extend TLSAssistant 's capabilities by *(i)* improving the webserver coverage; *(ii)* supporting more inputs (e.g., configuration files); *(iii)* automatize the mitigation process and further analyze experimentation's results by using statistical test, including co-factor analysis such as subject's experience, learning across tasks and more. As a second objective, we plan to use TLSAssistant to increase awareness and education in cybersecurity. A step in this direction is being made by integrating CVE identifiers, CVSS scores and modelling a series of attack trees [41], a hierarchical representation on how each attack can be mounted and which security properties it violates. Finally, we also plan to make TLSAssistant's source code freely available for anyone who wants to contribute to this project.

Acknowledgments. The authors would like to thank IPZS for the collaboration on the development of the authentication solution based on the CIE 3.0 carried out in the context of the joint laboratory DigimatLab between FBK and IPZS.

A   Post-questionnaire

Table 7 shows the content of the post-experiment survey questionnaire mentioned in Sect. 6.2. It deals with object clarity of the tasks, cognitive effects of the treatments on the behaviour of the subjects and perceived usefulness of TLSAssistant. The first set of questions (Q1–Q6) needs to be answered twice (one answer for each performed lab) while the remaining set only needs to be answered once as it refers to the overall session.

Table 7. Post-experiment survey questionnaire.

ID	Applies to	Question
Q1	Each lab	I had enough time to perform the tasks (1–5)
Q2	Each lab	I experienced no difficulty in patching the vulnerability given the report (1–5)
Q3	Each lab	How much time (in terms of percentage) did you spend looking at the TLS configuration code? (0, <20%, $\geq$20% and <40%, $\geq$40% and <60%, $\geq$60% and <80%, $\geq$80%)
Q4	Each lab	How much time (in terms of percentage) did you spend looking at online documentation on TLS vulnerabilities? (0, <20%, $\geq$20% and <40%, $\geq$40% and <60%, $\geq$60% and <80%, $\geq$80%)
Q5	Each lab	Provide some examples of online queries you used to search the vulnerabilities online (e.g., keywords used)
Q6	Each lab	Which steps did you take to perform the tasks? (e.g., run command Y, opened file X, ..)
Q7	Overall	Which report did you find more useful. (Report of Lab 1–2)
Q8	Overall	Which report did you find more easy to read. (Report of Lab 1–2)
Q9	Overall	Which report did you find more complex to understand. (Report of Lab 1–2)
Q10	Overall	The textual description of the mitigation is useful to complete the tasks (1–5)
Q11	Overall	The code snippet is useful to complete the tasks (1–5)
Q12	Overall	How did you use the code snippet? (Copy-pasted where needed, Typed manually where needed, Used to perform a web search)
Q13	Overall	Would you use TLSAssistant for your work? (Yes, No, Maybe)
Q14	Overall	Motivate your answer (to the previous question). (open question)
Q15	Overall	Do you know any tool that performs similar tasks? (open question)
Q16	Overall	Do you have any suggestion related to the tool usage? (open question)
Q17	Overall	Do you have any suggestion related to the amount of information provided by TLSAssistant's report? (open question)

B   Report snippet

To show the effort required by a system administrator in identifying the required mitigation, we show a snippet of the testssl's report (see Fig. 3). It contains the

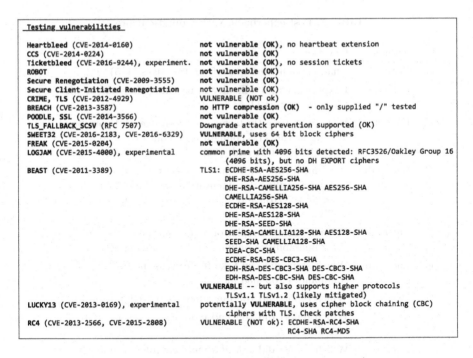

Fig. 3. testssl report snippet (Color figure online)

list of checked vulnerabilities matched with their presence in the analyzed TLS deployment. The status of each vulnerability is shown with a combination of a string (e.g.; "potentially vulnerable") and a color that represents the severity of the finding.

Not the shown snippet nor any other part of the report give any useful insight on how to actually mitigate the detected vulnerabilities.

References

1. Adrian, D., et al.: Imperfect forward secrecy: how Diffie-Hellman fails in practice. In: Proceedings of the 22nd ACM SIGSAC Conference on Computer and Communications Security (2015). https://doi.org/10.1145/2810103.2813707
2. AlFardan, N.J., Paterson, K.G.: Lucky thirteen: breaking the TLS and DTLS record protocols. In: IEEE Symposium on Security and Privacy, SP, pp. 526–540 (2013). https://doi.org/10.1109/SP.2013.42
3. Aviram, N., et al.: DROWN: breaking TLS with SSLv2. In: 25th USENIX Security Symposium (2016)
4. Bhargavan, K., Leurent, G.: Transcript collision attacks: breaking authentication in TLS, IKE and SSH. In: 23rd Annual Network and Distributed System Security Symposium, NDSS (2016)

5. Bhargavan, K., Leurent, G.: On the practical (in-)security of 64-bit block ciphers: collision attacks on HTTP over TLS and OpenVPN. In: Proceedings of the 2016 ACM SIGSAC Conference on Computer and Communications Security, Vienna, Austria, 24–28 October 2016 (2016). https://doi.org/10.1145/2976749.2978423

6. Blog, G.S.: SHA-1 Certificates in Chrome. https://security.googleblog.com/2016/11/sha-1-certificates-in-chrome.html

7. Bright, P.: Apple, Google, Microsoft, and Mozilla come together to end TLS 1.0. https://arstechnica.com/gadgets/2018/10/browser-vendors-unite-to-end-support-for-20-year-old-tls-1-0/

8. Cartwright, M.: Book Review: Experimentation in Software Engineering: An Introduction. By Wohlin, C, Runeson, P., Höst, M., Ohlsson, M.C., Regnell, B., Wesslén, A. Kluwer Academic Publishers (1999). ISBN 0-7923-8682-5. Softw. Test. Verif. Reliab. (2001). https://doi.org/10.1002/stvr.230

9. Dell'Interno, M.: Carta di identitá elettronica. https://www.cartaidentita.interno.gov.it

10. Desnos, A.: Github: Androguard. https://github.com/androguard/androguard

11. Dormann, W.: Announcing CERT Tapioca 2.0 for Network Traffic Analysis. https://insights.sei.cmu.edu/cert/2018/05/announcing-cert-tapioca-20-for-netwo rk-traffic-analysis.html

12. Ducklin, P.: Boffins 'crack' HTTPS encryption in Lucky Thirteen attack. https://nakedsecurity.sophos.com/2013/02/07/boffins-crack-https-encryptionin-lucky-thir teen-attack/

13. Ducklin, P.: The SLOTH attacks: why laziness about cryptography puts security at risk. https://nakedsecurity.sophos.com/2016/01/08/the-sloth-attacks-why-laziness-about-cryptography-puts-security-at-risk/

14. Fahl, S., Harbach, M., Muders, T., Baumgärtner, L., Freisleben, B., Smith, M.: Why Eve and Mallory love android: an analysis of android SSL (in)security. In: Proceedings of the 2012 ACM Conference on Computer and Communications Security, pp. 50–61 (2012). https://doi.org/10.1145/2382196.2382205

15. Georgiev, M., Iyengar, S., Jana, S., Anubhai, R., Boneh, D., Shmatikov, V.: The most dangerous code in the world: validating SSL certificates in non-browser software. In: ACM Conference on Computer and Communications Security, pp. 38–49 (2012). https://doi.org/10.1145/2382196.2382204

16. Gluck, Y., Harris, N., Prado, A.: BREACH: reviving the CRIME attack. http://breachattack.com/

17. Green, M.: A Diversion: BEAST Attack on TLS/SSL Encryption. https://blog.cryptographyengineering.com/2011/09/21/brief-diversion-beast-attack-on-tlsssl/

18. Green, M.: Attack of the week: Logjam. https://blog.cryptographyengineering.com/2015/05/22/attack-of-week-logjam/

19. Green, M.: The Internet is broken: could we please fix it? https://blog.cryptographyengineering.com/2012/02/28/how-to-fix-internet/

20. Grimmett, J.: Encryption export controls (2001). http://www.au.af.mil/au/awc/awcgate/crs/rl30273.pdf

21. Group, N.W.: The TLS Protocol: Version 1.0. https://tools.ietf.org/pdf/rfc2246.pdf

22. Group, N.W.: The Transport Layer Security (TLS) Protocol: Version 1.1. https://tools.ietf.org/pdf/rfc4346.pdf

23. Group, N.W.: The Transport Layer Security (TLS) Protocol: Version 1.2. https://tools.ietf.org/pdf/rfc5246.pdf

24. Group, N.W.: Transport Layer Security Protocol Compression Methods. https://tools.ietf.org/pdf/rfc3749.pdf

25. Group, O.W.: OAuth 2.0 Mutual TLS Client Authentication and Certificate Bound Access Tokens. https://tools.ietf.org/pdf/draft-ietf-oauth-mtls-10.pdf
26. IETF: The Transport Layer Security (TLS) Protocol: Version 1.3. https://tools.ietf.org/pdf/rfc8446.pdf
27. IETF: Transport Layer Security (TLS) Renegotiation Indication Extension. https://tools.ietf.org/pdf/rfc5746.pdf
28. IMPERVA: Attacking SSL when using RC4. https://www.imperva.com/docs/HII_Attacking_SSL_when_using_RC4.pdf
29. Kolybabi, M., Lawrence, G.: ssl-enum-ciphers. https://nmap.org/nsedoc/scripts/ssl-enum-ciphers.html
30. Lyon, G.: Nmap: the Network Mapper. https://nmap.org
31. Microsoft-Inria: Triple Handshakes Considered Harmful: Breaking and Fixing Authentication over TLS. https://www.mitls.org/pages/attacks/3SHAKE
32. Möller, B., Duong, T., Kotowicz, K.: This POODLE Bites: Exploiting the SSL 3.0 Fallback. https://www.openssl.org/~bodo/ssl-poodle.pdf
33. NIST: CVE-2012-4929. https://nvd.nist.gov/vuln/detail/CVE-2012-4929
34. NowSecure: Fully Validate SSL/TLS. https://books.nowsecure.com/secure-mobile-development/en/sensitive-data/fully-validate-ssl-tls.html
35. Poligrafico e Zecca dello Stato Italiano. https://www.ipzs.it
36. Pornin, T.: What is DROWN and how does it work? https://security.stackexchange.com/a/116140/186367
37. Qualys: SSL Pulse. https://www.ssllabs.com/ssl-pulse/
38. Qualys, I.: SSL Server Rating Guide. https://github.com/ssllabs/research/wiki/SSL-Server-Rating-Guide
39. rbsec. https://github.com/rbsec/sslscan/releases/tag/1.11.11-rbsec
40. Robshaw, M.: Stream ciphers (1995). ftp://ftp.rsasecurity.com/pub/pdfs/tr701.pdf
41. Schneier, B.: Attack Trees. https://www.schneier.com/academic/archives/1999/12/attack_trees.html
42. SecurityLearn: SSL Attacks. http://www.securitylearn.net/tag/ssl-renegotiation-attack/
43. Services, A.W.: Alexa Top Sites. https://aws.amazon.com/alexa-top-sites/
44. Shannon, C.E.: Communication theory of secrecy systems*. Bell Syst. Tech. J. **28** (1949). https://doi.org/10.1002/j.1538-7305.1949.tb00928.x
45. Siles, R.: TLSSLed v1.3. http://blog.taddong.com/2013/02/tlssled-v13.html
46. Smyth, B., Pironti, A.: Truncating TLS connections to violate beliefs in web applications. In: 7th USENIX Workshop on Offensive Technologies, WOOT (2013)
47. Somorovsky, J.: Systematic fuzzing and testing of TLS libraries. In: Proceedings of the 2016 ACM SIGSAC Conference on Computer and Communications Security, CCS 2016, pp. 1492–1504 (2016). https://doi.org/10.1145/2976749.2978411
48. Vanhoef, M., Piessens, F.: RC4 NOMORE (Numerous Occurrence MOnitoring & Recovery Exploit). https://www.rc4nomore.com/
49. Weisstein, E.: Hash Function. http://mathworld.wolfram.com/HashFunction.html
50. Wetter, D.: /bin/bash based SSL/TLS tester: testssl.sh. https://testssl.sh
51. Young, C.: TLS Extended Master Secret Extension: Fixing a Hole in TLS. https://www.tripwire.com/state-of-security/security-data-protection/security-hardening/tls-extended-master-secret-extension-fixing-a-hole-in-tls/

Contributing to Current Challenges in Identity and Access Management with Visual Analytics

Alexander Puchta[1](✉), Fabian Böhm[2](✉) (iD), and Günther Pernul[2](✉)

[1] Nexis GmbH, Franz-Mayer-Str. 1, 93053 Regensburg, Germany
alexander.puchta@nexis-secure.com
[2] University of Regensburg, Universitätsstr. 31, 93053 Regensburg, Germany
fabian.boehm@ur.de, guenther.pernul@ur.de

Abstract. Enterprises have embraced identity and access management (IAM) systems as central point to manage digital identities and to grant or remove access to information. However, as IAM systems continue to grow, technical and organizational challenges arise. Domain experts have an incomparable amount of knowledge about an organization's specific settings and issues. Thus, especially for organizational IAM challenges to be solved, leveraging the knowledge of internal and external experts is a promising path. Applying Visual Analytics (VA) as an interactive tool set to utilize the expert knowledge can help to solve upcoming challenges. Within this work, the central IAM challenges with need for expert integration are identified by conducting a literature review of academic publications and analyzing the practitioners' point of view. Based on this, we propose an architecture for combining IAM and VA. A prototypical implementation of this architecture showcases the increased understanding and ways of solving the identified IAM challenges.

Keywords: Identity and access management · Identity management · Visual Analytics

1 Introduction

Identity and access management (IAM) has become a vital component of modern companies as it enables the management of identities and grants access to necessary resources. IAM also assures compliance with regulations like SOX [41] or Basel III [3]. To achieve this, IAM systems consist of manifold policies, processes and technical solutions [13]. The core of IAM are identities like employees and their access rights to resources maintained within the system. Besides human identities, new technologies like the Internet of Things (IoT) require the integration of technical identities (e.g. sensors and machines) into IAM [27]. Thus, the number of elements maintained in the system is constantly rising. This will ultimately lead to an identity explosion where a vast amount of heterogeneous

© IFIP International Federation for Information Processing 2019
Published by Springer Nature Switzerland AG 2019
S. N. Foley (Ed.): DBSec 2019, LNCS 11559, pp. 221–239, 2019.
https://doi.org/10.1007/978-3-030-22479-0_12

identities has to be managed in a single system. This results in numerous problems to be addressed in the next years to ensure IAM systems remain an effective part of companies' IT landscapes.

A solution for those problems needs an effective way to manage and analyze the huge quantity of information with often thousands of identities and hundreds of thousands of entitlements. To decide whether information about an identity is wrong or redundant access rights are assigned to it, the knowledge of domain experts with experience and deep understanding of an enterprise's individual IAM landscape is needed. In this work we investigate how this domain knowledge can be integrated into an IAM landscape by leveraging Visual Analytics (VA) as VA is one of the central methods to include domain experts' knowledge and utilize their feedback [11]. In order to reach this goal, this work investigates three research questions:

- **RQ-1:** What are current and upcoming key challenges within IAM to be solved by integrating domain knowledge?
- **RQ-2:** How can VA be integrated into an existing IAM architecture and which steps are necessary?
- **RQ-3:** What could an exemplary VA solution for IAM look like and which challenges could be solved?

By answering these research questions our work focuses on two main contributions. We provide a list of challenges for current and future IAM. This list is an outcome of a structured analysis taking both academic and practice viewpoints into consideration. We also demonstrate how VA can be applied helping to integrate domain knowledge in tasks to identify IAM anomalies and possible erroneous configurations (e.g. over-authorization or wrong identity attributes). Therefore, we develop a prototypical visualization designed in cooperation with experienced IAM practitioners.

The remainder of this work is structured as follows. Section 2 introduces some background on IAM systems as well as related work regarding the integration of VA into IAM. Next, Sect. 3 follows a structured, two-fold approach to identify current challenges for IAM system as seen from academia and practice to answer *RQ-1*. An architecture to integrate VA into IAM (*RQ-2*) as well as a corresponding proof-of-concept visualization (*RQ-3*) are presented in Sect. 4. The benefits of this prototype regarding the identified challenges are highlighted with exemplary use cases in Sect. 5. Section 6 concludes our work and highlights possible future research directions as well as current limitations.

2   Background and Related Work

In this chapter we define key concepts of IAM and introduce related work regarding the integration of VA into IAM.

2.1 Background

IAM consists of two main fields which are managing identities and granting them access to resources. According to Pfitzmann and Hansen [33] an identity is a subset of attributes uniquely identifying a person. An identity is either real or exists as a digital identity like profiles in social media. Real and digital identities are often linked, and a real identity may own multiple digital personas. However, in the following we assume each entity to have exactly one digital identity as the scope of this work is limited to a single company's context. Currently, IAM regards employees, contractors or customers as identity because they all need to have access to certain resources [45]. In addition to humans having digital identities, technical equipment like machines or sensors are entities which need access to resources, too. Thus, these technical identities also are relevant for maintaining them within an IAM [12].

Digital identities in an IAM are managed from their creation to their deletion when not needed anymore. During this life cycle, access control is used to provide access to applications, data or other information [35]. Enterprises often employ role-based access control (RBAC) in order to grant access [37]. In RBAC, roles are utilized to bundle single access rights and consequently assigned to identities. On the contrary, attribute-based access control (ABAC) leverages identities' attributes and predefined access policies for dynamic access management [16].

To maintain landscapes with thousands of identities, enterprises employ IAM systems which are able to support the identity life cycle and provide identities with the correct entitlements. Besides that, modern IAM systems offer a variety of other functionalities (e.g. Single Sign-on) which are not detailed any further in this work.

2.2 Related Work

There are some existing publications applying visual representations for IAM problems. The earliest integration of VA to the best of our knowledge is the "role graph model" by Nyanchama and Osborn [32]. It is based on RBAC and is used to optimize existing roles for a company. In addition to that, several authors propose a matrix-based approach to visualize users and their entitlements [5,28]. Based on that, VA can be applied to identify suitable roles or outliers with extensive entitlement assignments. Recently, Morisset and Sanchez introduced a tool to visualize ABAC policies [30].

These approaches are focusing mostly on interactive visual techniques for Access Control. To the best of our knowledge there is no existing work taking Identity Management into consideration to build a more cohesive visual solution. Therefore, we try to fill this gap by identifying general IAM challenges where domain knowledge of experts is needed to solve them. After identifying those challenges, we build a prototypical visual approach to demonstrate how domain experts can be integrated.

3 IAM Challenges

This section defines current or future IAM challenges where domain knowledge of human experts may play a vital role. They can serve as a starting point to deduce requirements for any type of solution trying to tie experts and IAM systems closer together. In Sect. 4 we introduce a proof-of-concept visual solution to tackle some of the herein defined challenges.

For identifying the challenges, existing academic literature as well as practitioners experience within the field of IAM are taken into consideration. We are aware that there are far more challenges than the five proposed by us. However, based on the results of our structured analysis and the domain knowledge of practitioners, we chose the most relevant ones with respect to the necessity to integrate domain experts. An *IAM challenge* in the context of this work is a current or future problem with the need to be solved for IAM. Challenges already being tackled or focusing only on parts of an IAM system (e.g. access control) are not considered in this work. Neither do we consider problems where the inclusion of domain expert knowledge is not vital. To identify challenges, we follow a structured approach introduced in the Sect. 3.1.

3.1 Approach for Identifying Challenges

We derive current challenges following a structured approach depicted in Fig. 1. We ensure to include both the scientific and the practitioners' view as IAM is an active research field as well as a highly relevant topic in enterprises.

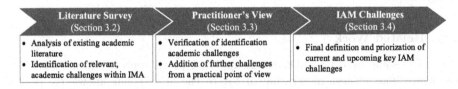

Fig. 1. Approach for defining the key IAM challenges.

During the literature survey we analyze existing academic literature published in the last ten years regarding IAM and respective challenges or problems. This scientific viewpoint allows us to derive a first set of IAM challenges. However, IAM is highly business-driven and there are numerous practical approaches outside the academic world. Thus, we also include the perspective of practitioners.

The goal of this second analysis is twofold. We verify the academic IAM challenges but also identify further challenges not yet considered by scientific literature. Three different sources of information are leveraged in order to minimize subjectivity of different business opinions:

1. Analyst reports and surveys from the IAM industry

2. Interviews with IAM consultants with 3 to 15 years of experience
3. Interviews with companies applying IAM solutions.

In a last step, we integrate all inputs from the analysis into five IAM challenges. The list of identified challenges is not exhaustive for the IAM field of research. Our work is focusing only on current challenges that can strongly benefit from integrating experts' knowledge.

3.2 Literature Survey

In order to identify relevant literature, we follow a structured approach by defining keywords to review relevant IAM literature. As we are defining challenges for the entire IAM system we only take resources into consideration which are dealing either with *"identity and access management"* or specific problems and challenges within *"identity management"* or *"access management"*. We transform these phrases into suitable search terms[1,2] and applied them to the dblp computer science bibliography[3]. Dblp is a service indexing relevant academic journals and proceedings of peer-reviewed conferences from computer science. Searching dblp results in a feasible number of results with a high suitability. Therefore, we can ensure to get only relevant academic publications. Dblp serves as a quality gate for our scientific analysis as it returns a manageable amount of entries compared to other engines like Google Scholar with nearly 10.000 results for the second search term. We manually filter the results based on title, abstract, and key sections to remove findings not mentioning any challenges or problems.

We apply a second, more unstructured search to identify additional relevant entries. In this step we include further academic databases (IEEE XPlore, ACM, Google Scholar) to find additional literature not listed within dblp. This results in a total of 19 academic publications mentioning or clearly defining relevant challenges for IAM. We group the identified problems and define the first four challenges (cf. *C1* to *C4* in Sect. 3.4).

3.3 Practitioner's View

We now conduct a business analysis to include the practitioners' point of view. Information received in this process step is often hard to generalize as it reflects subjective opinions. However, by including various sources of information we try to overcome this deficit. In the business analysis we look at reports from specialized IAM analysts namely KuppingerCole[4]. This company is focused on IAM and technologies around that sector and thus has accumulated valuable knowledge in this area [26,40,43]. Additional input is generated by Gartner [9], Forrester [7] and IDG Research Services [20].

---

[1] *identity—access management challenge/problem.*
[2] *identity-and-access-management.*
[3] https://dblp.uni-trier.de/.
[4] https://www.kuppingercole.com.

Furthermore, we conduct interviews with three different IAM consultants with several years of practical experience in the field of IAM projects. Besides that, four companies already applying IAM solutions are inquired regarding possible challenges. While the four previously defined challenges are verified throughout the interview, a fifth one ($C5$) arises as a current problem of IAM from a business viewpoint.

Table 1. Results of literature survey on ten years of academic work.

Source	Year	C1	C2	C3	C4	C5
Hovav and Berger [15]	2009	x	x			
Mahalle et al. [27]	2010	x	x	x		
Bandyopadhyay and Sen [2]	2011	x	x	x		
Jensen [22]	2012		x	x	x	
Kanuparthi et al. [23]	2013	x	x		x	
Fremantle et al. [12]	2014	x		x		
Xiong et al. [46]	2014		x			
Hummer et al. [18]	2015	x		x	x	
Kunz et al. [24]	2015			x	x	x
Hummer et al. [19]	2016	x		x	x	x
Moghaddam et al. [29]	2017		x			
Servos and Osborn [38]	2017		x			
Asghar et al. [1]	2018		x			
Damon et al. [8]	2018	x		x		
Hummer et al. [17]	2018	x		x	x	
Indu et al. [21]	2018	x	x	x		
Nuss et al. [31]	2018	x		x		
Povilionis et al. [34]	2018		x		x	
Kunz et al. [25]	2019			x	x	x

3.4 IAM Challenges

Within this section the identified IAM challenges are described in detail. A mapping of all relevant academic publication to the challenges is provided in Table 1. Table 2 maps the results of our analysis with practitioners to the challenges.

Challenge 1 - Identification of All Relevant Identities (C1): For current and future IAM systems the identification of all relevant identities may sound like a simple task. However, especially in practical application it is not. One of the major reasons for this is the integration of various types of identities

Table 2. Analysis results from practitioners' view.

Source	Year	C1	C2	C3	C4	C5
IDG Research Services [20]	2017	x		x		
KuppingerCole and CXP Group [26]	2017	x				
Tolbert [43]	2017	x		x		
Diodati et al. [9]	2018	x		x		
Small [40]	2018	x		x	x	
Cser and Maxim [7]	2018	x	x	x		
Interviews (IAM consultants)	2019	x	x	x	x	x
Interviews (Companies applying IAM)	2019	x		x	x	x

into IAM. Currently, mainly employee and contractor identities are maintained in an IAM system. A recent trend, customer IAM or shortly CIAM, strives to add customer identities into these systems as well [7]. Additionally, the Internet of Things requires integrating even more identities, mostly technical ones [31]. Furthermore, numerous IT systems are not even connected to IAM. Nevertheless, such systems also contain various identities with the need to be identified for IAM in order to prevent identities not being centrally manageable. These trends hinder IAM to establish a central view of all relevant identities. However, this view is vital for any further analysis to be done within IAM (e.g. identification of unnecessary accounts or entitlements).

Challenge 2 - Privacy Within IAM (C2): As modern IAM systems offer a centralized view on nearly all employees, contractors and even costumers including their attributes the need for privacy arises. Especially business solution power users like IAM administrators can easily retrieve personal information from the identities. Based on our practical experience this could be a simple mail address but may also uncover more sensitive information like wage brackets or entitlement usage information. In order to protect this information in compliance with regulations, privacy mechanisms are needed to grant access to such information only when necessary and for authorized users. This challenge is mainly focused by scientific research and not by practitioners at the moment. However, as the European General Data Protection Regulation (GDPR) came into effect in 2018, it certainly will have an impact on the business sector of IAM. Please note that this challenge is limited to the application of privacy mechanisms on IAM systems and does not include the application of IAM systems for enhancing GDPR compliance within companies.

Challenge 3 - Heterogeneity of Various Identities (C3): As there are various identities within an IAM system, they are not identical. In fact, they differ quite a bit as identities consist of various attributes (e.g. first name, department). Considering *C1*, it gets clear that not all identities have the same kind of attributes. Technical and human identities are likely to have a completely different set of attributes. For example, technical devices do not have a first name, but instead have an attribute indicating their software version. This, on the one hand, rises a technical challenge to integrate this variability of identities into one underlying data set for IAM. In addition, IAM mechanisms like provisioning of entitlements still need to be working for all of these identities. On the other hand, it also hinders the analytic part of IAM as domain experts need to browse through an enormously large, heterogeneous database. By applying VA, domain experts could be supported as various attributes can be displayed in a more accessible way than in currently deployed table-based reports.

Challenge 4 - Data Quality and Data Management (C4): When it comes to attributes and other data existing in IAM, data quality and the underlying data management in IAM system needs to be considered. Attributes are often manually entered by different people; thus, wrong or inconsistent values are very likely to occur. For example, the current business location of an employee may be added by HR employees. If the employee moves to another department of an enterprise, the location also needs to be changed. Manual processes for attribute modifications exacerbate data quality issues as one can forget to adjust the location attribute. Therefore, IAM mechanisms like provisioning of entitlements based on the attribute *location* might fail. Additionally, wrong attribute values limit the possibilities of IAM analytics. Although an approach to improve attribute quality management was lately introduced [25], algorithms can only detect anomalies but can neither confirm nor reject whether it is a real data error. To do so, domain experts are needed, and VA can be highly beneficial to support related decisions by integrating domain expert feedback.

Challenge 5 - Transformation from Role-Based IAM to Attribute-Based IAM (C5): Challenge 5 was identified during the interviews with IAM consultants as it is not explicitly defined as an upcoming challenge in academic literature. It comprises the enterprise IAM transformation from a role-based approach to an attribute-based one. As mentioned before, enterprises mainly depend on an RBAC approach. However, this can lead to an increasing number of existing roles and requires increasing effort regarding role management [10]. In order to overcome these limitations, ABAC can be applied [16]. However, as this is a fundamental change of approach for IAM companies have to consider various factors (e.g. processes, technologies and policies [13]). Changes needed for this transformation are therefore not limited to access control, but existing research is mainly focused on the transformation of the access control model [36,47]. To the best of our knowledge there is no overarching approach how an enterprise IAM can be transformed from a role-based approach to an attribute-based one.

Tables 1 and 2 compare the results and show that *C1*, *C3*, and *C4* are found in both worlds and can easily be identified as relevant IAM challenges. Privacy in IAM and therefore, *C2*, is mainly embraced by academic literature and not explicitly mentioned in the business sector. *C5* is not described explicitly in academic literature but only mentioned very shortly by 3 articles. We identified this challenge by conducting interviews with IAM consultants and companies.

4 Applying Visual Analytics to IAM

Any of the previously identified challenges can benefit from including domain experts and their knowledge. VA has proven its capabilities to help integrate domain expert knowledge in complex and data-intensive cyber security tasks throughout the last years [6,44]. Additionally, decision makers can be supported with VA by making highly technical data sources more accessible. Therefore, we argue that leveraging concepts from VA to solve the identified challenges in IAM is a reasonable approach. As described in Sect. 2, there is some existing work that has shown the feasibility and utility of VA in the context of IAM. However, none of the challenges identified in Sect. 3 has been explicitly tackled with visual approaches yet. We try to fill this gap as we describe the architecture and design of our new visualization approach. The visualization design cannot support all the identified challenges as they are far too different in requirements. However, our approach shows how heterogeneous information about human and technical identities can be integrated into a single visual representation. The resulting view allows identifying existing identities (c.f. C1) and their structures (c.f. C3) as well as users can detect problems regarding data quality (c.f. C4).

The visual representation is designed and implemented in close cooperation with IAM practitioners which were also part of our interviews during the challenge identification. By including them in development, we ensure that the representation that is helpful for practical use. The participating experts are IAM consultants working for numerous clients and with years of experience in practical work with IAM projects. While the current visual tool is at a proof-of-concept stage, we are planning to continue our fruitful cooperation with these experts to develop a solution that can be used in their day-to-day work. Our cooperation also allowed for the development of the prototype based on the adaption of anonymized real-world identity data from a medium-sized company in the manufacturing sector with around 1.200 employees.

The underlying architecture for our prototypical application is depicted in Fig. 2 and its main components - *Data Sources*, *Data Preparation*, and *Data Visualization* - are described in more detail throughout the following sections. This architectural design is based on the Information Visualization Pipeline [4] which is a widely accepted structural design concept for any interactive visualization approach. The applied architecture shows how identity-related information can be collected and integrated from different sources and how the information needs to be prepared for VA concepts supporting domain experts. The identification of different data sources and their integration into a single, displayable data set are a starting point for any visual representation of identity data. Therefore, the main part of our architectural design, the *Data Preparation*, demonstrates how visual representations in general can be integrated into an existing IAM structure. The operations executed during the *Data Integration Engine* and the *Data*

Transformation step need only small adjustments for varying *Data Sources*. The last part of the architectural design, the *Data Visualization*, demonstrates how VA can contribute to the focused challenges by introducing an exemplary visualization of identity information. Interaction in this step is crucial as it ensures that experts can adjust the view for their personal needs and explore the data based on their own preferences to gain insight.

4.1   Data Sources

Our current proof-of-concept tool collects information about identities from three main data layers. Although the company representing the use case has a central IAM system with a role-based access control mechanism, not all information about the existing identities is fed into it. Only partial information from the *Application Layer* is integrated into IAM, while other applications, like the company's Active Directory (AD) to manage windows accounts, are not connected to it. Therefore, information from these systems needs to be collected separately. Technical identities representing IoT devices are currently not integrated into IAM but rather maintained separately (IoT-layer). The wide variety of different data sources storing information about the companies' identities and the missing integration of this information into IAM are the main reasons for the challenges we focus in this prototype. It becomes increasingly hard for any company to keep track of its identities when the information about them is so spread out. Additionally, the different information systems store the available data in different formats or data models. Furthermore, it is very important for any company to keep the quality of their identity information at a high level. Spread out data makes this very difficult, especially when data is maintained redundantly in different repositories.

Additional data sources can be plugged easily into our architecture via the *Data Integration Engine*. Our proof-of-concept system works with one source from each layer. This number of data sources is already enough to demonstrate

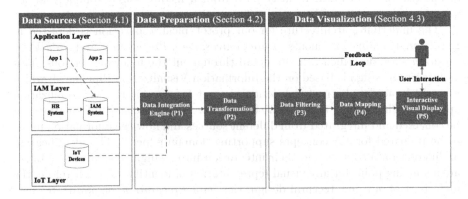

Fig. 2. Architecture for IAM Visual Analytics.

how VA helps to leverage experts' domain knowledge in the context of the aforementioned challenges as is demonstrated in Sect. 5.

4.2   Data Preparation

The main purpose of this part of the architectural design is to integrate and normalize the data from the sources into a single data model and format. Additional
fields are added and calculated in this step. The resulting data is structured as
a single table containing all relevant and necessary information about the identities. This step is essential for further visual display as it defines the level of
detail available to the users. The operations applied to the data in this step are
dynamic and can be changed whenever different information is of interest or new
data sources are plugged into the architecture.

Data Integration Engine (P1): This part of the architecture is responsible
for collecting data relevant from the data source and integrating them into a
single cohesive data set. Our proof-of-concept work extracts CSV data from all
data sources. However, each CSV export contains a different set of attributes.
To preserve the information about the source of a data set, we annotate the
data with a flag depicting the source. Additionally, we add a field describing
whether the identity is a human or a technical identity. In our conceptual setting, this identity type is mainly dependent on the data source. For example,
identities extracted from IAM are automatically annotated to be *"Human"* as
only employees or costumers are integrated into IAM. In the same way, identities extracted from the IoT layer are annotated to be *"Technical"* identities.
The cohesive data set is built as a union of the three data sets depicted in Fig. 2:
ApplicationLayer $\cup$ *IAMLayer* $\cup$ *IoTLayer*.

Data Transformation (P2): After integrating all available data sources into a
single, high-dimensional table, this data set is structured as needed for the visualization in this phase. This part of the architecture applies a variety of transformations. These include splitting a single field into multiple fields, replacing
values in a specific field, calculating additional fields based on existing information. The result of this step is a cohesive data set containing all relevant and
necessary information.

4.3   Data Visualization

This last part of the architecture is responsible for creating the interactive visual
representation with the subset of the data selected by the domain expert. The
interactions available to the users enable exploratory work with the visualization
and the identification of inconsistencies, miss-configurations as well as structures
and dependencies in the set of identities.

Data Filtering (P3): The interactive filtering assures the efficiency of the following steps and guarantees the expressiveness of the resulting view for the user. The subsequent *Data Mapping (P4)* can be CPU-intensive for very large data sets and, therefore, the input for this step needs to be as small as possible. It only contains the fields (columns) the user wants to see. The interactive selection of fields relevant for the user and the early integration of this interaction into the architectural design ensures that only relevant data is passed to the subsequent components. Up to this point the proposed architecture is generalized and can be applied to various visualization approaches within IAM. However, the *Data Mapping (P4)* and the *Interactive Visual Display (P5)* are highly dependent on the visualization technique selected for a specific VA solution. Therefore, the following considerations are specific to our exemplary solution.

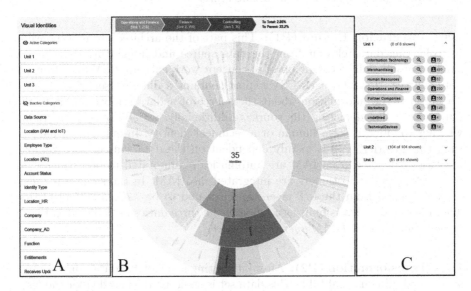

Fig. 3. Screenshot of the prototype available under http://bit.ly/iam-vis. Please note that the current version of the tool is only working in Google's Chrome Browser.

Data Mapping (P4): This phase in the architecture maps the filtered identity information into a dynamic hierarchical data structure which is necessary for the prototype to visualize the data correctly. We will not elaborate this data structure any further as it is specific for the proof-of-concept visualization and is prone to change for different visualization types.

Interactive Visual Display (P5): Before we are able to build a visual representation for the data at hand, it is necessary to choose a suitable visualization technique. This technique needs to be capable of displaying the dimensions and structure of the underlying data properly. For our prototypical visualization the technique must be able to represent multi-dimensional and hierarchical data. While there is a number of techniques (e.g. tree diagrams, circle packing, sunburst diagrams, or treemaps) which fulfill this requirement [39], each technique has its own advantages and disadvantages. It mainly comes down to the use case as well as the subjective preferences of the users which technique is most suitable. We used design sketches of the different suitable visualization techniques to interview IAM domain experts about their preferred visual representation. These interviews resulted in the *sunburst diagram* to be the most preferable technique to apply in the proof-of-concept application. However, any of the mentioned as well as a number of other techniques might be suitable, too.

The sunburst diagram displays a hierarchy using a series of concentric circles. Each ring itself corresponds to a level of the hierarchy. Therefore, underlying data structure is similar to a tree where the root node is depicted by the central ring and outermost circles represent the leaves of the tree-like structure. The sunburst's rings are sliced up and divided based on their hierarchical relationship to the parent slice. Therefore, the sunburst highlights structural, hierarchical relationships while being more scalable than other hierarchical visualization types. Figure 3 depicts the main view of our proof-of-concept application consisting of three main parts.

In (A) experts can drag-and-drop the boxes for the corresponding fields in the data set they want to be depicted in the sunburst diagram between two main lists. Each box represents an IAM employee attribute of the normalized data set. Exemplary fields which are contained in the proof-of-concept are the "Organisational Unit", "Data Source", or the number of entitlements of an identity. The upper list holds the currently active (i.e. displayed) fields and the lower one the inactive attributes. The first element in the list of active elements serves as the root element (innermost circle in the sunburst) when constructing the hierarchical data set. Accordingly, the second active attribute is displayed as the next outer circle. Logically, the last active element is included in the Sunburst diagram as the outermost circle.

The central part of the view is dedicated to the sunburst diagram (B). Each segment of the circles (attributes) depicts a single characteristic of an attribute in relative size to all existing identities. Hovering a segment brings up the number of identities depicted by this segment. The relative frequency with respect to the number of current root element (innermost circle) and the path to the hovered segment are displayed on top of the visual representation. Left-clicking a segment allows zooming in on this particular representation of an attribute. This improves the readability of specific hierarchy levels in very granular sunburst displays. When zoomed into a segment, clicking the white area in the middle of the sunburst diagram brings the zoom one hierarchy level upwards. Right-clicking a segment brings up a dialog. This dialog holds a table with identities in the rows and all attributes available for them in the columns. The identities displayed in the table are dependent on the clicked segment of the sunburst diagram as only the identities whose attributes fulfill the path to the segment are shown in the table. The table allows filtering and sorting of the currently displayed identities. In the dialog identities can also be reported for further analysis if necessary.

(C) holds the description of the sunburst as a dynamic list containing all currently visible circles (attributes) and the respective visible ring segments (attributes values). Clicking the magnifier for a list element zooms in into the segment representing this element. A click on the counter badge brings up the table with the identities included in corresponding node in the hierarchy. Within this table identities can be marked for further analysis by using a *"Report"*-button for each identity in the details table-view (e.g. after identification of an anomaly), thus, providing the possibility for integration of domain expert feedback into other applications. However, further functionality beyond this notification is out of scope for this work and needs to be implemented in a following version of the prototype.

5   Exemplary Use Cases

The current prototypical implementation[5] of our visualization for IAM was developed in co-creation with experts as suggested by Staheli et al. [42]. We regularly conducted semi-structured interviews with the participating practitioners to ensure that the implementation fits their needs and requirements. This section explicitly highlights how the visual display can support domain experts. We therefore go through several problems and inconsistencies based on one use case and identified by IAM experts while exploring the data. These had not been noticed before applying the visualization.

As the different problems only become evident in the sunburst diagram with different actively visualized attributes, we added predefined scenarios of the sunburst to our publicly available version of the prototype. Using the drop-down menu in the top right corner, we provide a video showcasing each of the following subsections. We would recommend to look at the corresponding video for each subsection in order to grasp the connection between the IAM problem and the sunburst visualization for identification of the inconsistency.

The exemplary use cases are based on the data set from a manufacturing company with 1.200 employees mentioned in Sect. 4. The company recently introduced an IAM system and connected the HR system as well as some minor applications. However, the Active Directory (AD) is currently not under IAM control because of its complexity as it was one of the company's first IT system growing for two decades. Therefore, some employees are missing an AD account while some AD accounts from former contractors and employees are still active. These orphan accounts are not identified via the IAM system, but they are still active and can be used for malicious activities (cf. Sect. 5.1).

Furthermore, the company made some investments in automating specific process tasks. Thus, two assembly machines and some automated users were

---

[5] The prototype is available under http://bit.ly/iam-vis. Please note that the current version of the tool is only working on Google's Chrome Browser.

integrated within the AD and were provisioned by an AD administrator. However, there was no communication with the IAM department and, therefore, no access management or integration into the IAM system took place. This results in technical identities with excessive entitlements. As the company is not experienced with such technical identities, the risk for failures (e.g. deletion of data) resulting from misconfiguration is high (cf. Sect. 5.2).

During configuration and assignment of a location to the technical identities some flaws regarding the existing location attribute values were detected. As entitlements shall be assigned automatically within the new IAM system based on a policy, the identification and correction of these values is highly relevant. Otherwise, identities with an incorrect value for their location attribute are not assigned enough entitlements (cf. Sect. 5.3).

5.1 Identities Not Managed Within a Central IAM (C1, C4)

As stated before, some identities within the company are not integrated in the central IAM system. Identifying these is a hard task considering the spread-out information. Our approach integrates applications not connected to the IAM system. Taking a look at the *"Data Source"* attribute the sunburst shows in which layer the respective data originates. Identities in the *"IAM Layer"* segment of the diagram are collected directly from IAM. However, another 17 identities are not managed by the IAM system. Three of them are maintained in the *"AD Layer"* while 14 are gathered from the *"IoT Layer"*. Taking a look at the details view of those 14 identities brings out that they are technical devices. Adding another circle for the *"Identity Type"* to the Sunburst allows an analyst to see that none of the identities within the *"IAM Layer"* are technical devices. So obviously the company has not integrated its technical identities into the central IAM system.

5.2 Identities with an Unusual Number of Entitlements (C3)

Another use case needing the attention of domain experts are identities with anomalous high number of entitlements. The Sunburst Diagram facilitates the identification of relevant entities and a decision how to proceed. Displaying the *"Entitlements"*, *"Identity Type"*, and the *"Function"* a small set of identities becomes visible having more than 76 entitlements. This seems conspicuous as most of the entities in the company have 0 to 25 roles assigned to them. Zooming into the segment with 76 to 100 entitlements a technical device attributed with function *"Support"* becomes visible and an identity from the company's customer *"Brandmark"* has an anomalous number of entitlements. These findings do not indicate an error per se, but it might be necessary to carry out further analyses. By browsing through the Sunburst domain experts are enabled to find various of similar cases. Any identity which might be over-authorized has to be examined, if all the entitlements are still needed (e.g. via recertification). If not, this indicates a serious security breach as identities having excessive permissions are legitimately allowed to access classified resources.

5.3 Poor Data Quality in IAM Data (C4)

The Sunburst diagram allows for identifying data quality flaws via several means. A first possibility is to compare similar attribute fields which originate from different data sources. Exemplary for this used when comparing *"Location (AD)"* and *"Location (IAM and IoT)"*. Information about locations of identities is administered in both the application layer and the IAM layer. Usually the IAM layer which contains the HR should be the master system for attributes like the location. After analyzing the data, some quality issues included in this system become apparent. An example is visible by zooming to the value *"Berlin"* of the *"Location (AD)"* attribute. The IAM layer has in fact 3 different attributes for identities having this value in the AD system, namely the correct value *"Berlin"* but also *"BER"* and *"10249 Berlin"*. Presumably, this value was recorded manually by the HR employees resulting in inconsistent data.

6 Conclusion

The complexity of modern IAM systems is constantly rising (e.g. increasing number of identities, further IAM mechanisms). Therefore, new challenges emerge. Within this work we showed that VA can be integrated into IAM in order to solve some of them. To achieve this, we initially identified five central challenges through a review of academic literature and analysis of the experience of practitioners (*RQ-1*). Thereby, we discovered two challenges especially connected to the identification and management of identities. Furthermore, we expect more challenges within the topics *Privacy* and *Data Quality*. Besides that, there will be the future challenge to transform role-based IAM into an attribute-based architecture for enterprises. We do not claim that our list of IAM challenges is exhaustive. However, we focused especially on problems where the integration of domain expert knowledge is vital. We detected some additional challenges but excluded them as they are not in the scope of the paper (e.g. inclusion of trust management in IAM, identity as a service, compliance with regulations).

Based on these challenges we identified VA as a possible solution as it enables enterprises to integrate domain expert knowledge and utilize their feedback to solve upcoming IAM challenges. We proposed an architecture how IAM by leveraging concepts from VA in order to answer our previously defined *RQ-2*. Additionally, we implemented proof-of-concept visualization according to our architecture and based on real world data (*RQ-3*). By applying VA, we have shown that problems tightly connected to the defined IAM challenges can be identified. However, the implementation should be regarded as a first example how the architecture can be implemented and as proof that VA can support enterprises to solve central IAM challenges. Other visualization techniques might be applied to solve another subset of our identified challenges.

After proposing an architecture for integration of VA and a first proof-of-concept implementation we want to focus further on the process to choose a suitable visualization for the *Interactive Visual Display* component. Additionally, we want to introduce further VA implementations to solve the remaining

IAM challenges. Afterwards, we can orchestrate the single implementations to an overarching coordinated view [14].

Acknowledgment. This research was supported by the Federal Ministry of Education and Research, Germany, as part of the BMBF DINGfest project (https://dingfest.ur.de).

References

1. Asghar, M., Backes, M., Simeonovski, M.: PRIMA: privacy-preserving identity and access management at internet-scale. In: Proceedings of the 2018 IEEE International Conference on Communications, pp. 1–6. IEEE Computer Society (2018)
2. Bandyopadhyay, D., Sen, J.: Internet of things: applications and challenges in technology and standardization. Wirel. Pers. Commun. **58**(1), 49–69 (2011)
3. Basel Committee on Banking Supervisions: Basel III: International Framework for Liquidity Risk Measurement, Standards and Monitoring (2010)
4. Card, S.K., Mackinlay, J.D., Shneiderman, B. (eds.): Readings in Information Visualization: Using Vision to Think. Morgan Kaufmann, Burlington (1999)
5. Colantonio, A., Di Pietro, R., Ocello, A., Verde, N.: Visual role mining: a picture is worth a thousand roles. IEEE Trans. Knowl. Data Eng. **24**(6), 1120–1133 (2012)
6. Crouser, R., Fukuday, E., Sridhar, S.: Retrospective on a decade of research in visualization for cybersecurity. In: Proceedings of the 2017 IEEE International Symposium on Technologies for Homeland Security, pp. 1–5. IEEE (2017)
7. Cser, A., Maxim, M.: Forrester - Top trends shaping IAM in 2018 (2018)
8. Damon, F., Coetzee, M.: The design of an identity and access management assurance dashboard model. In: Tjoa, A.M., Raffai, M., Doucek, P., Novak, N.M. (eds.) CONFENIS 2018. LNBIP, vol. 327, pp. 123–133. Springer, Cham (2018). https://doi.org/10.1007/978-3-319-99040-8_10
9. Diodati, M., Farahmand, H., Ruddy, M.: Gartner - 2019 planning guide for identity and access management (2018)
10. Elliott, A., Knight, S.: Role explosion: acknowledging the problem. In: Proceedings of the 8th International Conference on Software Engineering Research and Practice, pp. 349–355 (2010)
11. Federico, P., Wagner, M., Rind, A., Amor-Amorós, A., Miksch, S., Aigner, W.: The role of explicit knowledge: a conceptual model of knowledge-assisted visual analytics. In: Proceedings of the 2017 IEEE Conference on Visual Analytics Science and Technology (2017)
12. Fremantle, P., Aziz, B., Kopecký, J., Scott, P.: Federated identity and access management for the internet of things. In: Proceedings of the 2014 International Workshop on Secure Internet of Things, pp. 10–17. IEEE Computer Society (2014)
13. Fuchs, L., Pernul, G.: Supporting compliant and secure user handling - a structured approach for in-house identity management. In: The Second International Conference on Availability, Reliability and Security (ARES 2007), pp. 374–384. IEEE (2007)
14. Heer, J., Shneiderman, B.: Interactive dynamics for visual analysis. Queue **10**(2), 30 (2012)
15. Hovav, A., Berger, B.: Tutorial: identity management systems and secured access control. Commun. Assoc. Inf. Syst. **25**(1), 1–42 (2009)
16. Hu, V.C., et al.: Guide to attribute based access control (ABAC) definition and considerations. In: NIST Special Publication (2014)

17. Hummer, M., Groll, S., Kunz, M., Fuchs, L., Pernul, G.: Measuring identity and access management performance - an expert survey on possible performance indicators. In: Proceedings of the 4th International Conference on Information Systems Security and Privacy, pp. 233–240 (2018)
18. Hummer, M., Kunz, M., Netter, M., Fuchs, L., Pernul, G.: Advanced identity and access policy management using contextual data. In: Proceedings of the IEEE International Conference on Availability, Reliability and Security, pp. 40–49. IEEE Computer Society (2015)
19. Hummer, M., Kunz, M., Netter, M., Fuchs, L., Pernul, G.: Adaptive identity and access management - contextual data based policies. EURASIP J. Inf. Secur. **2016**(1), 1–19 (2016)
20. IDG Research Services: Studies Identity- & Access-Management 2017 (2017)
21. Indu, I., Anand, P.M.R., Bhaskar, V.: Identity and access management in cloud environment: mechanisms and challenges. Eng. Sci. Technol. Int. J. **21**(4), 574–588 (2018)
22. Jensen, J.: Federated identity management challenges. In: Proceedings of the 2012 IEEE International Conference on Availability, Reliability and Security, pp. 230–235. IEEE Computer Society (2012)
23. Kanuparthi, A., Karri, R., Addepalli, S.: Hardware and embedded security in the context of internet of things. In: Proceedings of the 2013 ACM Workshop on Security, Privacy & Dependability for Cyber Vehicles, pp. 61–64. ACM (2013)
24. Kunz, M., Fuchs, L., Hummer, M., Pernul, G.: Introducing dynamic identity and access management in organizations. In: Jajodia, S., Mazumdar, C. (eds.) ICISS 2015. LNCS, vol. 9478, pp. 139–158. Springer, Cham (2015). https://doi.org/10.1007/978-3-319-26961-0_9
25. Kunz, M., Puchta, A., Groll, S., Fuchs, L., Pernul, G.: Attribute quality management for dynamic identity and access management. J. Inf. Secur. Appl. **44**, 64–79 (2019)
26. KuppingerCole, CXP Group: State of organizations - does their identity & access management meet their needs in the age of digital transformation? (2017)
27. Mahalle, P., Babar, S., Prasad, N.R., Prasad, R.: Identity management framework towards internet of things (IoT): roadmap and key challenges. In: Meghanathan, N., Boumerdassi, S., Chaki, N., Nagamalai, D. (eds.) CNSA 2010. CCIS, vol. 89, pp. 430–439. Springer, Heidelberg (2010). https://doi.org/10.1007/978-3-642-14478-3_43
28. Meier, S., Fuchs, L., Pernul, G.: Managing the access grid - a process view to minimize insider misuse risks. In: Proceedings of the 11th International Conference on Wirtschaftsinformatik, pp. 1051–1065 (2013)
29. Moghaddam, F., Wieder, P., Yahyapour, R.: A policy-based identity management schema for managing accesses in clouds. In: Proceedings of the 8th International Conference on the Network of the Future, pp. 91–98. IEEE Computer Society (2017)
30. Morisset, C., Sanchez, D.: VisABAC: a tool for visualising ABAC policies. In: Proceedings of the 4th International Conference on Information Systems Security and Privacy. Newcastle University (2018)
31. Nuss, M., Puchta, A., Kunz, M.: Towards blockchain-based identity and access management for internet of things in enterprises. In: Furnell, S., Mouratidis, H., Pernul, G. (eds.) TrustBus 2018. LNCS, vol. 11033, pp. 167–181. Springer, Cham (2018). https://doi.org/10.1007/978-3-319-98385-1_12
32. Nyanchama, M., Osborn, S.: The role graph model and conflict of interest. ACM Trans. Inf. Syst. Secur. (TISSEC) **2**(1), 3–33 (1999)

33. Pfitzmann, A., Köhntopp, M.: Anonymity, unobservability, and pseudonymity—a proposal for terminology. In: Federrath, H. (ed.) Designing Privacy Enhancing Technologies. LNCS, vol. 2009, pp. 1–9. Springer, Heidelberg (2001). https://doi.org/10.1007/3-540-44702-4_1

34. Povilionis, A., et al.: Identity management, access control and privacy in integrated care platforms: the PICASO project. In: Proceedings of the 2018 International Carnahan Conference on Security Technology, pp. 1–5. IEEE Computer Society (2018)

35. Samarati, P., de Vimercati, S.C.: Access control: policies, models, and mechanisms. In: Focardi, R., Gorrieri, R. (eds.) FOSAD 2000. LNCS, vol. 2171, pp. 137–196. Springer, Heidelberg (2001). https://doi.org/10.1007/3-540-45608-2_3

36. Sandhu, R.S.: The authorization leap from rights to attributes: Maturation or chaos? In: Proceedings of the 17th ACM Symposium on Access Control Models and Technologies, pp. 69–70. ACM (2012)

37. Sandhu, R.S., Coyne, E.J., Feinstein, H.L., Youman, C.E.: Role-based access control models. Computer **29**(2), 38–47 (1996)

38. Servos, D., Osborn, S.L.: Current research and open problems in attribute-based access control. ACM Comput. Surv. **49**(4), 1–65 (2017)

39. Severino, R.: The data visualisation catalogue (2019). https://datavizcatalogue.com/index.html. Accessed 21 Feb 2019

40. Small, M.: Kuppingercole report - advisory note - big data security, governance, stewardship (2018)

41. SOX: Sarbanes-Oxley Act of 2002, pl 107–204, 116 stat 745 (2002)

42. Staheli, D., et al.: Visualization evaluation for cyber security. In: Proceedings of the 2014 IEEE Symposium on Visualization for Cyber Security, pp. 49–56. ACM (2014)

43. Tolbert, J.: Kuppingercole report - advisory note - identity in IoT (2017)

44. Wagner, M., Rind, A., Thür, N., Aigner, W.: A knowledge-assisted visual malware analysis system: design, validation, and reflection of kamas. Comput. Secur. **67**, 1–15 (2017)

45. Windley, P.J.: Digital Identity: Unmasking Identity Management Architecture (IMA). O'Reilly Media Inc, Newton (2005)

46. Xiong, J., Yao, Z., Ma, J., Liu, X., Li, Q., Ma, J.: PRIAM: privacy preserving identity and access management scheme in cloud. KSII Trans. Internet Inf. Syst. **8**(1), 282–304 (2014)

47. Xu, Z., Stoller, S.D.: Mining attribute-based access control policies from RBAC policies. In: Proceedings of the 10th International Conference and Expo on Emerging Technologies for a Smarter World. IEEE (2013)

Analysis of Multi-path Onion Routing-Based Anonymization Networks

Wladimir De la Cadena[1]([envelope]), Daniel Kaiser[1], Asya Mitseva[1],
Andriy Panchenko[2], and Thomas Engel[1]

[1] University of Luxembourg, Esch-sur-Alzette, Luxembourg
{wladimir.delacadena,daniel.kaiser,asya.mitseva,thomas.engel}@uni.lu
[2] Brandenburg University of Technology, Cottbus, Germany
andriy.panchenko@b-tu.de

Abstract. Anonymization networks (e.g., Tor) help in protecting the privacy of Internet users. However, the benefit of privacy protection comes at the cost of severe performance loss. This performance loss degrades the user experience to such an extent that many users do not use anonymization networks and forgo the privacy protection offered. Thus, performance improvements need to be offered in order to build a system much more attractive for both new and existing users, which, in turn, would increase the security of all users as a result of enlarging the anonymity set. A well-known technique for improving performance is establishing multiple communication paths between two entities. In this work, we study the benefits and implications of employing multiple disjoint paths in onion routing-based anonymization systems. We first introduce a taxonomy for designing and classifying onion routing-based approaches, including those with multi-path capabilities. This taxonomy helps in exploring the design space and finding attractive new feature combinations, which may be integrated into running systems such as Tor to improve users' experience (e.g., in web browsing). We then evaluate existing implementations (together with relevant design variations) of multi-path onion routing-based approaches in terms of performance and anonymity. In the course of our practical evaluation, we identify the design characteristics that result in performance improvements and their impact on anonymity.

1 Introduction

In modern society, people disclose a large quantity of digital traces via the Internet. Hence, privacy is attracting more and more attention and has become a serious concern. *Anonymization* is a basic technical means for achieving privacy. Despite the variety of approaches proposed for anonymous communication, only a few have reached widespread deployment. Currently, Tor [9] is the most popular low-latency anonymization network designed for TCP-based applications, serving more than two million daily users[1]. The main objective of Tor is to hide

---

[1] https://metrics.torproject.org/userstats-relay-country.html, October 2018.

© IFIP International Federation for Information Processing 2019
Published by Springer Nature Switzerland AG 2019
S. N. Foley (Ed.): DBSec 2019, LNCS 11559, pp. 240–258, 2019.
https://doi.org/10.1007/978-3-030-22479-0_13

the identities (i.e., IP addresses) of users who communicate through the Internet. To start a connection via Tor, the user runs local software, an *onion proxy* (OP), and creates a virtual tunnel, referred to as a *circuit*, to the destination over three nodes, known as *onion relays* (ORs) [8]. The ORs are run by volunteers who determine the amount of bandwidth they are willing to share. Depending on their position on the circuit, the ORs are denoted as *entry*, *middle*, and *exit*. Via a Diffie-Hellman key exchange, the user negotiates a distinct symmetric key with each OR on the circuit. The symmetric keys are used to encrypt the actual user data in multiple layers of encryption [8]. While forwarding user traffic, each OR on the circuit removes (or adds, depending on the direction) a layer of encryption. This ensures that none of the ORs on the circuit knows both the source and the destination of a connection at the same time. Along a circuit, user traffic travels encapsulated in fixed-size units referred to as *cells*.

Due to the diverse resource capabilities of ORs and their dynamic nature—anybody can join the network by running an OR or leave the network at any time—Tor suffers from both high congestion and latency. This often leads to significant delays for users which, in turn, may discourage them from using the network. Since the strength of anonymity provided by Tor strongly depends on the number of users, the protection of Tor clients utilizing the network is weakened by any user leaving the network. Therefore, performance improvements are necessary to make the system more attractive for both new and existing users. This will further improve the security of all users due to the increased anonymity set.

In response to this, a significant amount of research has focused on optimizing Tor's performance by improving its circuit processing [4,36,38], transport mechanisms [29,39,41], and relay selection algorithms [2,33,42], analyzing relay recruiting techniques [12,20,21], and adopting throttling methods [22] to reduce the load on the network. However, none of this work has investigated the performance benefits of multiple, disjoint paths used at overlay level when transmitting user data for a single Tor client. Although a few works [3,44] have suggested concrete approaches to deploying multi-path techniques in Tor, their evaluations are limited by unrealistic and outdated conditions.

In this paper, we present an up-to-date review of existing multi-path approaches particularly designed for Tor and similar onion routing-based low-latency anonymization systems. By conducting experimental evaluations at different scales, we analyze the state-of-the-art multi-path anonymization techniques in terms of the performance gain and anonymity implications of each approach. Our contribution is two-fold:

1. We provide a systematic survey of currently-existing multi-path approaches for Tor and other similar onion routing-based anonymization systems as well as techniques that allow adding multi-path capability. To this end, we introduce a taxonomy for onion routing-based low-latency designs with a focus on multi-path approaches and classify the existing related works accordingly.
2. We conduct a comprehensive evaluation to compare these approaches in terms of both performance and anonymity. Based on the results from our evalua-

tion and our theoretical analysis, we discuss which design choices should be considered to achieve a desired set of properties in new systems.

2   Related Work

To improve the performance of the Tor network, a significant amount of research has focused on exploring a variety of relay selection algorithms, e.g., by trying to avoid congested ORs [42], considering the geographical location [2] or bandwidth [34,35] of chosen ORs. Another group of works [10,27,29] criticizes the transport design applied by Tor, i.e., circuits from several users are multiplexed through a single TCP connection between two ORs. This may slow down the performance of interactive circuits. In response to this, several works evaluate advanced circuit scheduling mechanisms [36,38], propose improved congestion control algorithms [4,29] or even replacing the underlying transport protocol [26,41] to optimize the utilization of available bandwidth in Tor. In contrast to our study, these works did not evaluate the effect of multi-path techniques in Tor. Nevertheless, these proposals complement our work and their coexistence can further improve the performance and harden the security of Tor.

Karaoglu et al. [23] propose a multi-path routing scenario which emulates the operation of multi-path TCP [13]. Here, the Tor client is responsible for splitting and sending the traffic through multiple disjoint circuits to a web server which, in turn, is required to merge the received data. Thus, the authors do not require any modification in the core Tor network. However, Karaoglu et al. consider only a unidirectional scenario, in which the client uploads a file to the web server. Furthermore, the authors do not make any comparison with existing state-of-the-art multi-path approaches proposed for Tor or other onion routing-based anonymization systems. Last, but not least, Ries et al. [30] compare different low-latency anonymization networks with respect to their usability and the level of anonymity that they provide. Unlike our work, there is no evaluation of the applicability of multi-path techniques within these anonymization networks.

3   Multi-path in Anonymization Systems

Using multiple paths in anonymization systems has been also considered in previous theoretical analyses, simulations, and non onion routing approaches. The objectives pursued by those works were: passive attack resilience [11,32], multi-path as a means of anonymity [24], and performance improvements [23,33]. However, only three systems have been fully developed and implemented as multi-path onion routing-based approaches. Two of these, Conflux [3] and mTor [44], are extensions to vanilla Tor that adapt its traffic management design to utilize multiple circuits; the third, MORE [25], comprises a multi-path design over UDP where each cell travels along a different circuit. To our knowledge, there is no fully-developed multi-path approach that is both UDP-based and uses, as Tor does, fixed circuits per data transfer. For a more comprehensive analysis of standard transport protocol (UDP, TCP)-based multi-path approaches, we consider closing this gap in the design space to be necessary and so added multi-path support to UDP-OR [41] as a further contribution; we refer to the result as mUDP-OR. We chose enhancing UDP-OR because it is fully-developed and relies on standard transport protocols (see Sect. 4). The remainder of this section describes the multi-path onion routing-based systems analyzed and evaluated in this paper.

3.1   Conflux

In this design (see [3]), the OP builds multiple circuits with the same *exit* OR. Once those circuits are created, the OP sends a cell with a random nonce towards the *exit* OR as an identifier of the multi-path structure. To send each cell, the OP and the *exit* OR, known as *end-points*, select one of the multiple circuits according to its congestion, which is estimated as the time interval between the 100th cell being sent, and the corresponding *sendme*[2] being received. Cells that arrive out-of-order to the *end-points* are merged and sorted using a 4-byte sequence number included in the cell's payload. Conflux presents results from an implementation that supports only two circuits. For our analysis, we have enhanced the Conflux's design in order to support m circuits.

3.2   mTor

Here (see [44]), the multi-path structure and cell merging procedure is similar to Conflux. However, *end-points* choose one of the multiple circuits according to its current stream-level window[3] value. The *end-point* drops cells to the circuits in a first-in-first-out manner, while their stream window is greater than zero.

3.3   MORE

In MORE (see [25]), it is required that the client participates as OR within the network (peer-to-peer network). To send data, the client OR captures TCP data via a TUN device[4] and encapsulates it in cells, which will each be sent across a different circuit. This means that no initial circuit establishment takes place, but that each cell travels along its own randomly-chosen path. To guarantee reliability of cells traveling along different routes, MORE takes advantage of the TUN device's functionality and provides an IP overlay service for tunneling TCP data. In this sense, a multi-path layer TCP session exists between sender and receiver. To discover each cell's route, an intermediary OR onion-decrypts and reads the corresponding successor node from the header. To reduce the computational cost

---

[2] Cell used for flow control: an *end-point* sends it to acknowledge the arrival of 100 cells within a circuit, or 50 cells within a stream.

[3] As part of the flow control, the stream-level window caps the maximum amount of cells to 500 per stream at any moment.

[4] A TUN device is a virtual kernel network interface that works as a bridge between the user and kernel spaces.

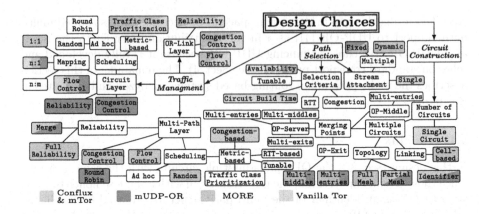

Fig. 1. Taxonomy of design choices for onion routing-based approaches

of re-setting up a cryptographic context for each cell, MORE uses elliptic curve cryptography (ECC). While using one circuit for each cell increases the resilience against traffic analysis attacks, it also considerably reduces performance.

3.4 mUDP-OR

Here (see [41]), the multi-path structure and circuit identification is performed in a manner similar to Conflux. However, ORs in a circuit communicate with each other using the UDP transport protocol. This circuit is used for tunneling TCP application data. Instead of encapsulating complete TCP segments, an *end-point* builds cells, appending to the header the necessary TCP fields (e.g. sequence numbers) to reconstruct a TCP packet at the other *end-point*. This TCP virtual connection is realized by setting up a SOCKS proxy in the *exit* OR, and establishing a virtual tunnel from a virtual TUN device in the OP. We implemented two strategies to dispatch cells into the circuits. In the first, the *end-point* chooses the circuit in a round robin (RR) manner with a configurable number of cells per circuit. In the second, the *end-point* randomly chooses through which circuit the next cell will be sent. We leverage the existing circuit-layer TCP session to merge cells arriving from different circuits. In this sense, the existing virtual end-to-end TCP connection is agnostic to the circuit(s) used.

4   Classifying Design Choices

In this section we introduce a hierarchical taxonomy for classifying and discussing onion routing design choices. The top level classes of our taxonomy comprise *traffic management*, *path selection*, and *circuit construction*; Fig. 1 illustrates our taxonomy. We focus on the multi-path aspects and the effect of adding multi-path capabilities. Based on the structure of our taxonomy, we classify and discuss the multi-path OR approaches introduced in the previous section.

4.1 Traffic Management

The *traffic management* class comprises design choices which are concerned with transmitting data over already-established circuits in an anonymization overlay network; specifically regarding providing a TCP-like end-to-end service and scheduling decisions. This class is a key element of designing OR approaches and significantly affects performance. It also has an effect on anonymity, as feedback mechanisms might leak information, allowing for fingerprinting attacks [29].

We classify traffic management into *OR-link layer*, *circuit layer*, and *multi-path layer*. These layers are intertwined, as their combination must provide the same service as a direct TCP connection, namely reliability, congestion control, and flow control. Inter-layer dependency causes some issues, the most prominent of which is cross-circuit interference [42]. In general, cross-circuit interference is a consequence of OR-link layer connection artifacts affecting virtually independent circuits, because several circuit-layer connections share the same OR-link.

OR-Link Layer: The *OR-link layer* comprises the transport connection between ORs. We classify the *OR-link layer* design according to which of reliability, congestion control, and flow control it incorporates. Tor uses TCP on the OR-link layer, realizing reliability, congestion control, and flow control on this layer. Since Tor multiplexes all circuit segments over a single OR-link layer connection (TCP connection) between ORs and TCP mechanisms are agnostic to these circuits, it is subject to cross-circuit interference; specifically, because of shared I/O buffers and congestion control. Shared I/O buffers are a problem because segments are taken out of the shared TCP buffer on a first-come-first-served basis, no matter which circuit they are associated with. This leads to high latency for all circuits in the presence of high-throughput circuits that congest the shared TCP I/O buffer. This, in turn, may render interactive sessions using a low-throughput circuit over the same TCP connection unusable, as there is no means for prioritizing an interactive session.

Congestion control causes TCP connections to be throttled in the case of a congestion event[5]; thus, if a congestion event occurs related to a single circuit, all circuits over the same TCP connection are throttled. Even without congestion control, reliability[6] would cause cross-circuit interference because the recovery from packet loss in one circuit would also affect all other circuits sharing the same TCP connection.

Two classes of solutions addressing Tor's cross-circuit interference have been proposed; firstly, dedicating a TCP connection to each circuit segment [5]; and secondly, using a simple transport protocol, e.g., UDP [41]. Conflux and mTor are Tor extensions that add the multi-path layer while inheriting this weakness of Tor. mUDP-OR and MORE both use UDP as a transport protocol, avoiding cross-circuit interference. However, this countermeasure leads to aggressive traffic[7], which might congest the network. This issue has been addressed in [39].

---

[5] A congestion event might, e.g., be a packet loss.

[6] The realization of reliability is typically intertwined with congestion control.

[7] Traffic sent at high rates even in case of network congestion.

A multi-path based mitigation technique for cross-circuit interference on the OR-link layer, which to our knowledge has not yet been discussed, would be the use of multi-path TCP [13] as a transport protocol. Since multi-path TCP handles scheduling among the various TCP sub-streams on the transport layer, it is not suited to circuit-aware scheduling. Still, having several TCP sub-streams would lower the risk of cross-circuit interference while potentially multiplexing several circuits over a single connection hiding them in an anonymity set. However, especially in congested networks, having several TCP connections also increases the aggressiveness of traffic [40].

Circuit Layer: The *circuit layer* comprises a single overlay connection between an OP and an *exit* OR. As with the OR-link layer, we classify the *circuit layer* design by which of reliability, congestion control, and flow control it incorporates. The Tor circuit layer protocol [8] does not implement reliability, since it is already provided by TCP at the OR-link layer. It provides flow control with a fixed-size-window-based mechanism and no congestion control. Reliability methods do not benefit from inter OR-link or inter-layer communication and thus should be realized on one layer exclusively. Flow control and congestion control can benefit from inter OR-link and inter-layer interaction [39], and thus may be (partially) realized on several layers. Both having a fixed-size window for flow control and not providing congestion control have been identified as the major performance limiting factors of Tor [6]. Prioritization of interactive connections on the circuit-level has been proposed by Tang et al. [36] as a mitigation technique for cross-circuit interference, making interactive connections more responsive.

Conflux and mTor also inherit the properties of Tor for the circuit layer. mUDP-OR tunnels TCP, meaning the onion proxy and the exit node have a virtual TCP connection; thus, mUDP-OR provides all of flow control, congestion control, and reliability on the circuit layer. MORE is an overlay IP service where TCP data can be tunneled, making it part of the same class as mUDP-OR. The advantage of both mUDP-OR and MORE is being able to avoid cross-circuit interference. However, the OP-to-exit feedback loop for congestion control and reliability realization is very long and therefore not responsive. If a packet is dropped on the first circuit segment, this packet loss is detected at the end of the last circuit segment and the notification of this event needs to travel all the way back. The same problem occurs for adapting the TCP congestion window. Further, because mUDP-OR tunnels kernel-level TCP, the feedback across the whole circuit allows OS fingerprinting attacks [29].

A further property we use to classify the *circuit layer* by is *circuit to OR-link mapping*. The *circuit to OR-link mapping* decides how circuit segments are mapped to connections between the corresponding pair of ORs. Realizations comprise (1) $n : 1$, where all circuit segments between a pair of ORs are multiplexed over one transport connection, (2) $1 : 1$, where each circuit segment is mapped to a dedicated transport connection, and (3) $n : m$, where several circuit segments between a pair of ORs are multiplexed over a set of transport connections. While (1) may suffer from cross-circuit interference (e.g., when reliability is provided) but offers the best anonymity properties, (2) prevents cross-circuit inference but may allow passive attackers to infer which circuit a given packet is associated with, which in turn might allow association with the sender. A compromise is provided by (3) which reduces cross-circuit interference while still hiding packets in an anonymity set. Tor implements strategy (1), which is inherited by Conflux and mTor. mUDP-OR also implements this strategy. MORE implements strategy (2) and further uses a new circuit for each (set of) cell(s). The multi-path TCP based solution described above is an example of (3).

We also classify the *circuit layer* by its *circuit scheduling* method. If several circuits share a transport connection, cells associated with various circuits are multiplexed over this connection. Circuit-layer scheduling is concerned with how to choose which cell from various circuit-level output queues should be the next to be put into the transport-level output queue. We classify *circuit scheduling* methods into (1) *ad hoc*, and (2) *metric-based*. Ad hoc methods do not depend on a metric; subclasses are, e.g., (1a) *random*, where cells are randomly taken from input queues and put into the output queue, and (1b) *round robin*. *Metric-based* methods collect information about available circuits. This information is used to calculate a metric, based on which scheduling decisions are made. A subclass is (2a) *traffic class prioritization*, where specific traffic classes, e.g., traffic from an interactive connection, are prioritized. (1) is simple to implement and neither consumes additional computational power nor needs extra network messages. However, as shown in [6], (2) provides superior overall performance.

Prior to 2012, Tor used *round robin* as its scheduler. Then, an improved scheduler based on the recent circuit's activity was implemented [36]. Most recently, in 2017 a new scheduler called KIST [18] was introduced. It uses feedback from the kernel to prioritize the traffic of each circuit's queue. Conflux and mTor inherit this characteristic from Tor. mUDP-OR does not maintain circuit-level queues and therefore directly passes cells to the transport layer. Because MORE has a 1 : 1 mapping between circuit segments and OR-links, it too does not implement any circuit-level scheduling and leaves this task to the transport layer. Not having a circuit-level queue decreases feedback time and total queueing delay, but comes at the cost of not having the advantages of *circuit scheduling*.

Multi-Path Layer: The *multi-path layer* incorporates sets of circuits jointly building communication channels. We classify the *multi-path layer* design by which of congestion control and flow control it considers. While it is a feasible design choice for the multi-path layer to be agnostic to both flow control and congestion control, the realization of reliability for a multi-path approach always includes the multi-path layer. The subclasses of multi-path reliability are *merge* and *full reliability*. The former expects the underlying circuits to provide a reliable ordered stream of cells—either by realizing reliability on the OR-link layer or on the circuit layer—and merges cells coming from different circuits. The latter collects all packets from the associated circuits and fully implements reliability. Having reliability on the multi-path layer allows for sending control information on less-congested circuits to reduce feedback time.

While Tor does not offer multi-path capabilities, both Conflux and mTor can be seen as multi-path extensions to vanilla Tor. As Tor already provides reliability and congestion control on the OR-link layer and flow control on the circuit layer, both solutions apply the *merge* strategy on the multi-path layer. Since mUDP-OR is a multi-path extension of UDP-OR, which already provides a means for anonymizing a reliable connection, mUDP-OR adds *merge* on top of the circuit layer provided by UDP-OR. MORE sends cell(s) over a different unreliable circuit; thus, *full reliability* is performed at the *multi-path layer*.

Another multi-path layer design choice is *multi-path scheduling*. While *circuit scheduling* decides from which circuit-level queue the next cell is put into the transport-level queue, *multi-path scheduling* decides over which circuit a given cell should be sent. The classes of scheduling algorithms, however, are the same as for *circuit scheduling*. New subclasses are (2b) *congestion-based*, where cells are sent through less congested circuits, (2c) *round trip time (RTT) based*, where circuits with lower RTT are prioritized, and (2d) *tunable*, which is a tunable combination of the other subclasses. As multi-path layer scheduling allows for congestion control which, in turn, leads to more even utilization of circuits, it also helps in mitigating cross-circuit interference. Both Conflux and mTor implement *congestion-based* scheduling. While Conflux's scheduling strategy has a very long feedback loop (see Sect. 3), mTor implements a more responsive method based on the stream-level receive window size. Still, in absolute terms, the feedback loop is long. The mTor scheduling algorithm improves the throughput of bulk transfers while not negatively affecting interactive sessions. The default scheduler used in mUDP-OR is *round robin*. MORE is special in this case, as it creates new circuits on the fly for each cell and sends cells over the respective newly-created circuit. Thus, it depends on path selection and circuit construction discussed in the following subsections. The scheduling itself is therefore ad hoc, because a cell is scheduled to the only available circuit at a given point in time.

Multi-path TCP [13] could be used not only on the OR-link layer, but also on the circuit and multi-path layers, tunneling multi-path TCP's sub-streams on the circuit layer and using its scheduling and merging strategy on the multi-path layer. While this solution has the advantage of using an established protocol, it comes with little flexibility for adapting it to be a Tor transport. Such a solution should *not* use TCP at the OR-link layer as this would lead to TCP over TCP throttling effects [37].

Summarizing the realization of TCP functionality, all approaches directly use TCP and do not introduce custom designs. Both Conflux and mTor use TCP on the OR-link layer, mUDP-OR uses TCP on the circuit layer, and MORE uses TCP[8] on the multi-path layer. Like Tor, Conflux and mTor add only a simple flow control mechanism on the circuit layer. More sophisticated approaches tailored to anonymization overlay networks (see, e.g., [39]) have not as yet been used in the context of multi-path onion routing.

---

[8] MORE uses TCP when anonymizing a reliable service. Because MORE provides an IP service on the overlay, it can also be used without providing TCP functionality at all.

4.2   Circuit Construction

This design class comprises the considerations for building the path(s) that the OP will employ. The only subclass of *circuit construction* is the *number of circuits* required by the OP for exchanging data. The subclass *single circuit* is valid for Tor, since only one circuit is required by the OP for a data transfer. If *multiple circuits* are required, the design choice needs to specify where the *merging/splitting points* are. This in turn defines how many ORs per position (*entry*, *middle*, or *exit*) can compose a circuit. This design choice influences anonymity, performance and implementation complexity. Conflux, mTor, and mUDP-OR enlarge the bandwidth capacity of the last hop by building extra middle-to-exit connections. From the anonymity perspective, using multiple *entry* ORs may improve the resilience against some attacks (see Sect. 6). None of the considered approaches merge on a *middle* OR; this scheme would represent a more complex implementation but at the same time an easier deployment in the network, since there are fewer requirements for starting a *middle* OR in Tor [1].

Another class refers to the *topology* formed by the selected ORs. Conflux, mTor, and mUDP-OR form a *partial mesh*, since each *entry* OR communicates with one *middle* OR. MORE tends to form a *full mesh* as the number of sent cells increases.

Lastly, the *linking* subclass refers to the mechanism to associate/save several circuits as a singular structure upon their creation. In Conflux, mTor and mUDP-OR, multiple circuits are referred by an *end-point* under a common identifier exchanged via a control cell. This type of *linking* comprises the subclass *identifier*. The other subclass, *cell-based*, is used by MORE. Here, paths are not linked in the construction process, but their cells will be grouped during the data transmission based on their header. This *linking* class is strongly related to the scheduling from the multi-path layer, and choosing it properly results in faster multi-path build times, and a more secure multi-path structure.

4.3   Path Selection

Preemptively, more than the required circuits can be built before streams are attached to them. This design choice determines which of the built path(s) will be next used for the data transfer. Once the path(s) are selected, the OP sends cells based on the *traffic management* design choices.

The subclass *selection criteria* determines which parameter(s) must be considered for defining which circuit(s) will be employed. In Tor, after discarding circuits with slow build times, the newest available is chosen. Other parameters such as RTT, congestion, or a tunable combination of these may be also considered. The subclass *stream attachment* comprises special choices for multi-path approaches. In contrast to Tor, where the stream will be directly attached to a single circuit, multiple circuits allow this attachment to be *fixed*, when the set of selected circuits does not change after they are chosen, or to be *dynamic* when the set of selected circuits may change during the data transfer. When the set of selected circuits changes as dynamically as in MORE, this design choice determines the *multi-path layer* scheduling.

To sum up, the top level classes of the taxonomy address the design choices to be considered before user data is sent (*path selection* and *circuit construction*), and for the data transmission itself (*traffic management*). In our evaluation, we identify the effects of the design choices employed by the analyzed approaches.

5   Performance Evaluation

In this section, we evaluate each approach within two scenarios: on an isolated private local network, and in a larger network using the NetMirage[9] emulator.

5.1   Private Local Network Experiment

We use this experiment to understand the differences between all designs without external influence. In a local network we set up seven ORs, one client for measurements, four web servers, and up to 30 clients to generate load on the employed circuit(s). Three metrics are reported on the client: TTFB (Time to First Byte), and download times (DT) for HTTP web (320 KiB) and bulk (1 MiB) requests[10]. Furthermore, on each OR the CPU usage was periodically logged. Considering that there are no congestion effects from other sources in an isolated network, we evaluated each approach with the round robin multi-path scheduler. This also ensures that multiple circuits will be equitably used. Effects of congestion-based schedulers are evaluated in the second scenario.

Multi-path Circuits and Load Balancing: In the left columns of Table 1 we present the average CPU load on each OR for the maximum number of clients. For multiple circuits, we present results for only one of them, since values in others are similar. It is clearly observable that the load assigned to each OR is decreased by using multiple circuits simultaneously. Furthermore, it is noticeable that translating the reliability and congestion control tasks to the *end-points* (in mUDP-OR and MORE) results in a higher load on them. We also observe that *entry* ORs are more loaded than others (except by MORE) in the circuit due to the cryptographic operations performed.

Client Performance: Performance metrics are presented in the right columns of Table 1. As an expected consequence of the dynamic stream attachment in MORE, its clients experience the slowest download times, making it unfeasible to complete data transfers in many cases (e.g., for bulk downloads). We confirm that a UDP-based approach such as mUDP-OR responds faster than a TCP-based approach. In contrast to Conflux and mTor, our multi-path enhancement to UDP-OR did not produce the desired improvements due to the still-existing very long feedback for retransmissions and acknowledgments.

---

[9] https://crysp.uwaterloo.ca/software/netmirage/.

[10] Values presented with 95% confidence and based on 200 repetitions.

Table 1. CPU usage percentage on the onion routers, and performance metrics for the maximum number of clients.

Approach	Paths	CPU usage on the OR			Client performance metrics		
		Entry	Middle	Exit	TTFB[s]	DT Web [s]	DT Bulk[s]
Conflux	1	57.87 ± 2.16	49.85 ± 1.82	31.74 ± 1.15	0.027 ± 0.005	0.059 ± 0.003	0.113 ± 0.008
	2	33.22 ± 2.25	31.01 ± 2.21	31.51 ± 2.22	0.016 ± 0.002	0.052 ± 0.0005	0.082 ± 0.002
	3	27.24 ± 1.94	22.78 ± 1.56	32.94 ± 2.28	0.014 ± 0.003	0.049 ± 0.0007	0.079 ± 0.0024
mTor	1	57.87 ± 2.16	49.85 ± 1.82	31.74 ± 1.15	0.027 ± 0.005	0.059 ± 0.003	0.113 ± 0.008
	2	25.10 ± 2.21	23.19 ± 1.73	32.51 ± 2.26	0.012 ± 0.002	0.055 ± 0.001	0.079 ± 0.003
	3	13.45 ± 0.97	14.62 ± 0.85	32.90 ± 2.27	0.017 ± 0.003	0.049 ± 0.0007	0.081 ± 0.002
mUDP-OR	1	88.31 ± 1.51	86.89 ± 1.48	71.33 ± 1.24	0.002 ± 0.0003	0.031 ± 0.0018	0.076 ± 0.005
	2	36.75 ± 2.51	32.67 ± 2.79	73.48 ± 3.09	0.002 ± 0.0001	0.034 ± 0.0017	0.081 ± 0.001
	3	22.73 ± 3.11	24.45 ± 3.27	75.30 ± 3.21	0.0024 ± 0.0001	0.035 ± 0.001	0.113 ± 0.014
MORE	1	6.73 ± 0.14	6.93 ± 0.19	75.92 ± 2.34	0.014 ± 0.004	1.41 ± 0.007	N.A
	2	6.65 ± 0.13	6.93 ± 0.2	70.09 ± 2.24	0.014 ± 0.004	1.37 ± 0.19	N.A
	3	6.57 ± 0.11	6.52 ± 0.08	68.19 ± 2.48	0.016 ± 0.006	1.39 ± 0.014	N.A

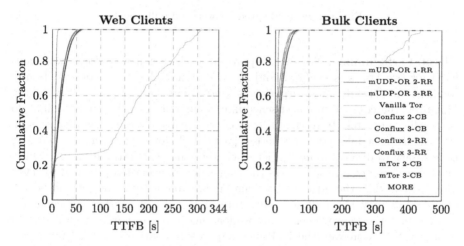

Fig. 2. Time to first byte for web and bulk clients

5.2 Larger-Scale Experiment

Currently, the Shadow [19] tool is widely used for large-scale Tor simulations. However, due to lack of support for some required functions (e.g., TUN devices) in Shadow, we opted for the NetMirage tool for building a common testbed. We based our experiment on the PlanetLab and Tor topologies included in version 1.12.1 of the Shadow simulator. It consisted of 303 nodes distributed all over the world, where we set up 206 web clients, 22 bulk clients, 14 exit nodes, 59 non-exit nodes and 14 web servers. Web clients performed successive downloads of 320 KiB data, waiting randomly from 0 to 20 s between each download, and bulk clients downloaded 1 MiB sequentially without pausing.

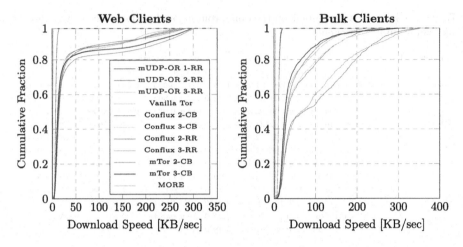

Fig. 3. Download speed for web and bulk clients

Client Performance: For every approach, each design variation[11] was emulated for two hours (see Figs. 2 and 3). We observe that, in a congested environment, mUDP-OR only outperforms other approaches in the TTFB metric. Moreover, the congestion-based scheduling techniques of mTor and Conflux do not profit completely from the utilization of multiple circuits; this may explain why the RR scheduler performs better, particularly for bulk downloads. Thus, it is necessary to develop a more efficient circuit congestion estimation procedure. Since Tor does not directly access the congestion information provided by TCP for each *OR link*, the estimations done in the circuit layer are not completely reliable and may not represent the state of the circuit at that moment. We observe that the improvements in downloading data are more advantageous to bulk transfers. Moreover, the TCP-based approaches outperform the UDP-based ones in terms of download speed for nearly all the 228 clients.

Network Scalability: In this experiment we incrementally introduced up to 228 clients (10% bulk and 90% web clients) and measured TTFB and download speed for each iteration (see Fig. 4). The fast first response of mUDP-OR is clearly advantageous within a congested network; however, clients of Conflux and mTor download data faster. We observe that the download speed for all approaches stabilizes to its minimum value when around 140 clients are present. After this point, differences between all approaches remain constant. We notice that using RR for multi-path scheduling scales better, due to the equitable usage of network resources. It is also noticeable that congestion-based mechanisms perform better in a lightly-congested environment; this reinforces the intuition

---

[11] Two design variations were evaluated: the number of paths (labeled as 1,2,3) and the *multi-path* scheduler (labeled as RR for round robin and CB for congestion based).

that the employed congestion estimation techniques are not fully precise. We refrain from comparing MORE in this regard due to its poor performance.

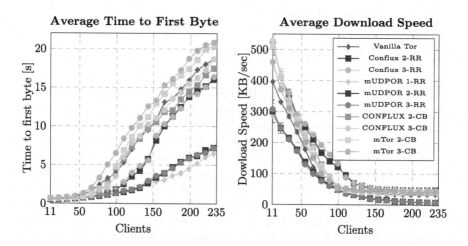

Fig. 4. Performance metrics for different number of clients

5.3   Design Recommendations

From the performed evaluations we identify that any design (single or multi-path) based on UDP, provides a fast first response. This feature comes however, at the cost of a degraded performance. If fast download speed is desired (e.g., in web browsing), the design should use the TCP protocol on the OR-link layer together with an effective congestion-based multi-path scheduler. If the objective is to ease the burden on ORs, the round robin scheduler ensures an equitable traffic distribution. If performance is not of the essence—for instance in non-time-sensitive applications like messaging or microblogging—even higher anonymity can be achieved by systems with the characteristics of MORE.

6   Anonymity Analysis

In this section, we address the anonymity implications produced by using multiple paths in the context of the evaluated approaches.

6.1   Client Multi-path Circuits Compromise

A circuit becomes compromised if an attacker gains control over both its edges. An adversary that controls a fraction of entry and exit nodes (f_g and f_x), can compromise any single circuit with a probability $P(c) \approx f_g f_x$ [3,7]. For multi-path clients employing m *entries* and one *exit* OR, this expression becomes $P_m(c) \approx f_x(1 - (1 - f_g)^m)$. This expression is valid for all evaluated approaches;

however, for MORE, an adversary must compromise many more than m circuits to fully affect one client, which means that this approach provides higher levels of anonymity. Even though $P_m(c) \geq P(c)$, this difference is negligible even in the presence of a powerful attacker[12].

6.2   Using Multiple Entry Onion Routers

To make the probability of de-anonymization vanishingly small, Tor clients try to choose the same *entry* OR from the priority-ordered *primary* list[13]. Since a multi-path client uses m entries, they should be taken from a *primary* list of minimum size m. In order to evaluate the anonymity implications, we leverage the framework presented in [17] together with metrics and adversary models presented in [15]. Two adversary models are considered for a client using m entries, the first determines that a client is compromised if at least one entry is controlled, which may be valid for confirmation and correlation attacks [14]. The second, defines a compromised client if and only if all m entries are controlled, which may be valid for *website fingerprinting* attacks [16,28,31,43]. Both models are valid for designs that assign streams to a fixed set of circuits (Conflux, mTor and mUDP-OR). For systems with dynamic stream attachment (MORE), the models are valid during the usage interval of the circuits. Using consensus data from 2015, we simulated 500,000 clients and a high-resource adversary controlling 10% of the overall entry bandwidth. Figure 5 shows the mean compromise rate (CR) of 50 simulations. We notice that, for the second adversary model, the CR decreases exponentially with each additional entry. Conversely, if one from m entries is enough to compromise a client, they become around twice as vulnerable.

Lastly, we analyze the *guard fingerprinting* attack[14], where using multiple entries decreases the mean anonymity set size $(\overline{A})$. Currently, each Tor client shares its *entry* OR with on average another 1,000 users $(\overline{A} = 1000)$. If clients used m paths, $\overline{A}$ would drastically decrease to $\frac{2 \times 10^6}{\binom{2000}{m}}$. Using the Tor source code, we simulated the creation of *primary* lists for 83,000 clients. For $m = 1$, we experimentally obtained $\overline{A} = 112$, while for $m = 2$ roughly 90% of clients had a unique pair of entries, and the user with the largest anonymity set shared its entries with another 14 users. For this attack, the dynamism of MORE is also favorable, because all clients tend to use all nodes as entries. Thus, $\overline{A}$ converges to its upper limit (the total number of clients).

To sum up, the anonymity advantages of using multiple *entry* ORs, together with the presented performance gains, are compelling reasons to enhance systems such as Tor with multi-path capabilities. The main constraint is the fact that

---

[12] An adversary controlling 20% of the total bandwidth (ca. 60 Gbit/s in Tor) is considered as very powerful; in this case $P(c) \approx 4\%$ and $P_m(c) \approx 9.8\%$ for $m = 3$.

[13] From all ORs with entry flags listed in the consensus, by default, a client filters three ORs from several lists, choosing the first that is reachable as its *entry* OR.

[14] Guard ORs refer to the entry ORs regularly selected by a client. This set of nodes may be used as a fingerprint for de-anonymizing a client.

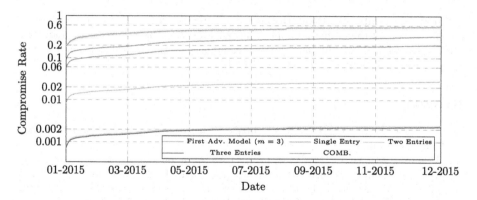

Fig. 5. Fraction of compromised clients (Compromise Rate) during one year: Single, two, and three entries refer to the second adversary model. In the scenario labeled as COMB, 60% of the clients use a single entry, 20% two entries and 20% three entries.

using multiple entries is not considered in the latest Tor specification, however future research directions [17] aim to give more flexibility in this regard.

7   Conclusions and Future Work

Onion routing-based approaches (e.g., Tor) can leverage multi-path capabilities as a means of enhancing the users' experience through performance improvement. To investigate these capabilities, we have presented a comprehensive analysis and evaluation of multi-path onion routing approaches regarding their design choices and realizations. By using the proposed taxonomy, we presented important guidelines to be followed not only for future multi-path onion routing-based designs, but also for other types of anonymization systems.

For future multi-path designs, greater performance improvements are expected if the current congestion estimation mechanisms can be refined to reflect the actual transport layer congestion into the multi-path layer. Furthermore, other aspects such as anonymity and load balancing should be taken into consideration when designing the multi-path circuit structure and scheduling mechanisms. We notice that for some attacks (e.g., *guard fingerprinting*) a considerable modification in the current node selection strategy is needed to guarantee a level of anonymity. Meanwhile, for other attacks such as *website fingerprinting* a quantitative analysis of their impact is required.

In future work we also plan to address cross-circuit interference, which is a significant problem in Tor, with mitigation techniques that often affect anonymity. We plan to analyze trade-offs between using different subsets of the mechanisms that TCP offers on the OR-link layer and specifically look into alternative congestion control methods. We want to improve performance while still avoiding network congestion, and also protect anonymity by not introducing end-to-end feedback and so opening additional attack vectors.

Acknowledgments. We thank the anonymous reviewers for their useful comments, and all authors of the evaluated approaches for providing their source code and assistance. This research was funded by the Luxembourg National Research Fund (FNR) within the CORE Junior Track project PETIT.

References

1. The Tor Relay Guide (2018). https://trac.torproject.org/projects/tor/wiki/TorRelayGuide
2. Akhoondi, M., Yu, C., Madhyastha, H.: LASTor: a low-latency as-aware tor client. In: Symposium on Security and Privacy (S&P). IEEE, San Francisco, May 2012
3. AlSabah, M., Bauer, K., Elahi, T., Goldberg, I.: The path less travelled: overcoming Tor's bottlenecks with traffic splitting. In: De Cristofaro, E., Wright, M. (eds.) PETS 2013. LNCS, vol. 7981, pp. 143–163. Springer, Heidelberg (2013). https://doi.org/10.1007/978-3-642-39077-7_8
4. AlSabah, M., et al.: DefenestraTor: throwing out windows in Tor. In: Fischer-Hübner, S., Hopper, N. (eds.) PETS 2011. LNCS, vol. 6794, pp. 134–154. Springer, Heidelberg (2011). https://doi.org/10.1007/978-3-642-22263-4_8
5. AlSabah, M., Goldberg, I.: PCTCP: per-circuit TCP-over-IPsec transport for anonymous communication overlay networks. In: Proceedings of the 2013 ACM SIGSAC Conference on Computer & Communications Security. ACM, Berlin, November 2013
6. AlSabah, M., Goldberg, I.: Performance and security improvements for tor: a survey. ACM Comput. Surv. (CSUR) **49**, 32:1–32:36 (2016)
7. Das, A., Borisov, N.: Securing anonymous communication channels under the selective DoS attack. In: Sadeghi, A.-R. (ed.) FC 2013. LNCS, vol. 7859, pp. 362–370. Springer, Heidelberg (2013). https://doi.org/10.1007/978-3-642-39884-1_31
8. Dingledine, R., Mathewson, B.: Tor protocol specification (2018). https://gitweb.torproject.org/torspec.git?a=blob_plain;hb=HEAD;f=tor-spec.txt
9. Dingledine, R., Mathewson, N., Syverson, P.: Tor: the second-generation onion router. In: 13th conference on USENIX Security Symposium. USENIX Association, San Diego, CA, USA, August 2004
10. Dingledine, R., Murdoch, S.: Performance improvements on Tor or, why Tor is slow and what we're going to do about it (2009). http://www.torproject.org/press/presskit/2009-03-11-performance.pdf
11. Feigenbaum, J., Johnson, A., Syverson, P.: Preventing active timing attacks in low-latency anonymous communication. In: Atallah, M.J., Hopper, N.J. (eds.) PETS 2010. LNCS, vol. 6205, pp. 166–183. Springer, Heidelberg (2010). https://doi.org/10.1007/978-3-642-14527-8_10
12. Ghosh, M., Richardson, M., Ford, B., Jansen, R.: A TorPath to TorCoin: proof-of-bandwidth altcoins for compensating relays. In: 7th Workshop on Hot Topics in Privacy Enhancing Technologies (HotPETs), Amsterdam, Netherlands, July 2014
13. Handley, M., Bonaventure, O., Raiciu, C., Ford, A.: TCP extensions for multipath operation with multiple addresses. RFC 1654 (1995)
14. Hayes, J.: Traffic confirmation attacks despite noise. In: Understanding and Enhancing Online Privacy Satellite Workshop of NDSS, San Diego, CA, USA, April 2016
15. Hayes, J., Danezis, G.: Guard sets for onion routing. Proc. Priv. Enhancing Technol. **2015**(2), 65–80 (2015). https://content.sciendo.com/view/journals/popets/2015/2/article-p65.xml

16. Hayes, J., Danezis, G.: k-fingerprinting: A robust scalable website fingerprinting technique. In: 25th USENIX Conference on Security Symposium. USENIX Association, Austin, August 2016
17. Imani, M., Barton, A., Wright, M.: Guard sets in Tor using AS relationships. Proc. Priv. Enhancing Technol. **2018**(1), 145–165 (2018). https://content.sciendo.com/view/journals/popets/2018/1/article-p145.xml
18. Jansen, R., Geddes, J., Wacek, C., Sherr, M., Syverson, P.F.: Never been KIST: Tor's congestion management blossoms with kernel-informed socket transport. In: Proceedings of the 23rd USENIX Security Symposium. USENIX Association, San Diego, August 2014
19. Jansen, R., Hopper, N.: Shadow: running Tor in a box for accurate and efficient experimentation. In: Proceedings of the Network and Distributed System Security Symposium (NDSS). Internet Society, August 2012
20. Jansen, R., Johnson, A., Syverson, P.: LIRA: lightweight incentivized routing for anonymity. In: 21st Annual Network and Distributed System Security Symposium (NDSS), February 2013
21. Jansen, R., Miller, A., Syverson, P., Ford, B.: From onions to shallots: rewarding tor relays with TEARS. In: 7th Workshop on Hot Topics in Privacy Enhancing Technologies (HotPETs), Amsterdam, Netherlands, July 2014
22. Jansen, R., Syverson, P., Hopper, N.: Throttling Tor bandwidth parasites. In: 21st USENIX Conference on Security Symposium. USENIX Association, Bellevue, August 2012
23. Karaoglu, H., Akgun, M., Gunes, M., Yuksel, M.: Multi path considerations for anonymized routing: challenges and opportunities. In: 5th International Conference on New Technologies, Mobility and Security (NTMS), pp. 1–5. IEEE, May 2012
24. Katti, S., Cohen, J., Katabi, D.: Information slicing: anonymity using unreliable overlays. In: 4th USENIX Conference on Networked Systems Design and Implementation (NSDI). USENIX Association, Cambridge, April 2007
25. Landsiedel, O., Pimenidis, A., Wehrle, K., Niedermayer, H., Carle, G.: Dynamic multipath onion routing in anonymous peer-to-peer overlay networks. In: 50th Annual IEEE Global Telecommunications Conference (GLOBECOM), pp. 64–69. IEEE, Washington, DC, November 2007
26. Loesing, K., Murdoch, S., Jansen, R.: Evaluation of a libutp-based Tor datagram implementation. Technical report 2013-10-001, The Tor Project (2013)
27. Murdoch, S.: Comparison of Tor datagram designs. Technical report (2011)
28. Panchenko, A., et al.: Website fingerprinting at internet scale. In: 23rd Annual Network and Distributed System Security Symposium (NDSS). Internet Society, San Diego, February 2016
29. Reardon, J., Goldberg, I.: Improving Tor using a TCP-over-DTLS tunnel. In: 18th Conference on USENIX Security Symposium, pp. 119–134. USENIX Association, Montreal, August 2009
30. Ries, T., Panchenko, A., State, R., Engel, T.: Comparison of low-latency anonymous communication systems - practical usage and performance. In: Ninth Australasian Information Security Conference (AISC) (2011)
31. Rimmer, V., Preuveneers, D., Juarez, M., Goethem, T.V., Joosen, W.: Automated website fingerprinting through deep learning. In: 25th Annual Network and Distributed System Security Symposium (NDSS). Internet Society, San Diego, February 2018
32. Serjantov, A., Murdoch, S.J.: Message splitting against the partial adversary. In: Danezis, G., Martin, D. (eds.) PET 2005. LNCS, vol. 3856, pp. 26–39. Springer, Heidelberg (2006). https://doi.org/10.1007/11767831_3

33. Snader, R.: Path selection for performance- and security-improved onion routing. Ph.D. thesis, University of Illinois at Urbana-Champaign (2010)
34. Snader, R., Borisov, N.: A tune-up for Tor: improving security and performance in the Tor network. In: 16th Annual Network and Distributed System Security Symposium (NDSS), February 2008
35. Snader, R., Borisov, N.: Improving security and performance in the Tor network through tunable path selection. Trans. Depend. Secure Comput. **85**, 728–741 (2011)
36. Tang, C., Goldberg, I.: An improved algorithm for tor circuit scheduling. In: 17th ACM Conference on Computer and Communications Security, pp. 329–339. ACM, Chicago, October 2010
37. Titz, O.: Why TCP over TCP is a bad idea. http://sites.inka.de/bigred/devel/tcp-tcp.html. Accessed 16 Nov 2018
38. Tschorsch, F., Scheuermann, B.: Tor is unfair - and what to do about it. In: 36th Conference on Local Computer Networks. IEEE, Bonn, October 2011
39. Tschorsch, F., Scheuermann, B.: Mind the gap: towards a backpressure-based transport protocol for the tor network. In: 13th Usenix Conference on Networked Systems Design and Implementation (NSDI), pp. 597–610. USENIX Association, Santa Clara, March 2016
40. Tschorsch, F., Scheurmann, B.: How (not) to build a transport layer for anonymity overlays. In: Proceedings of the ACM Sigmetrics/Performance Workshop on Privacy and Anonymity for the Digital Economy. ACM, New York, June 2012
41. Viecco, C.: UDP-OR: a fair onion transport design. In: 1st Workshop on Hot Topics in Privacy Enhancing Technologies (HotPETS), Leuven, Belgium, July 2008
42. Wang, T., Bauer, K., Forero, C., Goldberg, I.: Congestion-aware path selection for Tor. In: Keromytis, A.D. (ed.) FC 2012. LNCS, vol. 7397, pp. 98–113. Springer, Heidelberg (2012). https://doi.org/10.1007/978-3-642-32946-3_9
43. Wang, T., Cai, X., Nithyanand, R., Johnson, R., Goldberg, I.: Effective attacks and provable defenses for website fingerprinting. In: Proceedings of the 23rd USENIX Security Symposium. USENIX Association, San Diego, August 2014
44. Yang, L., Li, F.: mTor: a multipath tor routing beyond bandwidth throttling. In: IEEE Conference on Communications and Network Security (CNS), pp. 479–487. IEEE, Florence, September 2015

Distributed Systems

Shoal: Query Optimization and Operator Placement for Access Controlled Stream Processing Systems

Cory Thoma[(⊠)], Alexandros Labrinidis, and Adam J. Lee

Department of Computer Science, University of Pittsburgh, Pittsburgh, USA
{corythoma,labrinid,adamlee}@cs.pitt.edu

Abstract. Distributed Data Stream Processing Systems (DDSPS) execute on transient data flowing through long-running, continuous, streaming queries, grouped together in query networks. Often, these continuous queries are outsourced by the querier to third-party computing platforms to help control the cost and maintenance associated with owning and operating such systems. Such outsourcing, however, may be contradictory to a data provider's access controls as they may not permit their data to be viewed or accessed by an unintended third party. A data provider's access controls may, therefore, prevent a querier from fully outsourcing their query. Current research in this space has provided alternative access control techniques that involve computation-enabling encryption techniques, specialized hardware, or specialized query operators that allow for a data provider to enforce access controls while still allowing a querier to employ a third-party system. However, no system considers access controls and their enforcement as part of the *query optimization step*. In this paper, we present Shoal, an optimizer that considers access controls as first class citizens when optimizing and distributing a network of query operators. We show that Shoal can generate more efficient queries versus the state-of-the-art, as well as detail how changes in access controls can generate new query plans *at runtime*.

1 Introduction

The ever-increasing and ever-changing size, speed, and availability of accessible data has led to the rise of new outsourced data processing paradigms. One such paradigm is Data Stream Processing handled by *Distributed Data Stream Processing Systems* (DDSPSs). A DDSPS handles data on-the-fly by executing on transient data with long-running continuous computations (queries), such as streaming operations, map-reduce functions, or user-defined functions, etc. These computations are often outsourced to third-party systems that handle data processing and execution.

Outsourcing computation is desirable for the querier as it provides them with cost savings. For instance, the querier need not maintain expensive hardware

© IFIP International Federation for Information Processing 2019
Published by Springer Nature Switzerland AG 2019
S. N. Foley (Ed.): DBSec 2019, LNCS 11559, pp. 261–280, 2019.
https://doi.org/10.1007/978-3-030-22479-0_14

and software platforms. Further, the cloud provider offers guarantees on uptime and service availability that a querier can rely on. Finally, the querier can take advantage of a cloud provider's ability to scale to meet demand, by allocating new resources or freeing up underused ones. When a querier contracts a third-party cloud service provider, they are often able to optimize their query to take advantage of different third-party offerings and pricing models. In doing so, they are able to improve the efficiency of their query for some measurable metric (e.g., latency, throughput, monetary cost) by taking advantage of location, current pricing, current load, and other factors *at runtime* by changing the placement of certain components (i.e., operators) of their queries. This allows queriers to freely move queries around to improve some aspect of the query's performance.

When a data provider dictates access controls over their streaming data, however, a querier may lose some of these freedoms. For instance, if a data provider authors an access control policy that removes a third-party altogether, the querier would lose the ability to execute any part of their streaming query on that provider. Similarly, if access controls are enforced using some cryptographic method (e.g., Polystream [31] or Streamforce [3]), the querier may experience degraded performance as different permissions require different encryption schemes, each incurring different overheads. In both cases, a querier stands to lose some of the benefits of hosting their queries on a third-party system, and may even be required to host the query themselves.

Queriers must consider access controls when trying to generate and optimize their queries. For instance, when a query is broken down into different operators, some operators may be able to execute *directly* on plaintext, but may have to execute on a querier-maintained machine, whereas others may be more costly (monetarily, in terms of latency, or otherwise) but can be executed on a third-party system. A querier must now be able to reason about and decide which implementation of an operation to choose given the accesses they have been provided. This implies that a querier must be able to enumerate potential operators and consider them *at query optimization time* to ensure that the most efficient query plan is derived given each data providers' access controls. Further, when a data provider changes their access controls, query networks may need to be updated.

Currently, DDSPS optimizers and related work have explored DDSPSs optimization in a limited scope. Some have simply focused on better utilization of the underlying computation hardware alone [17], while others have focused on the underlying network alone [7,13,27,29]. Several optimizers and systems have focused on the impact of data variability on the system in the presence of access controls [3,31] and in the impact of data stream rates and selectivities [2,6]. Currently, there is no system that focuses on optimizing queries based on the underlying access controls from different data-providers. Further, the closest related work focuses on the use of an *optimize-then-place* approach in which a user's query is first optimized for non-distributed execution and then post-processed for placement on distributed resources. Finally, related work with enforcing access controls in a DDSPS focuses on a single query, which may not suit a querier as they may query many data providers.

We present an optimizer that considers a querier's access privileges at optimization time to produce a high-quality *placement* and *ordering* of individual streaming operators. Our proposal, *Shoal*,[1] uses a dynamic programming algorithm to guarantee optimal placement and orderings for moderate-sized sets of streaming queries on a DDSPS, and includes a heuristic approach for larger query networks. Shoal combines the ordering and placement steps to take advantage of the underlying system by considering multiple orderings on various distributed computation infrastructures, and avoids the pitfalls of the optimize-then-place approach. To this end, we make the following contributions:

- We show the optimize-then-place approach to be a sub-optimal approach for computing operator placement in a DDSPS.
- We introduce the first cost model for distributed streaming queries that leverages parallelism inherent to the DDSPS and accounts for key sources of heterogeneity such as fluctuations and changes in the underlying data streams and the underlying system and network.
- We detail an optimization algorithm that can execute both at query initialization as well as when a change in the system is detected. This algorithm only optimizes the parts of any queries that are potentially affected by a change in access controls. By only considering parts that are affected by system changes, this online algorithm is able to quickly re-optimize and recover while maintaining an optimal solution.
- We run an extensive evaluation of our algorithms and compare to several baseline, state-of-the-art, and optimize-then-place algorithms. We show that our proposed framework can produce higher quality optimization and placement plans (up to 2.2x better) with reasonably low overheads, and further show these plans are of a higher quality when compared to related work.

The remainder of this paper is organized as follows. We present the system model in Sect. 2 and formalize our problem statement in Sect. 3. We describe our proposed approach in Sect. 4 and present results form our experimental evaluation in Sect. 5. Section 6 summarizes related work. We conclude in Sect. 7.

2   Background and System Model

This section overview the features Shoal considers when optimizing. We further detail our system model and overview access control frameworks and their affect on a DDSPS.

2.1   Background on DDSPSs

Shoal uses a common Distributed Data Stream Processing System (DDSPS) model. DDSPSs separate the data provider from the data consumer, and often

---

[1] A shoal is a heterogeneous group of fish that is organized to function towards a common purpose, typically of a safety or social nature.

separate the data processing machines as well. DDSPSs rely on long running, continuous computations that execute on transient data. Once a DDSPS has processed the transient data, results can be stored or forgotten depending on the computation being done.

DDSPSs can implement many different stream-processing paradigms such as relational data stream processing [2,4,5], MapReduce [19], and user-defined tasks. Relational data stream processing systems use continuous SQL-like queries that are comprised of streaming operations that execute a single task. Similar to traditional database management systems, some operations require large amounts of information to produce their result (e.g., a join or aggregation). Since data is transient in a DDSPS, data is grouped by *windows* and *slides*: a window represents how much data to keep (e.g., 100 tuples, 10 min, etc.) and a slide represents how often to update the querier (e.g., every 10 s, every 500 tuples, etc.).

The system components of a typical DDSPS are listed below.

- **Data Providers** provide streaming data (subject to access controls) to the system.
- **Sites** are third-party computational and storage platforms (such as Amazon EC2 or Microsoft Azure). They are tasked with the execution of streaming operations as well as forwarding data.
- **Data Consumers** author and submit streaming queries to the system. These queries can be of a variety of paradigms such as relational streaming continuous queries, map-reduce computations, or user-defined functions.

2.2   Access Controls

In a DDSPS, access control enforcement becomes difficult since the data providers can not control the propagation of their data after it is transmitted. The current literature on enforcing access controls in a DSMS can be grouped into two categories: trusted third-party enforcement and untrusted enforcement. Trusted third-party enforcement techniques work by trusting that a computational site will enforce access to their data on their behalf. Such systems either work with special operators [9–11] or by re-writing queries [20,23] so that access can be limited.

Systems that do not trust third-party enforcement will rely on cryptographically enforced access controls. Rather than forcing a querier to process data only after it has been decrypted, systems like PolyStream [31] and Streamforce [3] allow the data provider to use specialized computation-enabling encryption techniques to enable third-party computation for a querier *directly on encrypted data*. These systems, however, limit the expressiveness and accessibility of a queriers' potential query. In Streamforce, a querier may only access *integer* data via a *view-like* format, (i.e., only allowing filtering and aggregations on numeric data). PolyStream supports a richer set of query operations than Streamforce, but cannot support join or complex user-defined functions over streams from multiple providers. Furthermore, these systems also leak information about the underlying plaintext values, such as equality, relative partial ordering, or relationships between groups of tuples (e.g., the encrypted aggregate of some encrypted data). In either the fully-trusted third-party or the untrusted third-party scenario, access control enforcement comes with computational overheads that must be properly accounted for when optimizing and placing a query.

3   Problem Description

In this section, we detail the exact optimization problem addressed by our framework and define the different components of a Distributed Data Stream Processing System (DDSPS). We offer a description of our optimization approach and show the optimize-then-place approach to be suboptimal.

3.1   Problem Description

In order to properly define the problem being addressed here, we must first formalize the required components. There are three main components to optimizing and placing a data consumer's computation: *sites, operations, queries,* and *query networks.*

Definition 1. *Site: As introduced in Sect. 2, a site s executes operators. Interconnections between sites have* bandwidth *(in bits/s, where tuples can be of varying size) and* latency *(in ms) characteristics that we represent as $b(s_1, s_2)$ and $l(s_1, s_2)$ respectively. Sites are associated with the following properties:*

- *s.cap is the site's processing capacity (in cycles, translated to tuples/s).*
- *s.name is the site's name used for unique identification purposes.*
- *s.per is a set of permissions $\{<o, f>|$ site s can execute a physical operator o on field f\}.*

Definition 2. *Operation: An operation op is a set of operators that execute the same task via different physical implementations.*

- *op.type represents the action to be performed on a data stream (e.g., filter, projection, summation, top-k, etc.).*
- *op.args includes metadata about the operations such as the join condition or selection criteria.*
- *op.input represents the set of fields required for this operation to execute.*
- *op.output represents the set of fields in the output of this operation.*
- *op.id is a unique identifier for the operation.*

A typical operation can be a filter over someone's age, a join to match two streams, or an aggregation to find the maximum profit in a given window of time. Operations can be implemented using different techniques, represented as operators.

Definition 3. Operator: *The basic computational unit used in Shoal is an operator o, which has the following properties:*

- *o.s represents the expected or actual selectivity of the operation. Selectivities can be derived either by estimation, measurements during a warm-up period, or historical selectivity data.*
- *o.c represents the cost of the operation in terms of the latency for computing on one tuple. It can be calculated in a manner similar to the selectivities.*
- *o.site represents the site an operator has been assigned to.*
- *o.opId represents the ID of the operation (that is to say, the logical operator that this physical operator represents) that this operator implements.*
- *o.window represents the window size for a stateful operator either in tuples or ms, with a default of 0.*

Operators allow flexibility when implementing an operation. Consider the potential difference between a hash-join implementation of a join operation versus a merge-join implementation. Given the input rate and selectivity of each stream, it is highly likely that one join would outperform the other in terms of overall latency. An operation would have the merge-join and hash-join as potential operators, and each operator would have a cost that can be used to better optimize the query network.

Definition 4. Query: *A query is represented as a set q of operations that describe the query or task that a data consumer wishes to execute over a set of data streams.*

- *leaves(q) returns the set of operations that operate on a raw data stream (i.e., do not require the output of another operation to execute).*

Definition 5. Query Network: *A query network is represented as a set qn of queries that will execute within the DSMS.*

- *sinks(qn) returns the set of operations that return a result to a querier (i.e., the last part of any one query).*

Using a query, permissions, and a set of available sites as input, Shoal produces a *plan* as output:

Definition 6. Plan: *A plan $p = (V_o, E_o)$ where V_o is a set of physical operators and $E_o \subset V_o \times V_o$ is the edge set linking the outputs of one operator to the inputs of adjacent operators.*

Definition 7. Satisfiability: *A plan p satisfies a query network qn if:*

- $\forall op \in qn, \exists! \, o \in V_o \text{ s.t. } o.opId = op.id$
- $\forall o \in V_o, \exists! \, op \in qn \text{ s.t. } op.id = o.opId \wedge \, < o, o.metadata > \in o.site.per$
- $\forall o \in p, o.input \subseteq \bigcup_{o' \mid <o', o> \in E_o} o'.output$

That is, each operation in each query that comprises the query network has a unique operator in the plan and that each operator in the plan is the implementation of one operation in the query, and that each operation in the query is represented in the resulting network. Additionally, each operator's input must be part of the output of its immediate predecessor, and each operator must be permitted to execute on its assigned site.

To determine the relative quality of a given plan p, we use the following cost model.

Definition 8. *Cost:* *For a plan p of a Query Network qn, the cost of p, starting at the leaf node(s), is determined by:*

$$\max_{path \in \mathsf{Paths}(p)} \mathsf{pathCost}(path) \tag{1}$$

where Paths is the set of paths from leaf nodes to sink nodes. The expected input rate for each operator (starting from the initial input rate of the leaf node from the source stream) is:

$$\mathsf{ir}(o_i) = IR_{qn} * \prod_{op_j \in \mathsf{pathUpTo}(op_i)} op_j.s \tag{2}$$

The function $\mathsf{pathUpdTo}(o)$ for an operator o is the ordered subset of operators that precede o in the plan (as part of the same query). IR_{qn} is the maximum input rate of any leaf operator on the current path. The cost of a path is:

$$\mathsf{pathCost}(path) =$$
$$\sum_{op_i \in path} \max(op_i.c, op_i.window) + \mathsf{Latency}(op_i, path) * \mathsf{Penalty}(o_i) \tag{3}$$

The penalty is defined as:

$$\mathsf{Penalty}(o) = \begin{cases} 1, & \text{if } \mathrm{pr}(o, o.site) > \mathrm{ir}(o_i) \\ \frac{\mathrm{ir}(o)}{\mathrm{pr}(o, o.site)}, & \text{otherwise} \end{cases} \tag{4}$$

The function $\mathrm{pr}(operator, site)$ determines what the processing rate of a site would be with the operator o assigned to it. If this processing rate is greater than the input rate, then there is no penalty. If the processing rate is insufficient to handle the input rate, the plan is penalized by the input rate over the processing rate. Latency is computed as:

$$Latency(o, path) = \begin{cases} 0, & \text{for } o \in leaves(qn) \\ \mathrm{l}(op, op_{i-1}), & \text{for } op_{i>1} \end{cases} \tag{5}$$

Constrained by $s.cap$ and $bandwidth(op_i, op_{i-1})$.

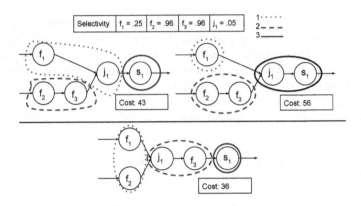

Fig. 1. Simple continuous query.

Definition 9. *Problem: Given a query network qn of queries, a set of Sites s, and a set of access control permissions per, produce a plan p that satisfies qn such that Eq. 1 is minimized.*

3.2 Optimize-then-place Approach

A reasonable first step solution to this optimization problem would be to separate optimization from operator placement. This would allow a placement algorithm to simply use an existing off-the-shelf optimizer and post-process the result for placement. This approach, however, can lead to a sub-optimal plan even if the query itself is fully optimized. Consider the following scenario for a continuous query optimizer, which we will use throughout the remainder of the paper to illustrate Shoal:

A simple query contains three filters (f_1, f_2, f_3), a join (j_1), and a projection (s_1) as depicted in the top of Fig. 1 as the result of an optimization step. Next consider the three sites available for placement, and assume the network cost is uniform. Each site has a capacity of 10 (unit-less for simplicity). The cost of each filter is 4, of the join 6, and of the project 2; each have selectivities shown in Fig. 1. Given these costs, either the join must be co-located with f_1 (the top of Fig. 1) or separated from all of f_1, f_2 and f_3, (the middle of Fig. 1). However, notice that the selectivity of f_1 combined with f_2 is .92, meaning that a tremendous amount of data is being sent over the network to the join. The selectivity of the join, however, is far lower at .05, meaning that a smaller amount of data is being produced. If the query was instead optimized so that f_3 were to follow the join, the overall network cost would be substantially reduced, resulting in a higher quality plan (78.8 ms vs. 67.8 ms with Eq. 1). This illustrates the need for the optimization and placement steps to be considered simultaneously.

Algorithm 1. DynamicProgramming

```
 1: DynamicProgramming(sc, perms, sites)
 2: optPlace= new Array(ArrayList(plan))
 3: for leaf ∈ leaves(sc) do
 4:    optPlace[0] = ∅                                    ▷ Initialize Empty list at level 0
 5:    for s ∈ sites do
 6:       for l ∈ operators(leaf) do
 7:          if s.cap ≥ l.c && (< l, l.metadata.field >∈ s.perms) then
 8:             l.site = s
 9:             optPlace[0].add(new plan(l))              ▷ Capacity kept per plan
10:    prunePlans(optPlace[0])
11: for lv = 1...|sc| do
12:    optPlace[lv] = ∅                                   ▷ Initialize Empty list at level lv
13:    for operation ∈ sc | operation.type = join do
14:       for plan1, plan2 ∈ optPlace[lv − 1] | (operation.input ⊆ (plan1.output ∩ plan2.output)) ∧
          operation.opId ∉ plan1.opIds ∧ operation.opId ∉ plan2.opIds do
15:          for s ∈ sites do
16:             for join ∈ allowableOps(operation) do
17:                if (updateCapacity(plan1, plan2, s) ≥ join.c) ∧ (< join, join.metadata.field >∈
                   s.perms) then
18:                   join.site = s
19:                   optPlace[lv].add(joinPlans(plan1, plan2, join))
20:       for plan ∈ optPlace[lv − 1] do
21:          for operation ∈ sc | (operation.type ≠ join)∧(operation.opId ∉ plan.opIds) ∧ (operation.
             input ⊆ plan.output) do
22:             for s ∈ sites do
23:                for o ∈ allowableOps(operation) do
24:                   if plan.s.cap ≥ o.c∧ < o, o.metadata.field >∈ s.perms then
25:                      o.site = s
26:                      optPlace[lv].add(combine(plan, o))
27:    prunePlans(optPlace[lv])
```

4   The Shoal Optimizer

In this section, we introduce our optimization and placement algorithms.

4.1   Online Optimization Approach

Given the long-running nature of continuous queries, there is a high chance (essentially a certainty) that a data provider's access controls will change over time, requiring re-optimization of the network of streaming queries currently deployed. When access controls change for any *one* data streaming operator, it could possibly have a ripple effect for other downstream operators as they may need to be moved to reallocate resources. To accommodate these changes, there are two possible approaches: *stop-the-world* and *on-the-fly*. The stop-the-world approach simply halts query execution and uses Algorithm 1 to re-optimize the query from the root nodes. This approach, however, can lead to large re-optimization times for larger query networks, and can end up doing repetitive work when a relatively small set of operations are affected by the change.

We introduce an *on-the-fly* approach to mitigate these overheads. The principle behind our on-the-fly approach is to execute Algorithm 1 from the operator that is first affected by the access control update relative to the data providers of the overall query network, which we will call the *first-impacted*. This requires the

---

Algorithm 2. Access Control first-impacted identifier.

```
 1: ACUpdate(Update u, Plan p)
 2: cld = sc.leaf                                    ▷ Operations not processed
 3: for o ∈ cld do
 4:   if o.input | u.protectedFields then
 5:     return p.levelOf(o)
 6:   else
 7:     cld.add(p.childrenOf(o))
 8:   cld.remove(o)
```
---

ability to determine the first-impacted, which depends upon the type of access control update that occurred.

Access Control. Algorithm 2 determines which operators are first-impacted by a change in access controls. It starts by adding operations that directly access raw data streams on Line 2 to the current query network, *cld*. These operators are then looped through on Line 3, and Line 4 determines if that operator accesses the data being protected by the new access control update. If so, this operation is the first-impacted and the algorithm determines its level by asking the plan for the level. If the operation does not access the protected data, its children are added to the *cld* set, and it removes itself from this set. This continues until the first-impacted is found. At this point, Algorithm 1 will execute on the *descendant* children of the first-impacted, as well as all operations at the same level and their descendants. Note that on Lines 16 and 23, we check for all allowable operations. Recall that one physical operation can be implemented by many physical operations (e.g., the querier may have sufficient access to query in plaintext local to their machine and further have access to use a computation-enabling encryption scheme such as an order preserving scheme on encrypted data in the cloud. This function enumerates the possible operators based on the current permissions of the querier. This leaves the already optimized operations and their ancestor operations intact from the previous plan, and re-optimizes the operations at and after the first-impacted's level, leading to less optimization time. The only alteration required for Algorithm 1 is the inclusion of the current plan from which to start, which is simply placed in the *optimalPlans* set and the Algorithm starts from Line 11 where the level is determined by traversing back to the leaf nodes.

4.2   Greedy and Hybrid Approaches

As with traditional dynamic programming optimizers, our algorithm could suffer from prohibitively large execution times for large or complicated query networks (explored further in Sect. 5). When query networks become too large or complex, we defer to a greedy approach. This approach simply considers one operator at a time and optimally places it. In the base case where each operator needs to be placed, the user defines a time threshold $t_{offline}$ for their optimization step. If the dynamic programming approach is expected to exceed $t_{offline}$, then the greedy approach is used. The online approach poses a different problem because there may be uncertainty in how costly an update may be to the system (i.e., the number of operators that need to be re-optimized).

The larger the number of operators that need to be considered, the greater the number of operators requiring re-optimization, and therefore the greater the cost of the update. In a system operating at or near capacity, online updates may end up hindering the quality of the result as some information may be lost during optimization, especially for costly updates on large overall query networks. To combat this problem, we use the greedy approach when updates are too costly relative to the system load. The greedy approach simply re-optimizes, placing each operator in the most optimal location, in a quick but likely non-optimal fashion.

The greedy approach lends itself nicely to distributed systems with heavy load where re-optimization needs to be quick to avoid losing data, but it will not produce plans of the same quality as the dynamic programming approach. To help a data consumer determine which to use, we propose a *hybrid* solution which automatically determines which approach to use given the current system state. The determination is based on three factors relative to the overall streaming query network submitted by the data provider: (a) buffer capacity, (b) processing time of a single streaming tuple (end-to-end), and (c) the input rate. When an update is deemed necessary, its cost c in seconds is determined by multiplying the number of operations needing to be re-optimized by the average amount of time to optimize one operator (based on the execution time of optimizing the entire query, or a running average). Then, the following equation is used to determine which algorithm to use:

$$uc = \left(\sum_i^{o \in p} (b_i * t_i) \right) + ir_o * c \tag{6}$$

where o is the operator in the plan p, b_i is the utilized buffer size of o, t_i is the processing time of o, ir is the input rate. If $c < uc$ then the dynamic programming approach is used, otherwise the greedy approach is used to minimize data loss.

4.3 Example

To help illustrate how Shoal optimizes a set of streaming queries, consider the following continuous query on a data stream that contains tuples with a timestamp, companyName, companyId, and the company profit, as illustrated in Fig. 2:

```
SELECT max(avg_profit), companyName
FROM (SELECT companyName, AVG(profit) as avg_profit
    FROM profitStream GROUP BY companyName EVERY 1m UPDATE 15s;)
GROUP BY companyName EVERY 1m UPDATE 15s;
```

This query requires five operations; a max (m), a projection (p), an average (a), and two group-by operations (g_1 and g_2) represented by circles in Fig. 2. Assume that the profit field is protected by a homomorphic encryption and the

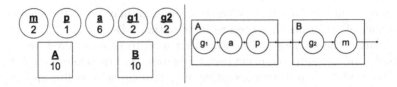

Fig. 2. Given a the set of operations and the sites A and B, Shoal optimizes and places the operations so that the first aggregation is placed on A with the projection reducing network load to the second aggregation operation placed on B.

others are plaintext. Further assume (for simplicity) that there are two sites A and B with capacities 10 and 10 respectfully, and a latency of 10ms between them (squares in Fig. 2).

Shoal starts with the operation a as it is the operation that accesses raw data. Since a homomorphic option exists for the aggregation, an operator executing a homomorphic scheme is put onto each site and the next round of dynamic programming is initiated. Further, plans are also added for random encryption and trusted machine processing. This aggregation requires a group-by operation, which can execute on the plaintext column for "company name". This operation is placed with the aggregation on each site, making each site's best plan having a cost of 8 which, along with other plans with varying physical operators, are kept for each site. Shoal then tests the remaining operations and determines that the projection p can be added to each site's best plan for a cost of 9. Plans are now kept for each site and for each physical operator, but the minimum plan score is 9. Note that this choice reduces the overall network load by eliminating all columns except the company name and the average. Shoal continues and determines that the maximum operation along with its group-by can not fit on either site and chooses them to operate on site B with site A keeping its previous plan. With all operators placed, the new plan resembles the one in the right half of Fig. 2.

5   Evaluation

To evaluate our optimizers, we decided to use relational continuous queries for the bulk of our experimentation.

Setup: For our evaluation, we limit Shoal to be used in a simple streaming system with data providers, data consumers, and data processing components. For our simulation, data is streamed from a laptop into Amazon AWS EC2 instances. Once data is processed, it is passed back to the laptop to act as the data consumer. We implement the streaming layer on the Apache Storm [30] framework. To keep the streaming layer simple, we use the most basic functionality of Storm where our data provider implements a *spout* and our data processing nodes implement *bolts* with no multi-threating or replication (i.e., a bolt just mimics a machine for our purposes). We use Storm *only* for the transport layer as it guarantees delivery and provides acking and nacking functionality. To simulate real-world streams, each stream is imposed with an artificial latency of 0–30 ms to emulate them being geographically separated.

Datasets: We use queries from the TPC-H [25] workload and modify them for use in a streaming system (e.g., aggregations use windows). We will explicitly call out any changes to the query we made, or if we use more than one query as part of the query network. We further segment data based upon a timestamp so that it is streamed into the system (in a pre-processing step) so that days are equivalent to minutes. All queries are referenced using the query number (e.g., q1 for TPC-H query 1) and the number of operators it translates to (e.g., q1(4) is TPC-H query 1, which has four operations).

Baseline Algorithms: In addition to our original and hybrid dynamic pro- graming algorithms, we chose three additional baselines for comparison: (1) *all- on-client*, where all of the operations run on one machine, (2) *first site*, where each operation is placed on the first site available, and switched to the next when either the site is at capacity or there is a conflict with access controls, and (3) *greedy*, where a plan is generated by greedily assigning each operation based on the best score.

5.1   Online Optimizer

This section evaluates our dynamic programing optimizer as compared to other baseline approaches. The cost of an update is based on how many operations in the query network are affected by the update, so we omit cases where the entire network was updated since it would degenerate to the basic case where each operator must be optimized.

Optimization Time. Given the cost of an update, this experiment determines the average optimizer execution time for the our dynamic programming approach as well as the baseline approaches.

Configuration: We combine queries in increasing size order (i.e., 1 query, 2 queries, 3 queries, up to 4 queries, or 8, 14, 28, and 45 operators). This provides four data points with an increasing size and number of sinks. All aggregation and join operations are given windows of 5 min (to directly use the date field in each relevant tuple). We trigger updates so that only a certain number of operators in each query are affected by the update. Each optimizer is then used to order and place the subsequent operations.

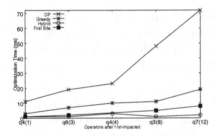
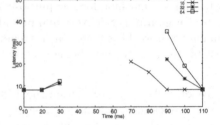

Fig. 3. Optimizer execution time for the online algorithm approaches.

Fig. 4. Recovery time caused by an access control update for different costs.

Results (Fig. 3): Here we can see the dynamic programming approach is the slowest. This optimizer execution time included the time to determine the first-impacted for each approach, for each query.

Takeaway: Although Shoal has the highest optimization time, it is still relatively low, especially when executing on a long-running continuous query in a network where the resulting plan quality is much more important.

Plan Quality. This experiment evaluates the overall plan quality of each approach in terms of latency (ms) for each updated plans. Again, we present both the expected latency, as well as the actual latency. Here, we include the hybrid approach to show when it may switch optimizers to reduce the overall impact of an update.

Configuration: Queries are executed for 10 min in total. There is a two-minute window for the initial query, after which an access control update is presented. The query is then updated and the remainder of the time is spent monitoring the updated query. The results presented below are the quality (latency in ms) of the updated query network, as presented by the number of operators updated in the largest network.

Results (Figs. 5a and b): Our dynamic programming optimizer produces the best overall latencies for both expected and actual evaluations for the query network. The difference between the expected and the actual is roughly 10.2%, which indicates that Shoal can produce results that are close to the actual values. Notice that the hybrid approach chose to switch to the Greedy optimizer in the last update to the query network. This is due to the system being near capacity when the update occurred (roughly 2,500 tuples/s with a processing rate of roughly 2,615 tuples/s), and in the time to process a new query, the system could have lost data, so the hybrid algorithm chose to use the greedy optimizer.

Takeaway: Shoal produces higher quality plans when compared to the baseline.

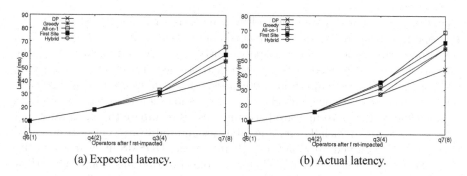

(a) Expected latency.   (b) Actual latency.

Fig. 5. Expected and Actual latency for Shoal on random data.

Recovery Time. When an update occurs, the system must determine how to re-optimize from first-impacted operation. This process takes time, and while it is processing, the query will still need to be executing. The time between the start of an update optimization and the normal execution of the resulting plan is the time it takes the system to recover from an update. In this experiment, we evaluate this *recovery time* for access control updates.

Configuration: For this experiment, we generated a 128-operator query network. Operations were selected from a random distribution of operations which included two-way joins, filters, summations, averages, projections, and decrypt-process-encrypt operations. Access control updates occur by specifying a specific change in access controls that target a specific operator such that the update cost remains consistent across the evaluation, and each update causes an increase in latency and a decrease in throughput (i.e., switch from plaintext to encrypted). A query is considered *recovered* once the latency has normalized back to a steady value.

Results (Fig. 4): When an update occurs to an access control policy (Fig. 4), the data consumer may lose access as indicated by the unreported latency values. Once the query has been resumed, the larger updates cause a large spike in latency that takes more time to recover from, as expected. Note the processing time of each update also increases, but the recovery time is more-or-less the same (20–30 ms). This shows that the new queries can handle the increased workload to make-up for the lost work and then maintain a new latency rather quickly.

5.2   Comparison to the State-of-the-Art

We now evaluate the quality of the plans produced by Shoal versus other operator placement approaches, namely Pietzuch et al. [24] and Srivastava et al. [29].

Algorithms: Pietzuch et al. [24] propose a solution that focuses on placing operators in a large-scale distributed network using a latency metric. Their optimizer takes a query plan and places it using a two-step algorithm: first a Virtual

Operator Placement step and then a Physical Operator Placement step. The virtual operator placement step considers all operators in a query and places them based on a cost space. This cost space consists of a decentralized view of the network from a single node's perspective and focuses on the latencies between potential sites. There is also a load dimension that can ensure that a single site does not become overwhelmed. Their approach allows for access control updates by allowing operators to migrate between sites. To compare to our work, we fix the cost space by artificially creating latencies and data rates between potential sites (i.e., the assumed information gathered by the DDSMS in their work) and then allow it to adjust over time. The main optimization function used in their work is to minimize the following formula:

$$\sum_{l \in L} DR(l) * LAT(l) \tag{7}$$

Where l is the link between two nodes, $DR(l)$ is the data rate of that link, and $LAT(l)$ is the latency of that link.

Srivastava et al. [29] also reduce data transmission, but do so for localized networks. Their work focuses on using parts of the query itself, as well as the machines available for placement, to make a placement decision. Specifically, they focus on the selectivity of filtering operations, the cost associated with each operation, and the cost associated with sending a tuple through the network. In addition to the above costs, a join's cost is calculated using its selectivity and the cost per unit time for processing one tuple. The cost of a placement plan is therefore the sum of all of the nodes where the selectivities of upstream filters are multiplied by the cost of the current filter. Some filters are correlated and some are not, so the ordering decision comes from the commutative aspect and the overall cost comes from minimizing the cost of the filter and join orderings. To compare with our work, we again assume an artificially created latency and use the same operators' costs and latencies across all approaches.

Configuration: For our comparison, we use multiple queries over a fixed number of sites. We use 5 sites, each connected to each other with an initial latency randomly selected from a range of 5–500 ms. Each query is comprised of between 4 and 128 operations selected as either filters (selection operations) or joins, with plaintext data. Since [24] requires an initial query plan, we use Shoal with a single site and sufficient capacity to generate a non-distributed query plan. Finally, each filter is given a selectivity randomly selected from the set $\{.1,.2,....,.9\}$. To gather information on actual latencies, each query was executed for a total of five minutes for each approach.

Results (Figs. 6a & b): As depicted in Figs. 6a and b, Shoal produces plans with better expected and actual latency. As before, the expected and actual are within an average of 8%, however the Pietzuch et al. approach is more predictable since its expected is on average only 4% different from the actual value. Shoal is able to outperform the other approaches because it attempts to find an optimal solution that takes into account the parallelism inherent to a distributed system by preferring plans that allow work to be done on multiple devices simultaneously. The Pietzuch et al. approach relies on an optimize-then-place approach and missed better filter orderings, which becomes more apparent as queries grow larger. The Srivastava et al. approach does consider ordering, but does not consider the parallelism inherent to a distributed system and would often serialize sets of operations that could have otherwise been done in parallel.

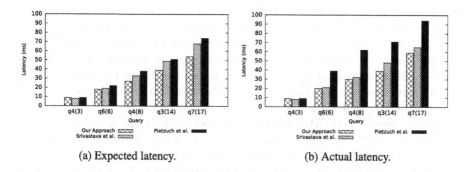

(a) Expected latency.   (b) Actual latency.

Fig. 6. Expected and Actual latency for Shoal on random data.

Takeaway: By considering ordering and placement at optimization time, as well as taking advantage of parallelization inherent to the distributed system, Shoal can out-preform other state-of-the-art optimizers in terms of end-to-end latency.

6   Related Work

Stream processing has been rigorously studied in the literature to include novel systems such as Aurora [1], Borealis [2], and Twitter Herron [18]. For traditional database applications, the focus for operator placement in distributed database systems usually focuses on replication, sharding, or scalability [12,14,15,28]. The PAQO [16] optimizer focuses on placing operators in a distributed database system so that one entity does not learn the underlying intension of the query. For data steaming systems, operator placement is of a larger concern since queries are long-running and operators are expected to consume resources for long periods of time while possibly fluctuating in their required resource utilization. The contributions in [8] explore the general problem of operator placement on heterogeneous computational platforms for DDSMs, and propose a linear programming model to place operators. Their approach processes placement in a separate step from optimization, which can lead to suboptimal results (cf. Sect. 3).

Huang et al. [17] fit operators onto sites by calculating the execution time of an operation and place it based on the capacity of each site, using end-to-end delay and throughput as the metrics. Thoma et al. [32] place operators in a DDSMS where queriers have the ability to control where operators are placed via a set of constraints. These constraints generally cover all aspects of the placement, but do not consider the access control policies of a data consumer. Operators placement using heuristics to optimize for end-to-end latency and network traffic have also been explored [7,13,27].

Finally, some related work has focused on the impact of enforcing access controls in a DDSPS. Enforcement systems such as FENCE [21,22] include the enforcement overheads in the optimization step by adding streaming operations that can be handled like any other operation, but do so without considering operator placement. Other systems will rewrite queries or alter streaming operators [9–11,23], while others focus on protecting a single system, such as Borealis [20]. These systems simply explore the overheads associated with access control enforcement and do not consider them at optimization time or during operator placement. Furthermore, these systems do not explore the tradeoff between different types of access control enforcement during optimization time, which is provided in ShoalSystems like PolyStream [31], and Streamforce [3], CryptDB [26] consider such tradeoffs, but do either do not operate in a distributed fashion (CryptDB), or do not consider them at optimization time.

Thus far current optimizers and systems have focused on a limited scope of characteristics within a DDSPS, mostly excluding access controls. Either they do not consider optimization and placement simultaneously, or they limit their approach to optimize solely for something like network, hardware, or other traditional metrics. Shoal provides a general cost model and dynamic programming algorithm that accounts for data provider's access control enforcement at query optimization time.

7   Conclusion

We present Shoal which considers access controls as first-class-citizens during query optimization. By simultaneously ordering and placing streaming query networks on a per-operator level, Shoal can guarantee optimal results through a dynamic programming algorithm. Further, Shoal reduces optimization time for updates based on changes in access controls by identifying the precise operators that need to be re-optimized and only optimizing from those points forward in an online fashion. Finally, we show that Shoal produces higher quality plans (up to 2.2x) versus the state-of-the-art optimizers, and does so while considering data provider's access controls.

Acknowledgements. This work was supported in part by the National Science Foundation under awards CNS–1253204 and CNS–1704139.

References

1. Abadi, D., et al.: Aurora: a new model and architecture for data stream management. VLDB **12**(2), 120–139 (2003)
2. Abadi, D., et al.: The design of the borealis stream processing engine. In: CIDR (2005)

3. Anh, D.T.T., Datta, A.: Streamforce: outsourcing access control enforcement for stream data to the clouds. In: ACM CODASPY, pp. 13–24 (2014)
4. Arasu, A., et al.: Stream: the Stanford data stream management system. Book chapter (2004)
5. Arasu, A., et al.: The CQL continuous query language: semantic foundations and query execution. VLDB J. **15**(2), 121–142 (2006)
6. Arasu, A., et al.: Stream: the Stanford data stream management system. In: Garofalakis, M., Gehrke, J., Rastogi, R. (eds.) Data Stream Management. Data-Centric Systems and Applications. Springer, Heidelberg (2016). https://doi.org/10.1007/978-3-540-28608-0_16
7. Backman, N., Fonseca, R., Çetintemel, U.: Managing parallelism for stream processing in the cloud. In: HOTCDP Workshop, pp. 1–5. ACM (2012)
8. Cardellini, V., et al.: Optimal operator placement for distributed stream processing applications. In: DEBS, pp. 69–80. ACM (2016)
9. Carminati, B., et al.: Enforcing access control over data streams. In: ACM SACMAT, pp. 21–30 (2007)
10. Carminati, B., Ferrari, E., Tan, K.L.: Specifying access control policies on data streams. In: Kotagiri, R., Krishna, P.R., Mohania, M., Nantajeewarawat, E. (eds.) DASFAA 2007. LNCS, vol. 4443, pp. 410–421. Springer, Heidelberg (2007). https://doi.org/10.1007/978-3-540-71703-4_36
11. Carminati, B., et al.: A framework to enforce access control over data streams. ACM TISSEC **13**(3), 28 (2010)
12. Cattell, R.: Scalable SQL and NoSQL data stores. ACM SIGMOD Rec. **39**(4), 12–27 (2011)
13. Chatzistergiou, A., Viglas, S.D.: Fast heuristics for near-optimal task allocation in data stream processing over clusters. In: CIKM, pp. 1579–1588. ACM (2014)
14. Corbett, J.C., et al.: Spanner: Google' globally distributed database. ACM Trans. Comput. Syst. (TOCS) **31**(3), 8 (2013)
15. Curino, C., et al.: Relational cloud: a database-as-a-service for the cloud. In: CIDR (2011)
16. Farnan, N., et al.: PAQO: preference-aware query optimization for decentralized database systems. In: ICDE (2014)
17. Huang, Y., et al.: Operator placement with QoS constraints for distributed stream processing. In: CNSM, pp. 1–7. IEEE (2011)
18. Kulkarni, S., et al.: Twitter heron: stream processing at scale. In: SIGMOD, pp. 239–250. ACM (2015)
19. Lee, K.-H., Lee, Y.-J., Choi, H., Chung, Y.D., Moon, B.: Parallel data processing with MapReduce: a survey. ACM SIGMOD Rec. **40**(4), 11–20 (2012)
20. Lindner, W., Meier, J.: Securing the borealis data stream engine. In: IEEE IDEAS, pp. 137–147 (2006)
21. Nehme, R., et al.: A security punctuation framework for enforcing access control on streaming data. In: ICDE, pp. 406–415 (2008)
22. Nehme, R.V., et al.: Fence: continuous access control enforcement in dynamic data stream environments. In: ACM CODASPY, pp. 243–254 (2013)
23. Ng, W.S., et al.: Privacy preservation in streaming data collection. In: ICPADS, pp. 810–815 (2012)
24. Pietzuch, P., et al.: Network-aware operator placement for stream-processing systems. In: ICDE, pp. 49–49. IEEE (2006)
25. Poess, M., Floyd, C.: New TPC benchmarks for decision support and web commerce. ACM SIGMOD Rec. **29**(4), 64–71 (2000)

26. Popa, R., et al.: CryptDB: protecting confidentiality with encrypted query processing. In: ACM SOSP, pp. 85–100 (2011)
27. Rizou, S., et al.: Solving the multi-operator placement problem in large-scale operator networks. In: ICCCN, pp. 1–6. IEEE (2010)
28. Shute, J., et al.: F1: a distributed SQL database that scales. VLDB 6(11), 1068–1079 (2013)
29. Srivastava, U., Munagala, K., Widom, J.: Operator placement for in-network stream query processing. In: SIGMOD, pp. 250–258. ACM (2005)
30. StormProject: Storm: distributed and fault-tolerant realtime computation (2014). http://storm.incubator.apache.org/documentation/Home.html
31. Thoma, C., et al.: Polystream: cryptographically enforced access controls for outsourced data stream processing. In: SACMAT, vol. 21, p. 12 (2016)
32. Thoma, C., Labrinidis, A., Lee, A.J.: Automated operator placement in distributed data stream management systems subject to user constraints. In: ICDEW, pp. 310–316. IEEE (2014)

A Distributed Ledger Approach to Digital Twin Secure Data Sharing

Marietheres Dietz$^{(\boxtimes)}$, Benedikt Putz, and Günther Pernul

University of Regensburg, Regensburg, Germany
{marietheres.dietz,benedikt.putz,guenther.pernul}@ur.de

Abstract. The Digital Twin refers to a digital representation of any real-world counterpart allowing its management (from simple monitoring to autonomy). At the core of the concept lies the inclusion of the entire asset lifecycle. To enable all lifecycle parties to partake, the Digital Twin should provide a sharable data base. Thereby, integrity and confidentiality issues are pressing, turning security into a major requirement. However, given that the Digital Twin paradigm is still at an early stage, most works do not consider security yet. Distributed ledgers provide a novel technology for multi-party data sharing that emphasizes security features such as integrity. For this reason, we examine the applicability of distributed ledgers to secure Digital Twin data sharing. We contribute to current literature by identifying requirements for Digital Twin data sharing in order to overcome current infrastructural challenges. We furthermore propose a framework for secure Digital Twin data sharing based on Distributed Ledger Technology. A conclusive use case demonstrates requirements fulfillment and is followed by a critical discussion proposing avenues for future work.

Keywords: Trust frameworks · Distributed systems security ·
Distributed ledger technology · Digital twin

1 Introduction

Hardly anything has revolutionized society as much as digitization. At its beginning, data from everyday life was captured and stored digitally. After reaching significant amounts of digital data, recent years have been devoted to gaining relevant insights into data by leveraging Big Data Analytics, Artificial Intelligence and so on. A next step in digitization is now emerging in the form of the Digital Twin (DT) paradigm.

The Digital Twin refers to a digital representation of any real-world counterpart, at most times an enterprise asset. Its core building blocks are asset-specific data items, often enhanced with semantic technologies and analysis/simulation

---

The first two authors have contributed equally to this manuscript.

S. N. Foley (Ed.): DBSec 2019, LNCS 11559, pp. 281–300, 2019.
https://doi.org/10.1007/978-3-030-22479-0_15

environments to explore the real-world asset digitally. The DT thus allows management of such an asset ranging from simple monitoring to autonomy. An essential part of the concept is the inclusion of the whole asset lifecycle. To integrate all lifecycle participants, the DT should provide comprehensive networking for its data, allowing it to be shared and exchanged [4].

Although the DT concept certainly advances digitization, it nevertheless poses new challenges in terms of IT security, especially in industrial ecosystems [10,18]. Most notably, security must be maintained during the exchange of DT data between different, non-trusting parties. For instance, consider the DT of a power plant. Synchronizing tasks between twins should uphold integrity to avoid manipulated operations on the power plant. Also, involved parties should not be able to read every shared data element (e.g. the manufacturer of the power plant need not know the plant's current status), resulting in confidentiality requirements. To the best of our knowledge, current DT frameworks do not permit secure data sharing. Bridging this gap, our work provides a framework introducing security-by-design in DT data sharing.

To achieve this goal, we consider Distributed Ledger Technology (DLT). DLT is the umbrella term for distributed transaction-based systems, shared among several independent parties in a network. Distributed Ledgers have built-in mechanisms for access control and asset management, including authentication and authorization mechanisms. We focus on permissioned distributed ledgers, which target enterprise usage by restricting access to fixed set of independent and semi-trusted participants. One of the main reasons for using a Distributed Ledger is disintermediation, replacing the need for trust in a third party or central operator through a replicated and integrity-preserving database. Inherent transparency and auditability are additional advantages over centralized solutions. Due to these properties, DLT is uniquely suited to solve the challenges of DT secure data sharing.

Accordingly, this work proposes a framework for secure DT data sharing across an asset's lifecycle and collaborating parties based on DLT. We contribute to the body of knowledge by offering a solution without a trusted third party (TTP) based on security-by-design. The remainder of this paper is organized as follows: Sect. 2 introduces the background of our work. Afterwards, we proceed to the description of the current problems in DT data sharing and name the resulting requirements for secure DT data sharing (Sect. 3). In Sect. 4, we provide a framework for secure DT data sharing for multiple parties based on DLT. To show practical relevance and the functionality of our framework, a use case is provided in Sect. 5. In Sect. 6, we evaluate our approach in terms of fulfillment of the stated requirements. To conclude, Sect. 7 sums up the main contributions and gives an outlook for future work.

2   Background

At present, the *Digital Twin* phenomenon is still in its infancy. Nevertheless, implementation and design of this concept are addressed to date, especially in the area of Industry 4.0. With strong focus on the industrial domain, the major part of research suggests DT implementation through AutomationML-formatted descriptive data of the real-world counterpart, e.g. [2,6,20]. The XML-based AutomationML (AML) format describes industrial assets and provides object-orientation for modeling the asset's physical and logical components [20]. Eckhart and Ekelhart [6] propose a framework for using a DT's simulation mode for security purposes such as pen testing. While these works focus on an initial development of a DT, the consideration of data sharing functions are still missing. However, exchanging data is vital for enabling the lifecycle integration and collaboration [4]. Our work builds on existing DT propositions, resulting in a concept that can be applied in a complementary way to enable secure DT data sharing.

Regarding DT data sharing, both the communication between lifecycle parties and the bidirectional communication between the DT and its real-world asset counterpart need to be considered. Bidirectional communication consists of the DT's instructions for the asset and the asset's status update for the DT. To uphold integrity among multi-domain DT models, Talkhestani et al. [21] offer a solution. They detect model changes by applying anchor points, and upon detection synchronize the DT while keeping model dependencies consistent. However, this includes drawbacks such as the manual creation of anchor points and reliance on a Product Lifecycle Management (PLM) system, while our solution offers platform-independence. Security aspects, such as the guarantee for all lifecycle partners to access the data while upholding confidentiality, are not considered to date, but integrated in our solution.

DT management is a form of enterprise asset management, which is one of the prime use cases of Distributed Ledgers [1]. Distributed Ledgers are able to track events and provenance information along an asset's lifecycle and increase transparency for all participants. For example, Litke et al. [12] studied the benefits of Distributed Ledgers for different actors in supply chain asset management, a research area closely related to DT asset management. In another study, Meroni and Plebani [14] investigate how the blockchain technology can be used for process coordination among smart objects. Smart objects are similar to DTs in that they are applied for monitoring physical artifacts. An issue with their proposed approach is that sensor data is also stored on the blockchain, which can be detrimental to performance and scalability. We consider this issue and provide a solution to overcome this obstacle.

3   Problem Statement

On the one hand, DTs should facilitate the access to asset information for different stakeholders along its lifecycle [17]. It is a task which enables feedback loops, while stepping towards a circular economy [3]. On the other hand, the involved parties do not necessarily trust each other, resulting in a confidentiality dilemma. A useful example is given in [13]: Two separate standalone DTs exist for a single device instance, one for the manufacturer and the other at the customer site – due to information security reasons. Additionally, current works state that enterprise infrastructures need to overcome the following obstacles to provide secure DT data sharing:

- application of different tools [13, 24]
- usage of various data formats [13]
- missing standards [4]
- broken information flow across lifecycle phases [13, 24]
- clarification of the ownership of information [13].

This calls for a holistic approach that provides confidentiality and integrity, two central security dimensions in networks [26].

3.1   Digital Twin Model

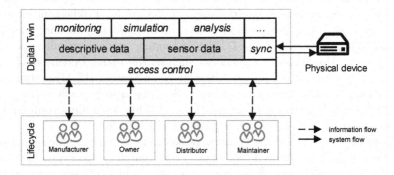

Fig. 1. Overview of the asset lifecycle participants interacting with the DT.

Figure 1 illustrates DT data sharing and an exemplary set of lifecycle stakeholders. The depicted DT model comprises different *capabilities* and two types of asset-specific data. *Descriptive data* refers to static properties of the device and infrequently changing state information. This data is mainly produced by users. *Sensor data* occurs frequently and should be available in near real-time. It is generated by sensors of the physical asset or in its proximity, which provide valuable information on the asset's environmental conditions. Moreover, data of both types needs to be synchronized with the physical counterpart. Therefore, the *sync* capability compares the state of the DT to its real-world counterpart and resolves possible discrepancies.

The *access control* capability provides authentication and authorization modules to enable data sharing of involved parties without hampering confidentiality. The *monitoring, simulation* and *analysis* capabilities represent advanced operations of the DT. Depending on the extent of the operations present in a DT, DT status data can be returned to the participant or the real-world counterpart's state can be modified.

The depicted information flows show how information about the physical device is gathered from and sent to the lifecycle parties. Generally, the type of data accessed and shared by the different lifecycle parties depends on the real-world twin, the parties' roles in its lifecycle and thus, the specified access control mechanisms in the DT. The system flows represent necessary bidirectional synchronization between the DT and its real-world counterpart as stated in Sect. 2. Both flows contribute to making the data sharing activities of the involved parties traceable. This enables feedback from the latest stages of the asset lifecycle to the earliest ones [17].

3.2   A Formal Basis for Secure Digital Twin Data Sharing

Although a methodological literature analysis to establish requirements is the state-of-the-art approach, it is currently not sensible to carry out with regard to our research focus. On the one hand, this is due to the fact that only a small number of publications exist. In addition, data sharing has not yet been a focus in DT literature to date. Moreover, security-by-design concepts have not been considered yet. Therefore, we establish a formally valid basis in order to create a uniform understanding of DT data sharing.   To derive the requirements, the

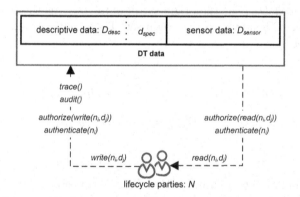

Fig. 2. Control flows for a single DT.

mechanisms to achieve the central goal of **secure DT data sharing** have to be examined in detail. Figure 2 illustrates the formal functions required to achieve this goal, which are also described hereafter.

DT Data: We see DT data twofold: At first, there is a set of descriptive data elements $D_{desc} := \{d_1, ..., d_m\}$ varying from documents to models or analytic outcomes. Its essential data element is the specification of the DT $d_{spec} \in D_{desc}$. The second set contains environmental, device-produced data, namely sensor data $D_{sensor} := \{d_1, ..., d_n\}$, whereby $D_{desc} \setminus D_{sensor}$.

Sharing: A finite set of lifecycle parties $N := \{n_1, ..., n_k | k \geq 2\}$ can share the respective data elements of D_{desc} (write operation) or access the data elements of D_{desc}, D_{sensor} (read operation). This results in the following necessary functions:

$write(n_i, d_j | d_j \in D_{desc})$ and $read(n_i, d_j | d_j \in D_{desc} \vee d_j \in D_{sensor})$.

$$\text{Note that } 1 \leq i \leq k \text{ as well as } j \begin{cases} 1 \leq j \leq m & \text{if } j \in D_{desc} \\ 1 \leq j \leq n & \text{if } j \in D_{sensor} \end{cases}.$$

Security-by-design: Security-by-design infers introducing security mechanisms at the very beginning of a system's design [22]. In terms of DT data and sharing security, data integrity and confidentiality mechanisms are of special interest. Confidentiality in terms of securing data from view of non-trusted third parties can be reached by access control mechanisms [19]:

authentication:
$$authenticate(n_i)$$

authorization:
$$authorize(read(n_i, d_j))$$
$$authorize(write(n_i, d_j))$$

Integrity of data can be achieved by auditability and traceability of write operations. Given D_{desc} as the origin set of data, D'_{desc} is the set of data after a data element d_j is added to the origin set. The following functions can cover integrity aspects:

auditability:

$$audit() : D_{desc} \rightarrow D'_{desc} \iff write(n_i, d_j) \quad \wedge$$

$$D_{desc} \nrightarrow D'_{desc} \iff \neg write(n_i, d_j)$$

traceability:
$$trace() : D_{desc} \rightarrow D'_{desc} \implies D_{desc} \circ D'_{desc}$$

Thereby, auditability guarantees that D_{desc} is transformed to D'_{desc} in case of an authorized write operation whereas other operations are not able to transform the data in any way. Traceability ensures that authorized writes of data elements and thus, transformations of D_{desc} to D'_{desc}, are chained up. In conclusion, data integrity is ensured as the data cannot be manipulated or tampered with in retrospect.

3.3 Requirements for Secure DT Data Sharing

To provide a sound solution for secure DT data sharing, the following requirements were derived from the formal basis and the aforementioned challenges identified in the literature analysis.

R1. Multi-party Sharing. To enable lifecycle inclusion, a vital characteristic of the DT paradigm [4], the multiple stakeholders N involved in the lifecycle have to be considered. As described in Fig. 1, parties can vary from manufacturer to maintainer. However, all involved parties are pre-registered and therefore determinable.

R2. Data Variety Support. At the heart of the DT lie the relevant digital artifacts D_{desc}, D_{sensor}, which vary from design and engineering data to operational data to behavioral descriptions [4]. Thus, different data types and data formats [13] need to be supported during data sharing. For instance, Schroeder et al. claim that using the semi-structured AutomationML format to model attributes related to the DT (d_{spec}) is very useful for DT data exchange [20]. In addition to semi-structured data, structured data (e.g. sensor tuples in D_{sensor}, database entries) and unstructured data such as human-readable documents can be asset-relevant and shared via the DT.

R3. Data Velocity Support. Often, DT data is distinguished between descriptive, rather static data, and behavioral, more dynamic data (see Fig. 1). The latter changes with time along the lifecycle of the real-world counterpart [20]: With each lifecycle stage the asset-related information evolves, resulting in different versions and a dynamic information structure [17]. Naturally, dynamic data includes sensor data D_{sensor} – which mostly refers to the actual state of the real-world counterpart [8]. While the infrequently changing data D_{desc} might not require high throughput, sensor and dynamic data D_{sensor} accrues in intervals ranging from minutes to milliseconds. Therefore, the solution must support high throughput and low sharing latency for efficient sharing of dynamic data – thus supporting data velocity.

R4. Data Integrity and Confidentiality Mechanisms. An important requirement is taking into account data security features, especially integrity and confidentiality. At first, this requirement aims at safeguarding data integrity to avoid wrong analytic decisions based on manipulated data. It can be ensured by $audit()$ and $trace()$ mechanisms. The second main security objective is to avoid confidentiality and trust problems while enabling multi-party participation. This calls for restricted data access dependent on the party through $authenticate()$ and $authorize()$ functions, while ideally keeping the effort for user registration low. Different levels of confidentiality should be possible for different data elements. For instance, D_{sensor} might need a lower level of protection than D_{desc}, as the latter might include sensitive corporate information such as blueprints. Detailed $authorize()$ functions, providing access-restrictions for each data element, can cover this aspect.

R5. Read and Write Operations. To interact with DT data, a DT data sharing solution must provide $read()$ and $write()$ data operations for the sharing parties. The allowance of operation modes for the data elements should be chosen carefully for each party to ensure **R4** (cf. Fig. 2).

Overall, we do not claim that these requirements are complete. There may be other requirements of importance, but regard these as essential for the following reasons. On the one hand, these requirements were found to be mentioned most often in the reviewed literature, while others were less frequently mentioned and are therefore considered of lower importance (see Sect. 6.2 for further explanation). On the other hand, the stated requirements were also the main focus in various practitioners reports (e.g. [9, 16, 25]) and during discussions with experts.

4 Solution Architecture

In order to develop a framework for secure DT data sharing, we first evaluate the suitability of DLT in Sect. 4.1. Afterwards, Sect. 4.2 explains the system architecture and Sect. 4.3 explains how the various data types are stored. Section 4.4 details the inclusion of the DT capabilities as part of the DLT solution. Finally, Sect. 4.5 explains the initial setup procedure for our framework.

4.1 Technology Selection

To develop a solution architecture, we first evaluate different data storage solutions' properties to select the technology best suited to fulfill the requirements.

A centralized solution could be created in the form of a portal, operated by a third party or the operator of the twinned device. This requires trust of the participating parties towards the portal maintainer, as the maintainer could manipulate data or revoke access to the DT for other parties. A distributed approach jointly operated by all participants could solve this trust issue. Distributed Ledgers represent such a distributed solution. They permit verifiable decentralized execution of business logic via *smart contracts*, ensuring that rules and processes agreed upon by the lifecycle participants are followed.

We evaluate the applicability of Distributed Ledgers to our DT data sharing requirements based on the blockchain applicability evaluation framework by Wüst and Gervais [28]. As illustrated in Fig. 1, there is a need to store various types of data as part of the DT state. Multiple parties interact with the twin during its lifecycle who do not fully trust each other. These writers are usually known in advance or change infrequently (i.e. the maintenance service provider changes). These characteristics lead to the choice of a public or private permissioned blockchain in the framework [28]. In our case, this choice depends on whether public auditability is required or not. While use-case dependent, we focus on private permissioned blockchains for the rest of this paper. If needed, public read-only access to blockchain data can be enabled during implementation for most permissioned blockchain frameworks (i.e. through a REST API).

4.2 System Architecture

The proposed DLT-based architecture for secure DT data sharing is shown in Fig. 3. Every participant runs three components: a node of a **Distributed Hash Table (DHT)**, a node of the **Distributed Ledger** and a **client application**. The DHT and Distributed Ledger make up the shared data storage, while the client application is responsible for the user interface and backend logic for retrieving and processing the data stored on the ledger and DHT. For owners

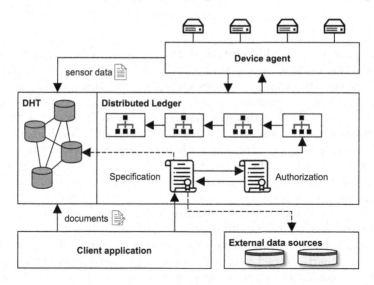

Fig. 3. DLT-based architecture for DT data sharing.

of twinned physical devices, a **Device agent** manages the physical devices and coordinates their interactions with the system. As part of operational technology, the Device agent functions as a bridge between the cross-organizational asset management system and the physical devices controlled by a single organization.

Data storage systems based on distributed ledgers have two ways of storing data: on-chain and off-chain [29]. On-chain storage is restricted to transactions and the internal state storage of smart contracts. Due to full replication of on-chain data, items larger than a few kilobytes in size need to be stored in a different, off-chain location. Using a traditional database would however result in a single point of failure or reintroduce a trusted party.

For this reason, we resort to a structured DHT for large data items. DHTs are distributed key-value stores, where all key-value pairs are mapped to one or more nodes. The DHT entries can be linked to the corresponding on-chain asset based on the DHT key hash. By storing the hash on the blockchain, integrity of the off-chain data can be verified after retrieving it from the DHT. To maintain confidentiality and availability, data stored on the DHT is encrypted, sharded and replicated. Correspondingly, an access control mechanism is needed to allow authorized parties to access the data. The k-rAC scheme illustrates how a DHT can implement the required functionality [11]. In k-rAC, access control is implemented using access control lists (ACL) stored along with each key-value pair on the DHT. We propose reusing the Distributed Ledger's public key identities for DHT authentication. A symmetric key is used for encryption, which is then available to authorized parties by encrypting it with their public key. The encrypted access keys are distributed with each data item's ACL. Manipulation of the ACL is prevented by requiring a quorum of $2k + 1$ nodes for write operations, where k is the number of tolerated malicious nodes.

4.3   Data Storage

There are two types of **descriptive data** that need to be stored by the system: a machine-readable specification and device-related unstructured data (i.e. human-readable documents). The **specification** includes a description of the device's hardware components as well as their functions. The DT's physical properties are derived from this specification. For our work we assume that AML is used to describe the physical asset. The AML specification is stored on the ledger in a modifiable way. This approach guarantees that updates to the device specification are observed by all parties. Distributed Ledgers can store complex modifiable state by using *smart contracts*. We thus refer to the resulting contract as the *specification contract*.

Unstructured data can be uploaded to the system and may subsequently be annotated or modified by other parties. Due to its size it cannot easily be parsed and stored in contracts. For this reason, it is stored off-chain and registered in the smart-contract with a document title and a hash of the contents. To update a document, a new version must be uploaded to the DHT and the smart contract reference updated. This ensures that changes to the documents are traceable.

Sensor data needs to be stored off-chain due to its frequent updates and the considerable amount of generated data. A history of the sensor data is kept to allow for further analysis, e.g. predictive maintenance or troubleshooting. The link to the on-chain data is established via a pointer to the off-chain storage location, stored on-chain in the specification contract. To avoid having to update the storage location hash every time new sensor data is uploaded to the DHT, we take advantage of *DHT feeds*. This concept is inspired by the Ethereum network's DHT Swarm [7]. In Swarm, a feed is defined by a feed manifest with a unique address on the network. The feed manifest's owner (i.e. the physical device) is the only user permitted to upload signed and timestamped data to this address. Any data format can be used and a history of uploaded data is kept. The DHT feed enables frequent sensor data sharing without having to update an on-chain reference. Based on the feed, the client application may compare sensor updates with expected values derived from the specification contract to detect anomalies. Additionally, there is no need for directly accessing the physical device, which may reside in a protected network. Instead, data updates are readily available on the DHT for authorized participants.

Many organizations also have additional internal data sources or microservices that provide structured data relevant to the Digital Twin. These data sources can be included in the twin by adding references (i.e. an URI) to the DT specification contract. This allows inclusion of legacy data sources and complex data which cannot easily be stored on a DHT (i.e. relational data). If the external data source requires authentication, it is the responsibility of the data source provider to ensure access rights for the DT ledger's identities.

Listing 1.1 shows a pseudocode representation of the data types stored in the specification contract. The syntax is inspired by Ethereum's Solidity smart contract programming language. All data stored on the contract is readable by all lifecycle participants. Besides general device metadata, the contract also includes a program call queue for interaction with the physical device's program interfaces (see also Sect. 4.4). Since smart contracts must be deterministic and thus cannot interact with files, the AML specification is stored in a string variable. This variable can later be parsed and modified, as illustrated in Sect. 4.4. Hash references to new original documents on the DHT are kept track of in the `documents` mapping. The hash serves as an identifier, while the `document` struct provides metadata. Updated versions of each document are stored in the `documentVersions` mapping. The componentID and corresponding feed reference of the sensor data stream on the DHT are stored in the `sensorFeeds` mapping.

```
/* metadata and specification*/
string deviceName
string deviceID
string deviceAML
string[] callProgramQueue

/* additional descriptive data */
struct Document {
    uint timestamp
    string description
    address owner
}

struct ExternalSource{
    string URI
    address owner
}

mapping(string=>Document) documents
mapping(string=>string[]) documentVersions
ExternalSource[] externalSources

/* sensor data */
mapping(string=>string) sensorFeeds
```

Listing 1.1. Data structures of the specification contract

```
/* descriptive data interfaces */
function addDocument(document)
function addDocumentVersion(string hash)
function removeDocument(string hash)

function addExternalSource(string URI)
function removeExternalSource(string URI)

/* sensor data interfaces */
function addSensorFeed(string componentID,
        string reference)

function removeSensorFeed(string componentID)

/* interaction with the specification */
function insertAML(string amlCode, string
        parentID, string afterID)
function removeAML(string ID)
function callProgram(string programName,
        string parameters[])
```

Listing 1.2. Function interfaces of the specification contract

4.4   Capabilities

We focus on the three capabilities required for accessing and publishing DT data: *DT interaction, access control* and *sync*.

DT interaction refers to the information flows in Fig. 1, which allow users to interact with the twin's data. The specification contract implements this functionality. It allows users to read and potentially modify the DT instance. The relevant interfaces that can be called with transactions are shown in Listing 1.2. New or updated references to documents may be appended by any authorized user. The same applies to external data sources and sensor feed references to the DHT. The specification can be manipulated by inserting or removing specific AML segments, which are identified by their ID. To determine the position of a new AML code segment in the AML document, the parent ID and the ID of the preceding element need to be passed as parameters. The twin's program

interfaces for setting device parameters can be accessed via `callProgram`. This
function checks authorization, finds the requested program in the AML specification and places it in a queue for the Device agent to retrieve. The agent then
forwards the program call to the device for execution.

The *access control* capability is responsible for authentication and authorization of user interactions with the DT data. For user authentication, accounts
are created on the blockchain and represented by their public key. An initial
solution could be provided by the framework's built-in identity management,
for example Hyperledger Fabric's Membership Service Provider (MSP) [1]. The
MSP lists the certificate authorities who may issue digital identities for the Distributed Ledger. The same identity can then be reused for authentication in
the DHT. Authorization is realized in a separate access control smart contract.
Any protected interaction with the Digital Twin is first authorized through that
contract. Such interactions are for example modifications of the twin's properties, like changing parameters or modifying its specification. A query from the
client application provides an identity to the specification contract, which then
interacts with the authorization contract to determine if the user is allowed to
perform the action. Authorization is then granted or denied based upon a stored
role-permission mapping. Accordingly, the contract's interfaces are based upon a
Role-based Access Control (RBAC) scheme. We do not describe the access control contract in detail here, as there are other works describing blockchain-based
access control schemes [5].

The *sync* capability requires regular interaction between the Device agent and
the Distributed Ledger. For synchronization, the Device agent pulls updates from
the real-world asset and uploads them to the off-chain DHT sensor data feed.
The Device agent monitors the ledger and pushes any modifications instructed
by committed on-chain transactions to the asset. The synchronization interval
depends on the use case.

Other DT capabilities like *monitoring, simulation* and *analysis* can be executed off-chain by interacting with the local copy of the ledger. Simulation or
analysis instructions and results can be shared on the ledger as documents. This
would allow other parties to verify the results, should they desire to do so.

4.5   Setup Process

Initially, each lifecycle participant sets up one network node running both a DHT
and a Distributed Ledger node. These serve as local replicas of ledger data and
access points for off-chain data. They may also be used for transaction-based
interaction with the smart contracts. Additionally, an identity provider must be
set up to allow federated identities from all participating organizations based on
public key certificates.

Once the network is set up, a Digital Twin instance can be created on the
ledger by the device owner. The manufacturer should first provide the AML file
to the owner, who then proceeds to set up a Digital Twin sharing instance on
the ledger. The client application provides the interface to upload the file and
create a smart contract based on it. Before uploading, the owner also needs to

specify the access rights associated with the various parts of the specification. Although use case dependent, sensible default values could be *write* access by owner and maintainer and *read* access by everyone else.

In this way, any number of Digital Twin instances can be created by the various parties on the network. Each instance is represented by a specification contract. Subsequent modifications take place via authorized on-chain transactions and are stored as part of the contract's internal state. As a result, auditing the twin is possible by (actively or retroactively) monitoring smart contract transactions for anomalies.

5   Use Case

This chapter intends to show how the theoretical framework developed in Sect. 4 is traversed in a use case. To begin with, the overall setting of the use case is described in Sect. 5.1, while the subsequent Sect. 5.2 iterates the use case through the solution architecture. At last, a summary is given, focusing on the automation degree in data sharing and the reading operation (Sect. 5.3).

5.1   Setting

The setting is chosen close to reality. The asset, the real-world counterpart to the DT, is a bottling plant, where bottles are filled with beverages. The parties involved in the asset lifecycle are a manufacturer, an owner, a maintainer of the bottling plant and an external auditor that audits the safety of our bottling plant. For our use case, we consider the following scenario: The bottles are flooding due to a broken sensor in the bottling plant. Consequently, the maintainer detects the damage and changes the broken sensor in the bottling plant.

This entails the following shared data interactions. At first, the specification of the plant needs to be updated by replacing the broken sensor's specification entry with the newly added sensor. Additionally, the new sensor's data stream has to be integrated in place of the old sensor stream. Other documents concerning the maintenance task might also be shared, such as a maintenance report.

While the maintainer is the only party sharing data in this scenario, the owner should also be updated on the state of the bottling plant. Furthermore, the manufacturer needs to be informed that the sensor is broken, so that an analysis of the time and circumstances can be conducted. This way relevant insights for future plant manufacturing can be gained. Additionally, the external auditor needs to access the information about the maintenance task to review the procedure in terms of safety compliance.

5.2   Framework Iteration

This use case triggers a specific logical order of events in the framework, which are highlighted in Fig. 4 and described hereafter. The framework first comes into play when the maintainer replaces the broken sensor.

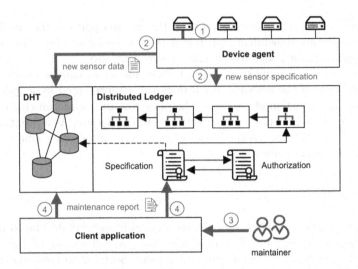

Fig. 4. Use case tailored architecture for DT data sharing.

1. All devices are connected with the **Device agent**, which registers the exchange of the broken sensor. Additionally, it gathers information about the new sensor.
2. Following the new sensor connection, the **Device agent** forwards the new incoming data stream of the sensor into the **DHT**. The location of the stored sensor stream in the **DHT** is registered by the **Device agent**.

    The **Device agent** then sends a transaction containing the new sensor specification to the **Distributed Ledger**. This transaction invokes the specification contract, resulting in several updates. First, the old sensor entry is removed and the new sensor specification given by the **Device agent** is added. Secondly, the storage location of the sensor stream on the **DHT** is added by a reference to the location. These three transactions concerning the specification are stored on the **Distributed Ledger**.
3. Having performed the maintenance task, the maintainer writes a maintenance report and pushes it onto the **Client application**.
4. The **Client application** adds the maintenance report by performing two actions. Firstly, it adds the report to the off-chain **DHT**. Secondly, it stores the reference to the **DHT** location of the report on the specification contract. Thereby, the location is added to the entry of the sensor specification.

5.3   Results

In a nutshell, the recognition of new sensor and the AML update with the new component is already accomplished by the **Device agent** without requiring human interaction. The new data stream is automatically forwarded to the **DHT** and the reference to the new storage location of the component's data stream is added to the specification contract. Additional unstructured non-specification data (e.g. the maintenance report) can be added manually. The **Client application** takes care of the necessary background work by inserting the file into the **DHT** and adding the respective storage reference into the specification contract.

All participating parties can view the latest transactions on the ledger – presented in a comprehensive way in the **Client application**. Advanced **Client applications** could also notify the user whenever an ledger update takes place.

Considering security, the advantages of this framework shine when compared to the alternative solution: A TTP could deliberately transfer shared information and know-how to rival enterprises. For instance, confidential sensor data or blueprints could be leaked to competitors, which may then deduce quality issues of the rival product. The service of the TTP could also be compromised by attackers, resulting e.g. in a violation of integrity so that the sharing parties receive inconsistent asset versions.

6 Evaluation

To evaluate our framework, Sect. 6.1 discusses the suitability of the framework in reference to the requirements. Finally, the results are discussed in Sect. 6.2.

6.1 Requirements Fulfillment

To sum up, our approach fulfills the requirements **R1–R5**. The following paragraphs explain how each requirement was addressed in our solution architecture.

R1. Multi-party Sharing. The main argument for using Distributed Ledgers is the involvement of multiple parties N who produce and consume data. Next to the ledger, our approach provides a client application for all parties that accesses the data on the ledger and the DHT. Therefore, our approach clearly fulfills **R1**.

R2. Data Variety Support. To enable the sharing of different data in various formats, our approach provides a central documentation and two storage options. The standardized asset description d_{spec} is included in the Distributed Ledger and serves as the basis of the DT within the specification contract. All other data of D_{desc} as well as the sensor data D_{sensor} are stored off-chain in the DHT. Moreover, each stored data element in DHT is registered in the central specification contract as a reference to the storage location of the data element. For instance, a sensor in the specification contract contains a reference to the storage position of its data stream in the DHT. Hence, **R2** is met.

R3. Data Velocity Support. Modern sensor data streams' frequency and volume exceed the performance characteristics of current Distributed Ledger frameworks. Since the data streams D_{sensor} do not describe main features of the DT (d_{spec}), they are stored off-chain in the DHT. This way, high throughput of D_{sensor} is supported, while the sharing latency is also kept low (seconds). The Distributed Ledger maintains verifiability by storing the hash reference to the data stream on the DHT in the specification contract. This ensures no loss in performance and data access through the DHT, supporting **R3**.

R4. Data Integrity and Confidentiality Mechanisms. With respect to data integrity, the Distributed Ledger attaches every new data element (*trace()*) and prevents manipulation of the data by replicating it among all involved parties. A manipulation would result in a version mismatch or loss of consensus and could be detected easily (*audit()*). The second storage component (DHT) also supports integrity by storing the respective hash values to the data. A manipulation of DHT data would also be detected by a mismatch between the hashes in the nodes (*audit()*). However, there remains the problem of adding non-valid data, which is a common issue in the area of DLT. Here, we rely on the parties' interest in sharing valid data and on mechanisms ensuring quality of input data that the respective responsible party applies.

In terms of data confidentiality, our approach ensures that the data is read only by authenticated and authorized parties. Authentication is ensured through lifecycle party login to the client application (*authenticate()*). Access control concerning the party and the data elements is realized through an ACL and encryption for off-chain data and an authorization smart contract for on-chain data (*authorize()*). In concrete terms, the ACLs specify access rights on a per-document basis, while the smart contract stores authorization information for all involved parties. Therefore, different confidentiality levels can be realized.

To conclude, our approach provides data integrity and confidentiality mechanisms (**R4**) – reinforcing data security in DT data sharing.

R5. Read and Write Operations. Read and write operations are managed through the Client application. For *read()* operations, the Client application fetches the requested data from the DHT and the ledger and presents the data in a comprehensive way adjusted for the demanding party. In case of a *write()* operation, the Client application triggers the right procedure to alter the smart contract with a transaction and uploads additional asset-relevant data beyond specification to the DHT. Consequently, our approach also fulfills **R5**.

6.2   Discussion

Keeping the requirements *variety* (**R2**) and *velocity* (**R3**) in mind, the question arises why data *volume* is not considered a requirement. As literature is currently not at consensus regarding the relevance of the Big Data feature *volume* [15] for Digital Twins, we consider explicit support for data volume to be non-necessary. Nevertheless, by storing documents off-chain, our approach can handle considerable amounts of data. Future implementations of our concept may conduct benchmark studies to explore scalability limits with regard to big data volumes.

It should be noted that our approach depends on multi-party participation. The more independent parties maintain the Distributed Ledger and DHT, the less vulnerable the data sharing is to manipulation. With regard to the access control capability, a decentralized identity management solution with a shared identity database could be an even more holistic, next-generation solution.

While we are aware that our approach currently lacks an implementation, we nevertheless believe that the use case shows suitability for practice. Future work will focus on implementing the framework. Here, challenges might include adjusting a DHT framework to support authorization and data feeds (although Swarm shows promise in this regard [7]), as well as selecting a suitable Distributed Ledger framework.

The Distributed Ledger and the concomitant smart contracts could also be handled in a different way. For instance, the AML could be transformed into classes and types in the smart contract, similar to the BPMN to Solidity transformation in [27]. However, the effort clearly outweighs the utility as AML is a very powerful standard allowing very complex descriptions. Moreover, not all of the hypothetically generated classes and functions might be needed. Plus, functions or classes might be newly added later on, which results in the need to re-create the smart contract as they are currently not represented in the smart contract. This clearly increases effort and downgrades utility.

Another issue is entailed by the possibility to directly alter variable values referring to an actual function in our current version of the ledger. For instance, consider a PLC device with various functions such as setting a conveyor belt's velocity (with an integer parameter). Without constraints, the changed velocity could exceed safety bounds. Safety threats like this one, be they malicious or accidental, need to be mitigated in a production system. Therefore, we suggest integrating safety and security rules as proposed in [6]. They could be integrated as part of the specification contract, with the Device agent checking conformance of program calls on synchronization.

With respect to the current problems hampering secure DT data sharing, our approach tackles the issues stated in Sect. 3 in the following ways:

- The usage of different tools that can be connected with our main data sharing approach (External data sources, Fig. 3) is possible (*application of different tools*)
- Our approach is tailored for the integration of data in multiple formats and variety as stated in Sect. 4.3 (*usage of various data formats*)
- An agreement only on the standard describing the asset (e.g. AML) is required to transform the main description of the asset into the specification smart contract, while other standardized or non-standardized data can still be shared via the DHT (*missing standards*)
- The proposed shared collaborative data basis is distributed among all involved parties and the information flow is universal across the lifecycle phases (*broken information flow across lifecycle phases*)
- The Distributed Ledger registers the data as well as the involved party sharing the data, while mechanisms such as access control (Authorization contract, Fig. 3) support confidentiality issues (*clarification of the ownership of information*).

To sum up, the major part of the identified issues in the literature referring to DT data sharing are diminished or solved by our approach.

7 Conclusion

DT data not only ties physical and virtual twin [23], it also enables integration of the whole asset lifecycle, which is essential for realizing the DT paradigm. Moreover, the exchange of asset-relevant data (DT data) is vital for achieving the effects of a feedback loop. Closing the feedback loop in turn favors the development of a circular economy.

However, maintaining data security becomes a major requirement when sharing DT data between multiple parties, especially as the parties do not necessarily trust each other. Our approach of applying DLT can clearly solve this issue and enable secure multi-party data sharing. It provides confidentiality through access control arranged by usage of a smart contract. Moreover, data integrity is implicitly supported through the immutability of the original data in the ledger.

To conclude, our approach fulfills the requirements **R1–R5** for secure DT data sharing. Nevertheless, there remain minor drawbacks that need to be addressed in future research (see Sect. 6.2). Our upcoming work will focus on implementing our theoretical concept to demonstrate its feasibility in practice.

References

1. Androulaki, E., et al.: Hyperledger fabric: a distributed operating system for permissioned blockchains. In: Proceedings of the Thirteenth EuroSys Conference, EuroSys 2018, pp. 30:1–30:15. ACM, New York (2018). https://doi.org/10.1145/3190508.3190538
2. Banerjee, A., Dalal, R., Mittal, S., Joshi, K.P.: Generating digital twin models using knowledge graphs for industrial production lines. In: Workshop on Industrial Knowledge Graphs, No. June, pp. 1–5 (2017). http://ebiquity.umbc.edu/paper/html/id/779/Generating-Digital-Twin-models-using-Knowledge-Graphs-for-Industrial-Production-Lines
3. Baumgartner, R.J.: Nachhaltiges Produktmanagement durch die Kombination physischer und digitaler Produktlebenszyklen als Treiber für eine Kreislaufwirtschaft. In: Interdisziplinäre Perspektiven zur Zukunft der Wertschöpfung (2018). https://doi.org/10.1007/978-3-658-20265-1_26
4. Boschert, S., Heinrich, C., Rosen, R.: Next generation digital twin. In: Proceedings of TMCE 2018, No. May (2018). https://www.researchgate.net/publication/325119950
5. Di Francesco Maesa, D., Mori, P., Ricci, L.: Blockchain based access control. In: IEEE Blockchain Conference 2018, pp. 1379–1386 (2018). https://doi.org/10.1007/978-3-319-59665-5_15
6. Eckhart, M., Ekelhart, A.: Towards security-aware virtual environments for digital twins. In: Proceedings of the 4th ACM Workshop on Cyber-Physical System Security - CPSS 2018, pp. 61–72 (2018). https://doi.org/10.1145/3198458.3198464
7. Ethereum Swarm Contributors: Swarm 0.3 documentation (2019). https://readthedocs.org/projects/swarm-guide/downloads/pdf/latest/
8. Glaessgen, E., Stargel, D.: The digital twin paradigm for future NASA and U.S. air force vehicles. In: 53rd AIAA/ASME/ASCE/AHS/ASC Structures, Structural Dynamics and Materials Conference (2012). https://doi.org/10.2514/6.2012-1818

9. Greengard, S.: Building a Better Iot (2017). https://cacm.acm.org/news/218924-building-a-better-iot/fulltext
10. ICS-CERT: Overview of cyber vulnerabilities. Technical report (2017). https://ics-cert.us-cert.gov/content/overview-cyber-vulnerabilities
11. Kieselmann, O., Wacker, A., Schiele, G.: k-rAC - a fine-grained k-resilient access control scheme for distributed hash tables. In: Proceedings of the 12th International Conference on Availability, Reliability and Security, ARES 2017, Reggio Calabria, Italy, pp. 1–43. ACM, New York (2017). https://doi.org/10.1145/3098954.3103154
12. Litke, A., Anagnostopoulos, D., Varvarigou, T.: Blockchains for supply chain management: architectural elements and challenges towards a global scale deployment. Logistics **3**(1) (2019). https://doi.org/10.3390/logistics3010005
13. Malakuti, S., Grüner, S.: Architectural aspects of digital twins in IIoT systems. In: Proceedings of the 12th European Conference on Software Architecture Companion Proceedings - ECSA 2018, pp. 1–2 (2018). https://doi.org/10.1145/3241403.3241417
14. Meroni, G., Plebani, P.: Combining artifact-driven monitoring with blockchain: analysis and solutions. In: Matulevičius, R., Dijkman, R. (eds.) CAiSE 2018. LNBIP, vol. 316, pp. 103–114. Springer, Cham (2018). https://doi.org/10.1007/978-3-319-92898-2_8
15. Negri, E., Fumagalli, L., Macchi, M.: A review of the roles of digital twin in CPS-based production systems. Procedia Manuf. **11**(June), 939–948 (2017). https://doi.org/10.1016/j.promfg.2017.07.198
16. Ovtcharova, J., Grethler, M.: Beyond the Digital Twin - Making Analytics come alive. visIT [Industrial IoT - Digital Twin], pp. 4–5 (2018). https://www.iosb.fraunhofer.de/servlet/is/81714/
17. Ríos, J., Hernández, J.C., Oliva, M., Mas, F.: Product avatar as digital counterpart of a physical individual product: literature review and implications in an aircraft. In: Advances in Transdisciplinary Engineering (2015). https://doi.org/10.3233/978-1-61499-544-9-657
18. Rubio, J.E., Roman, R., Lopez, J.: Analysis of cybersecurity threats in industry 4.0: the case of intrusion detection. In: D'Agostino, G., Scala, A. (eds.) CRITIS 2017. LNCS (LNAI and LNB), vol. 10707, pp. 119–130. Springer, Heidelberg (2018). https://doi.org/10.1007/978-3-319-99843-5_11
19. Sandhu, R.S., Samarati, P.: Access control: principles and practice. IEEE Commun. Mag. (1994). https://doi.org/10.1109/35.312842
20. Schroeder, G.N., Steinmetz, C., Pereira, C.E., Espindola, D.B.: Digital twin data modeling with automationML and a communication methodology for data exchange. IFAC-PapersOnLine **49**(30), 12–17 (2016). https://doi.org/10.1016/j.ifacol.2016.11.115
21. Talkhestani, B.A., Jazdi, N., Schloegl, W., Weyrich, M.: Consistency check to synchronize the Digital Twin of manufacturing automation based on anchor points. Procedia CIRP (2018). https://doi.org/10.1016/j.procir.2018.03.166
22. Tankard, C.: The security issues of the Internet of Things. Comput. Fraud Secur. **2015**(9), 11–14 (2015). https://doi.org/10.1016/S1361-3723(15)30084-1
23. Tao, F., Cheng, J., Qi, Q., Zhang, M., Zhang, H., Sui, F.: Digital twin-driven product design, manufacturing and service with big data. Int. J. Adv. Manuf. Technol. **94**(9–12), 3563–3576 (2018). https://doi.org/10.1007/s00170-017-0233-1
24. Uhlemann, T.H., Lehmann, C., Steinhilper, R.: The digital twin: realizing the cyber-physical production system for industry 4.0. Procedia CIRP (2017). https://doi.org/10.1016/j.procir.2016.11.152

25. Usländer, T.: Engineering of digital twins. Technical report, Fraunhofer IOSB (2018). https://www.iosb.fraunhofer.de/servlet/is/81767/
26. Voydock, V.L., Kent, S.T.: Security mechanisms in high-level network protocols. ACM Comput. Surv. (1983). https://doi.org/10.1145/356909.356913
27. Weber, I., Xu, X., Riveret, R., Governatori, G., Ponomarev, A., Mendling, J.: Untrusted business process monitoring and execution using blockchain. In: La Rosa, M., Loos, P., Pastor, O. (eds.) BPM 2016. LNCS, vol. 9850, pp. 329–347. Springer, Cham (2016). https://doi.org/10.1007/978-3-319-45348-4_19
28. Wüst, K., Gervais, A.: Do you need a blockchain? In: 2018 Crypto Valley Conference on Blockchain Technology (CVCBT), pp. 45–54 (2018). https://doi.org/10.1109/CVCBT.2018.00011
29. Xu, X., Pautasso, C., Zhu, L., Gramoli, V., Ponomarev, A., Tran, A.B., Chen, S.: The blockchain as a software connector. In: Proceedings - 2016 13th Working IEEE/IFIP Conference on Software Architecture, WICSA 2016, pp. 182–191. IEEE (2016). https://doi.org/10.1109/WICSA.2016.21

Refresh Instead of Revoke Enhances Safety and Availability: A Formal Analysis

Mehrnoosh Shakarami[(✉)] and Ravi Sandhu

Institute for Cyber Security (ICS), Center for Security and Privacy Enhanced
Cloud Computing (C-SPECC), Department of Computer Science,
University of Texas at San Antonio, San Antonio, USA
mehrnoosh.shakarami@my.utsa.edu, ravi.sandhu@utsa.edu

Abstract. Due to inherent delays and performance costs, the decision
point in a distributed multi-authority Attribute-Based Access Control
(ABAC) system is exposed to the risk of relying on outdated attribute
values and policy; which is the safety and consistency problem. This
paper formally characterizes three increasingly strong levels of consistency to restrict this exposure. Notably, we recognize the concept of
refreshing attribute values rather than simply checking the revocation
status, as in traditional approaches. Refresh replaces an older value with
a newer one, while revoke simply invalidates the old value. Our lowest
consistency level starts from the highest level in prior revocation-based
work by Lee and Winslett (LW). Our two higher levels utilize the concept
of request time which is absent in LW. For each of our levels we formally
show that using refresh instead of revocation provides added safety and
availability.

Keywords: ABAC · Refresh · Consistency · Safety · Availability

1 Introduction

In Attribute-Based Access Control (ABAC), access decisions are made based on
attribute values of subjects, objects and environment with respect to a given
policy. Attribute values for subjects and objects are typically provisioned by an
Attribute Authority (AA) and presented in credentials as name, value pairs. A
credential must be trustworthy, perhaps by a cryptographic signature or trusted
delivery. Attribute values are susceptible to change. Ideally the decision point
should know real-time values, which is practically impossible due to inherent
delays of distributed systems and performance costs. This can lead to granting
access when it should be denied (safety violation) or denying access when it
should be granted (availability violation). The longer the gap between updates
of credentials, the higher the risk of relying on stale attribute values.

© IFIP International Federation for Information Processing 2019
Published by Springer Nature Switzerland AG 2019
S. N. Foley (Ed.): DBSec 2019, LNCS 11559, pp. 301–313, 2019.
https://doi.org/10.1007/978-3-030-22479-0_16

In this paper we formally characterize three increasingly strong levels of consistency to restrict the exposure of the decision point to stale attribute values. For simplicity, we develop our formalism based on changing subject attribute values. Extension to changing object and environment attribute values is straightforward. Extension to policy changes is more subtle. Policy changes may require additional credentials come into play. While acquiring these additional credentials the policy may change again. In principle, this could lead to an infinite regress. In practice such an infinite regress is unlikely. Policies composed of multiple sub-policies specified by different authorities also raise issues of policy conflicts [5,17]. A formal treatment of policy changes is beyond our scope.

The closest prior work is by Lee and Winslett (LW) [15,16]. Our paper is inspired by LW but presents a completely new perspective by considering refresh instead of revocation. We build our levels on top of the highest consistency level of LW, recasting it in the refresh framework. Taking request time into account, we propose two higher consistency levels not available in LW.

Our main contribution is to develop a formal framework for safety, availability and consistency problems of ABAC systems, via introducing the refresh scenario instead of the traditional revocation check. As we will show, this enhanced possibility of getting a new value rather than an invalid response enhances safety and availability. We also define the concept of being `satisfactory` for an attribute value with respect to a policy, which is first introduced in our work to the best of our knowledge. Relying on the history of `satisfactory` attribute values, we introduce additional flexibility to grant access to authorized users.

The paper is organized as follows. A review of related work and a comparison to LW is given in Sect. 2. Section 3 documents our system model and assumptions. The formalism of our consistency levels along with guaranteed properties by each specification is given in Sect. 4. Limitations and practical implications are discussed in Sect. 5. Section 6 concludes the paper.

2   Related Work

There is a rich body of research work on consistency in distributed systems [1,2,8,20,24]. Many access control models are not completely compatible with distributed systems in that they are not deployed for such systems in the first place [10]. ABAC is well adjusted to distributed environments due to its flexibility and granularity. In this paper, we consider an ABAC model to be in place and define consistency as communicating credentials' updates as quickly as possible to the decision point. To the best of our knowledge there is very limited directly related research in this arena. Especially there is no work done toward utilizing the refresh operation to obtain recent information.

The closest to our research is LW in trust negotiation environments [14–16]. Another closely related research is on stale-safety which tries to safely uses stale attributes [12,13]. Although the problem is similar, it mainly differs from our work since it has been applied in a non-ABAC, single authority environment. Policy staleness of cloud transactions proposed in [11].

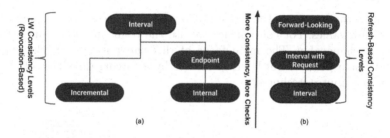

Fig. 1. (a) LW revocation-based levels [15,16] (b) Our refresh-based levels

Ciphertext-Policy Attribute-Based Encryption [3,4] is broadly applicable in decentralized multi-authority environments, but presents challenge to handle attribute revocation [22,26–28]. Moreover, it imposes a heavy performance burden which makes it impractical [6]. There are other researches concerning the policy consistency in distributed environments [6,11,18,29] focusing on cloud environments. In this paper, policy assumed to be known with high assurance. There are research works utilizing revocation in authenticated dynamic dictionaries [7,21,25], which enable dissemination of information from a secure central repository to multiple recepients.

Comparison to LW Model. LW presented the first organized work on consistency in trust negotiation systems. They proposed four consistency levels based on timeliness of credentials revocation checks. In common with our model they considered every credential to have its lifetime specified by start time and end time. While all levels in LW model utilize the notion of receive time of credentials, we are agnostic to it. We consider decision time as central and utilize it explicitly in all levels, whereas in LW it is explicit only in top two levels. Revocation check in LW is replaced with refresh in our model, as will be discussed in next section. An alternate formulation of LW without use of receive time is given in [23], which includes an additional level based on request time.

Figure 1-(a) shows the levels in LW which are partially ordered. We do not recommend using incremental and internal levels since in both cases decision point may use a credential which is known to be expired or revoked. Our proposed levels are shown in Fig. 1-(b) with a total order among levels. We set LW's highest level as the base level in our definitions. By taking request time into account, we propose two additional stronger levels of consistency. We provide further availability in our model by letting the decision point consider valid authorization should the current and cached values of relevant credentials be `satisfactory`, as defined in following sections.

3   System Model and Assumptions

We assume an ABAC authorization system in a distributed multi-authority environment. For a particular access request, there is a single decision point which determines whether or not the access is allowed by the access control policy based on attribute values. For convenience we use the terms attribute and attribute value interchangeably. The main focus of this paper is to limit the exposure of the decision point to outdated attributes by enforcing timeliness of checking subjects' attributes freshness.

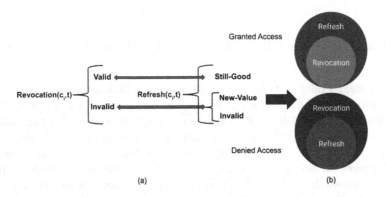

Fig. 2. (a) Revocation vs. Refresh (b) Comparing Grant vs. Deny

Table 1. Summary table of symbols

Symbol	Meaning
t_{req}	request time
t_d	decision time
c_i	i^{th} credential
t^i_{revoc}	actual revocation time of c_i (the AA always knows this time)
$t^i_{ref,k}$	time of k-th refresh of c_i
$t^i_{start,k}$	attribute start time of c_i after k-th refresh
$t^i_{end,k}$	attribute expiration time of c_i after k-th refresh
$kmax(t)$	latest refresh of c_i before time t (c_i is determined by context)
$val^i_{kmax(t)}$	the value of c_i after $kmax(t)$-th refresh
$t^i_{ref,kmax(t)}$	time of $kmax(t)$-th refresh of c_i
$t^i_{start,kmax(t)}$	attribute start time of c_i after $kmax(t)$-th refresh
$t^i_{end,kmax(t)}$	attribute expiration time of c_i after $kmax(t)$-th refresh

3.1   Refresh Vs. Revocation

Subject attributes might change during credential lifetime. A change could be a new value, a new lifetime or a premature revocation. In all cases the decision point needs to be updated about the latest changes of the attribute through either revocation or refresh. In revocation, AA would represent the current status of the credential as either `Valid` (no change) or `Invalid` (otherwise). However, with refresh AA can indicate the credential's status as `Still-Good`, `New-Value` or `Invalid`. `Still-Good` and `Invalid` correspond to `Valid` or `Invalid` in revocation scenario. `New-Value` reflects any change in credential's start time, end time or new value. So, `Invalid` status in revocation splits in two possibilities of `Invalid` and `New-Value` in refresh (see Fig. 2-a). Thereby, refresh can allow more accesses than revoke and deny fewer accesses (see Fig. 2-b).

Refresh function is defined as follows. T is the set of possible time stamps and C represents the set of all credentials in the system. Table 1 defines the symbols used in this definition and throughout the paper.

$$Refresh : C \times T \rightarrow \{Invalid, Still\text{-}Good, New\text{-}Value\} \tag{1}$$

$$Refresh(c_i, t) = \begin{cases} Invalid & \Longleftrightarrow (t \geq t^i_{end,kmax(t)}) \vee (t \geq t^i_{revoc}) \\ New\text{-}Value & \Longleftrightarrow (t^i_{start,kmax(t)} \neq t^i_{start,kmax(t)-1}) \\ & \vee (t^i_{end,kmax(t)} \neq t^i_{end,kmax(t)-1}) \vee (val^i_{kmax(t)} \neq val^i_{kmax(t)-1}) \\ Still\text{-}Good & \Longleftrightarrow (t^i_{start,kmax(t)} = t^i_{start,kmax(t)-1}) \\ & \wedge (t^i_{end,kmax(t)} = t^i_{end,kmax(t)-1}) \wedge (val^i_{kmax(t)} = val^i_{kmax(t)-1}) \end{cases}$$

Following example highlights the benefits provided by considering refresh rather than revocation. Although granting illegitimate access is considered as a greater risk in many systems, availability is also important in which a legitimate user should not be denied access.

Example 1. Authorization policy in a coding company grants read access to a project's code to managers and test engineers and read/write access to developers. Alice was a test engineer. But her role has changed to a developer in the same project. Subsequently she submits a write request to the decision point. In revocation, checking her cached role credential results in `Invalid` response since she is no longer a test engineer. So her request would be denied. In refresh, however, `New-Value` response along with a new credential asserting her new role would be returned and access would be granted, as it should be based on policy.

Claim. If a subject can proceed to utilize a requested access in a revocation scenario, it can proceed in a refresh scenario as well. But there are scenarios in refresh-based systems which let the subject proceed, whereas it would be denied in revocation-based systems.

Proof. If nothing changed about a required credential, revocation and refresh would return `Valid`/`Still-Good` respectively. So, the first part of the claim follows. For the second part, it is possible that a required credential has changed with respect to start/end time or the value. So AA response in revocation scenario will be `Invalid` which prohibits subject's access. However with refresh the response would be `New-Value`, so access would be granted (see Fig. 2).

3.2   System Assumptions

Without loss of generality, we suppose that the policy is stated in Disjunctive Normal Form (DNF), which is the disjunction of different conjuncts. The decision point tries to find the first conjunct which satisfies the desired level of consistency. This conjunct is called the `View` of the decision point at any specific time t with respect to the policy P which we denote as $V_{DP}^{P,t}$. We assume the decision point can instantaneously check the policy and identify the view.

Definition 1. *At any time t, we call the set of subject's attributes included in $V_{DP}^{P,t}$ as the relevant credentials.*

We make following assumptions in this paper.

1. Attributes do not change as the result of attribute credentials usage, that is we assume attributes to be immutable in sense of [19].
2. We will not utilize any expired credential. If any required credential is beyond its end time, decision point polls AA to get a new credential for the attribute.
3. We do not refresh any credential after it has been found to be `Invalid`.
4. There is one instantaneous decision time (t_d) and one instantaneous request time (t_{req}).
5. V_{DP}^{P,t_d} is the only view of our interest as described above.
6. If refresh returns a `New-Value` result, its start time cannot be prior to its previous start time, i.e., $t_{start,k}^i \geq t_{start,k-1}^i$.
7. AA will not return a credential along with `New-Value` which has not been started yet, so, $t_{ref,k}^i \geq t_{start,k}^i$.

4   Consistency Levels Formal Characterization

4.1   Preliminaries

Satisfactory Values. We define an attribute to be `satisfactory` if and only if its value fulfills the policy conditions. For instance if the policy requires the security level to be at least 3, any security level credential with the value greater than or equal to 3 is considered as `satisfactory`. Obviously the same credential may not be `satisfactory` with respect to another policy. We formally define `satisfactory` with respect to a policy P at the specific time t as follows.

Definition 2. *The view at time t has the structure $V_{DP}^{P,t} = \bigwedge_{1 \leq i \leq n} F(i)$ in which $F(i)$ is an atomic expression specifying required conditions for c_i's value. We define Sat as follows to determine satisfactory requirements for c_i's value.*

$$Sat_{c_i}^{P,t} = True \iff F(val_{kmax(t)}^i) = True \tag{2}$$

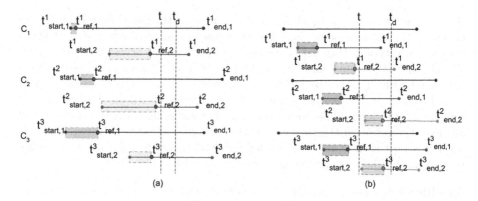

Fig. 3. Interval consistency

Freshness. We rely on the freshness concept in refresh scenario, compared to validity in revocation scenario. We formally define freshness via *Fresh* function as follows. When *Fresh* is used in a boolean expression, we understand $Fresh(c_i, t)$ to be *False* when its value is *Unknown*.

$$Fresh : C \times T \rightarrow \{\, True, False, Unknown \,\}$$

$$Fresh(c_i, t) = \begin{cases} True & \Longleftrightarrow (t^i_{start,k} \leq t \leq t^i_{ref,k}) \\ & \land (Refresh(c_i, t^i_{ref,k}) \neq Invalid) \\ Unknown & \Longleftrightarrow (t^i_{ref,k} < t < t^i_{end,k}) \lor (t \geq t^i_{ref,kmax(t)}) \\ False & \Longleftrightarrow [(t \geq t^i_{ref,k}) \land (Refresh(c_i, t^i_{ref,k} = Invalid))] \\ & \lor [t \geq t^i_{end,kmax(t)}]) \end{cases}$$

$$\tag{3}$$

Following example is used throughout the paper.

Example 2. In a company, project managers and testing engineers with the security level of at least 5 can access project's documents. The policy in DNF form is $P = [(role \in \{manager, engineer\}) \land (security\text{-}level \geq 5)]$. Bob is a project manager since January 1st to January 25th based on a refresh at January 15th. A refresh at January 21st shows his role has changed to testing engineer as of January 20th through March 20th. A refresh at January 15th shows his security level is 6 as of January 10th to March 20th. Another refresh at January 28th reveals security level has been downgraded to 4 since January 26th through March 20th.

We now introduce three levels of consistency taking both old and new values of relevant credentials into account. We provide specifications and consequent properties guaranteed by each level in the rest of this section.

4.2 Interval Consistency

At this level, it is required to find overlap of freshness intervals (simultaneous freshness) of relevant credentials before the decision time. In Example 2, suppose

Bob requests access to project documents on Jan 18th. Based on refresh results at Jan 15th, decision point finds simultaneous freshness of relevant credentials during Jan 10th-Jan 15th with satisfactory values. So the access will be granted. The stipulated overlap could be found for most recent refresh results of relevant credentials (Fig. 3-(a)) or by considering both old and new refresh results (Fig. 3-(b)). In these and subsequent figures if any refresh shown on the first line, it returns Still-Good while any other refresh returns New-Value. Moreover, in all cases the values of the three credentials are satisfactory. In Fig. 3-(a) the overlap is for the most recent refreshed values, whereas in Fig. 3-(b) the overlap is for a mix of the refreshed values, one new and two older.

Specification. Every credential has been refreshed at least once before the decision time and found to be fresh. Most recent values of all relevant credentials are satisfactory with respect to the policy. Any overlap of freshness intervals for the freshest/cached credentials is acceptable so long as the values are satisfactory.

$$Interval(V_{DP}^{P,t_d}) \iff (\exists t \le t_d)(\forall c_i \in V_{DP}^{P,t_d})$$

$$[\max_{\forall c_j \in V_{DP}^{P,t_d}} t_{start,kmax(t)}^j \le t_{ref,kmax(t)}^i < \min_{\forall c_i \in V_{DP}^{P,t_d}} t_{end,kmax(t)}^i$$

$$\land\ Fresh(c_i, t_{ref,kmax(t)}^i) \land Fresh(c_i, t_{ref,kmax(t_d)}^i) \land Sat_{c_i}^{P,t} \land Sat_{c_i}^{P,t_d} \qquad (4)$$

$$\land\ \max_{\forall c_i \in V_{DP}^{P,t_d}} t_{start,kmax(t_d)}^i < t_d < \min_{\forall c_i \in V_{DP}^{P,t_d}} t_{end,kmax(t_d)}^i]$$

Property 1. There is a time interval during which all relevant credentials were simultaneously fresh with satisfactory values with respect to the policy.

Proof. Based on Eq. (4), there exists a time (t) prior to the decision time at which the latest refresh of every relevant credential happens after all have been started and before any of them ends. This implies all credentials are simultaneously fresh during $[\max_{\forall c_i \in V_{DP}^{P,t_d}} t_{start,kmax(t)}^i, \min_{\forall c_i \in V_{DP}^{P,t_d}} t_{ref,kmax(t)}^i]$.

Corollary 1. if $t = t_d$, latest values of relevant attributes have freshness overlap.

Comparing with Revocation-Based Scenario. Based on the Claim in Sect. 3.1, revocation and refresh are the same in case of Valid and Still-Good responses from AA. But if the result is New-Value, the corresponding revocation result would be Invalid which denies the access. In Example 2, if Bob requests access to the project's documents on Jan 25th and decision point rechecks the credentials, although Bob's role has changed, he would get the access in the refresh scenario whereas he would be denied in revocation scenario.

4.3   Interval Consistency with Request Time

In first level, the decision point relies on what avails of previous refresh results for relevant credentials and access would be denied in case of any unrefreshed

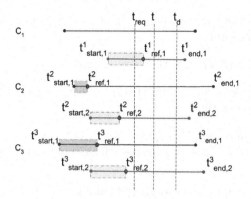

Fig. 4. Interval consistency with request time

credential. By considering the request time we could compensate for missing refreshes. In Example 2, if Bob requests for accessing project's documents on January 14th, the access would be denied at first level since there is no refresh result available for required credentials. At second level, the decision point refreshes the credentials after the request time and then checks the consistency requirements. Figure 4 shows similar example where the top credential is refreshed after request time.

Specification. Decision point refreshes any credential with missing refresh results after the request time. Afterwards, relevant credentials should satisfy the interval consistency (previous level) requirements.

$$
\begin{aligned}
IntervalWithReq(V_{DP}^{P,t_d}) &\iff (\forall c_i \in V_{DP}^{P,t_d})\ [t_{ref,kmax(t_{req})}^i \neq \bot \\
&\vee (\exists t_r\ t_{req} < t_r < t_d)\ Refresh(c_i, t_r)] \wedge Interval(V_{DP}^{P,t_d})
\end{aligned}
\tag{5}
$$

Proposition 1. *We assume the set of relevant credentials would not change during the short gap between request time and decision time, so, $V_{DP}^{P,t_{req}} = V_{DP}^{P,t_d}$. In other words the policy will not frequently change in the system.*

Property 1. There is a time interval during which all relevant credentials are simultaneously fresh. Possible lack of refresh would not unnecessarily deny access.

Proof. Use of the same requirement of $Interval(V_{DP}^{P,t_d})$ guarantees the same property of freshness overlap of relevant credentials. Any missing refresh results would be compensated after request time. It is possible that the gap between the request time and decision time does not last enough to compensate all the lacking information, but we consider it as an administrative setting which is out of scope for this paper to quantify.

Property 2. Every interval consistent view with request time satisfies the interval consistency requirements as well.

Proof. The proof is trivial since this level is defined based on interval level.

Property 3. An interval consistent view may deny access allowed by interval consistent with request time.

Proof. Since we do not consider request time in first level, there is no opportunity to compensate possible missing refreshes which could enable access.

Comparing with Revocation-Based Scenario. Considering the formal specification in Eq. (5), which is based on first level, the comparison is trivial. If refresh is substituted with revocation, system's availability would decrease as discussed in Sect. 3.1. The same situation may happen with regard to Example 2 as discussed in Sect. 4.2.

4.4 Forward-Looking Consistency

This level provides simultaneous freshness of all relevant credentials *after the request time*, considering both new and old credentials. Overlapping interval could either include the request time (Fig. 5-(a)) or not (Fig. 5-(b)). In Example 2, if Bob requests access to project's documents on Feb 1st his credentials would be refreshed afterwards revealing changes in role and security level leading to deny. However in previous levels an unauthorized access may be granted.

Specification. Any relevant credential has to be refreshed at least once after the request time. All relevant credentials have to be found simultaneously fresh at or after the request time.

$$
\begin{aligned}
ForwardLooking(V_{DP}^{P,t_d}) &\iff (\exists t\ t_{req} < t \le t_d)(\forall c_i \in V_{DP}^{P,t_d})[(t_{req} < t^i_{ref,kmax(t)}) \\
&\wedge (\max_{\forall c_i \in V_{DP}^{P,t_d}} t^i_{start,kmax(t)} \le t^i_{ref,kmax(t)} < \min_{\forall c_i \in V_{DP}^{P,t_d}} t^i_{end,kmax(t)}) \\
&\wedge Fresh(c_i, t^i_{ref,kmax(t)}) \wedge Fresh(c_i, t^i_{ref,kmax(t_d)}) \wedge Sat_{c_i}^{P,t} \wedge Sat_{c_i}^{P,t_d} \\
&\wedge \max_{\forall c_i \in V_{DP}^{P,t_d}} t^i_{start,kmax(t_d)} < t_d < \min_{\forall c_i \in V_{DP}^{P,t_d}} t^i_{end,kmax(t_d)}]
\end{aligned}
$$

$$(6)$$

Property 1. There is a time interval during which all relevant credentials are simultaneously fresh after the request time.

Proof. Based on Eq. (6), all relevant credentials are simultaneously fresh during $[\max_{\forall c_i \in V_{DP}^{P,t_d}} t^i_{start,kmax(t)}, \min_{\forall c_i \in V_{DP}^{P,t_d}} t^i_{ref,kmax(t)}]$. Part of this interval is located after the request time since refresh has been done after it.

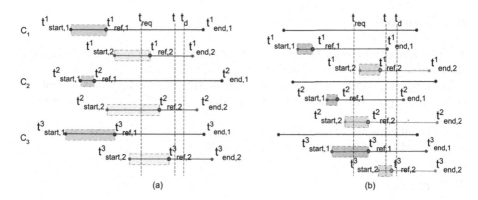

Fig. 5. Forward looking consistency

Property 2. Every forward-looking consistent view is interval consistent with request time as well.

Proof. Comparing Eqs. (5) and (6) shows forward-looking consistency is a restricted version of its preceding level, so the proof is trivial.

Property 3. Not every interval consistent with request time view is necessarily forward-looking as well.

Proof. At second level of consistency, only some credentials need to be refreshed after request time to compensate for lacking information. Whereas in forward-looking consistency, all have to be refreshed after the request time.

Comparing with Revocation-Based Scenario. Changing credentials in revocation scenario leads to hinder the access, whereas in refresh scenario, the New-Value in case of any changes would let the subject proceed. In Example 2, Bob's request to access project's documents at Jan 20th would be denied in a revocation-based scenario, however in refresh scenario access would be granted.

5   Limitations and Practical Issues

We presented three levels of consistency, where each higher level provides enhanced availability and safety at the cost of refreshing more frequently. We compared qualitative benefits of each level. Quantifying cost-benefit is highly implementation and application specific, and is beyond the scope of this paper. Furthermore, there are issues related to manage the risks inherent to applying ABAC in a distributed environment, since ABAC introduces new challenges in selecting appropriate trust models [9]. Finally, the formal correctness and appropriateness of the proposed criteria notwithstanding, the underlying information could be vulnerable to attack. The attack models would depend on the particular protocols and data structures used to implement credential transfer and refresh. As such they are out of scope for an abstract framework.

6   Conclusion

We formally characterize the safety and availability problem in multi-authority distributed ABAC systems. Our major contribution is to utilize the concept of refresh, which provides new attribute values rather than simply invalidating old ones. We propose three consistency levels which are totally ordered in strictness.

Acknowledgements. This work is partially supported by NSF CREST Grant HRD-1736209 and DoD ARL Grant W911NF-15-1-0518.

References

1. Adya, A.: Weak consistency: a generalized theory and optimistic implementations for distributed transactions. Ph.D. thesis, MIT (1999)
2. Bernstein, P.A., Goodman, N.: Concurrency control in distributed database systems. ACM Comput. Surv. **13**, 185–221 (1981)
3. Chase, M.: Multi-authority attribute based encryption. In: Vadhan, S.P. (ed.) TCC 2007. LNCS, vol. 4392, pp. 515–534. Springer, Heidelberg (2007). https://doi.org/10.1007/978-3-540-70936-7_28
4. Chase, M., Chow, S.S.: Improving privacy and security in multi-authority attribute-based encryption. In: ACM CCS (2009)
5. Cheminod, M., Durante, L., Valenza, F., Valenzano, A.: Toward attribute-based access control policy in industrial networked systems. In: IEEE WFCS (2018)
6. Garrison, W.C., et al.: On the practicality of cryptographically enforcing dynamic access control policies in the cloud. In: IEEE S&P (2016)
7. Goodrich, M.T., Shin, M., Tamassia, R., Winsborough, W.H.: Authenticated dictionaries for fresh attribute credentials. In: Nixon, P., Terzis, S. (eds.) iTrust 2003. LNCS, vol. 2692, pp. 332–347. Springer, Heidelberg (2003). https://doi.org/10.1007/3-540-44875-6_24
8. Harding, R., Van Aken, D., Pavlo, A., Stonebraker, M.: An evaluation of distributed concurrency control. Proc. VLDB Endow. **10**, 553–564 (2017)
9. Hu, V.C., et al.: Guide to attribute based access control (ABAC) definition and considerations. NIST SP 800-162 (2019)
10. Hu, V.C., Kuhn, D.R., Ferraiolo, D.F.: Access control for emerging distributed systems. IEEE Comput. **51**, 100–103 (2018)
11. Iskander, M.K., Wilkinson, D.W., Lee, A.J., Chrysanthis, P.K.: Enforcing policy and data consistency of cloud transactions. In: IEEE ICDCSW (2011)
12. Krishnan, R., Niu, J., Sandhu, R., Winsborough, W.H.: Stale-safe security properties for group-based secure information sharing. In: ACM FMSE (2008)
13. Krishnan, R., Sandhu, R.: Authorization policy specification and enforcement for group-centric secure information sharing. In: Jajodia, S., Mazumdar, C. (eds.) ICISS 2011. LNCS, vol. 7093, pp. 102–115. Springer, Heidelberg (2011). https://doi.org/10.1007/978-3-642-25560-1_7
14. Lee, A.J., Minami, K., Winslett, M.: Lightweight consistency enforcement schemes for distributed proofs with hidden subtrees. In: ACM SACMAT (2007)
15. Lee, A.J., Winslett, M.: Safety and consistency in policy-based authorization systems. In: CCS. ACM (2006)
16. Lee, A.J., Winslett, M.: Enforcing safety and consistency constraints in policy-based authorization systems. In: TISSEC. ACM (2008)

17. Lupu, E.C., Sloman, M.: Conflicts in policy-based distributed systems management. IEEE Trans. Softw. Eng. **25**, 852–869 (1999)
18. Myers, M., Ankney, R., Malpani, A., Galperin, S., Adams, C.: X.509 internet public key infrastructure online certificate status protocol-OCSP (RFC 6960)
19. Park, J., Sandhu, R.: The $UCON_{ABC}$ usage control model. In: ACM TISSEC (2004)
20. Perrin, M.: Distributed Systems: Concurrency and Consistency. Elsevier, Amsterdam (2017)
21. Reyzin, L., Meshkov, D., Chepurnoy, A., Ivanov, S.: Improving authenticated dynamic dictionaries, with applications to cryptocurrencies. In: Kiayias, A. (ed.) FC 2017. LNCS, vol. 10322, pp. 376–392. Springer, Cham (2017). https://doi.org/10.1007/978-3-319-70972-7_21
22. Sciancalepore, S., et al.: On the design of a decentralized and multiauthority access control scheme in federated and cloud-assisted cyber-physical systems. IEEE IoT J. **5**, 5190–5204 (2018)
23. Shakarami, M., Sandhu, R.: Safety and consistency of subject attributes for attribute-based pre-authorization systems. In: NCS. Springer, Heidelberg (2019)
24. Van Steen, M., Tanenbaum, A.S.: Distributed Systems (2017)
25. Tamassia, R., et al.: Independently verifiable decentralized role-based delegation. IEEE Syst. Man Cybern.-Part A: Syst. Hum. **40**, 1206–1219 (2010)
26. Yang, K., Jia, X.: Attributed-based access control for multi-authority systems in cloud storage. In: IEEE ICDCS (2012)
27. Yang, K., Jia, X.: Expressive, efficient, and revocable data access control for multi-authority cloud storage. IEEE Parallel Distrib. Syst. **25**, 1735–1744 (2014)
28. Yang, K., et al.: DAC-MACS: effective data access control for multiauthority cloud storage systems. IEEE Inf. Forensics Secur. **8**, 1790–1801 (2013)
29. Zahoor, E., Ikram, A., Akhtar, S., Perrin, O.: Authorization policies specification and consistency management within multi-cloud environments. In: Gruschka, N. (ed.) NordSec 2018. LNCS, vol. 11252, pp. 272–288. Springer, Cham (2018). https://doi.org/10.1007/978-3-030-03638-6_17

Source Code Security

Wrangling in the Power of Code Pointers
with ProxyCFI

Misiker Tadesse Aga[(✉)], Colton Holoday, and Todd Austin

University of Michigan, Ann Arbor, USA
{misiker,choloday,austin}@umich.edu

Abstract. Despite being a more than 40-year-old dark art, control flow attacks remain a significant and attractive means of penetrating applications. Control Flow Integrity (CFI) prevents control flow attacks by forcing the execution path of a program to follow the control flow graph (CFG). This is performed by inserting checks before indirect jumps to ensure that the target is within a statically determined valid target set. However, recent advanced control flow attacks have been shown to undermine prior CFI techniques by swapping targets of an indirect jump with another one from the valid set.

In this article, we present a novel approach to protect against advanced control flow attacks called ProxyCFI. Instead of building protections to stop code pointer abuse, we replace code pointers wholesale in the program with a less powerful construct – pointer proxies. Pointer proxies are random identifiers associated with legitimate control flow edges. All indirect control transfers in the program are replaced with multi-way branches that validate control transfers with pointer proxies. As pointer proxies are uniquely associated with both the source and the target of control-flow edges, swapping pointer proxies results in a violation even if they have the same target, stopping advanced control flow attacks that undermine prior CFI techniques. In all, ProxyCFI stops a broad range of recently reported advanced control flow attacks on real-world applications with only a 4% average slowdown.

Keywords: CFG mimicry attacks · CFI · Pointer proxy

1 Introduction

For more than four decades, control flow attacks, in which attackers force programs into executing code sequences not anticipated by the developer, have played an important role in the infiltration of vulnerable systems. These attacks are particularly attractive to attackers because they immediately give them the agency necessary to deploy attack payloads, leak important information, embed a rootkit, launch an additional attack such as privilege escalation, etc. As such, there has been much attention paid to reducing a system's vulnerability to control flow attacks.

© IFIP International Federation for Information Processing 2019
Published by Springer Nature Switzerland AG 2019
S. N. Foley (Ed.): DBSec 2019, LNCS 11559, pp. 317–337, 2019.
https://doi.org/10.1007/978-3-030-22479-0_17

Early measures to stop control flow attacks include StackGuard [16], data execution prevention (DEP) [1,2] and address space layout randomization (ASLR) [3]. However, subsequent attacks have skirted this defenses [27,34,35]. CFI [10] follows a principled approach to mitigating control flow attacks by enforcing the runtime execution path of a program to adhere to the statically determined CFG. It does these by checking if the target of an indirect jump is within a valid set of targets. However, prior proposed CFI solutions are either impractical or ineffective. Some, which strictly follow a program's CFG [28], have high overheads that render them impractical to production systems. Others attempt to reduce overheads by approximating the CFG with limited classes of targets (*e.g.*, two classes for function pointers and return addresses) [4,6,41,42], but these do not protect against control flow attacks that swap targets while remaining on the CFG [14,20,21,33].

In this work, we make the key observation that many of the vulnerabilities in control flow stem from the excessive power inherent in code pointers. To stop the tide of control flow attacks, we propose a novel approach to control flow integrity, called ProxyCFI, that replaces all code pointers in the program with *pointer proxies*. A pointer proxy is a unique random identifier (64-bits in our implementation), which represents a forward or backward control flow edge in the program. Consequently, all indirect jumps in the program (*e.g.*, returns and jumps-through-register) are replaced with multi-way branches that implement a direct jump to the address associated with the pointer proxy. As pointer proxies are a function of both the source and the target of an edge, swapping pointer proxies results in a violation even if they have the same target.

To ensure that all execution flows stay on the program CFG for even third-party ProxyCFI compliant code, a binary-level program verifier first validates at load-time that programs and libraries have CFGs that are fully discoverable, use only pointer proxies, and avoid all indirect jumps/returns. Finally, to thwart attacks based on binary analysis, the verifier re-randomizes pointer proxies at load time. In addition, the loader marks code sections unreadable, to protect from active-read attacks that gather pointer proxies using memory leaks.

Table 1. Comparison of Code Pointers to Pointer Proxies. Pointer proxies preserve program control integrity by reducing their capabilities. This table lists the differences in capabilities between code pointers and pointer proxies. Ultimately, it is the powerful nature of code pointers that enable many control flow attacks.

	Code pointers	Pointer proxies
Arithmetic allowed	Yes	No
Totally ordered	Yes	No
Trivial forgery attacks	Yes	No
Permit relative distance attacks	Yes	No
Replay attacks on returns and fptrs	Yes	Only from the same source address

More importantly, ProxyCFI has a number of powerful features to deter attacks that mimic legitimate control flow (*i.e.*, control flow attacks that seemingly remain on legitimate control flow edges), such as control flow bending (CFB) [14]. These attacks exploits the fact that existing CFI techniques allow executions to maliciously divert indirect branches if the target address is still in the valid set of targets. ProxyCFI thwarts this as a pointer proxy is unique to a particular source and target address which makes a pointer proxy used in one function context invalid in another even if they share the same target addresses.

Table 1 lists the comparative capabilities of traditional code pointers versus pointer proxies. As shown in the table, pointer proxies do not support arithmetic manipulation; thus, relative-address based control flow attacks, such as ASLR de-randomization attacks [34] would not be possible with pointer proxies. Moreover, pointer proxies are much more difficult to forge, since their assignment is not in anyway related to other pointer proxies, whereas pointer values often reveal much information through relative address distances to other code objects, facilitating relative address inspired attacks. Since pointer proxies are unique to a given function, return address copy attacks, such as the return-into-libc [37] and backward-edge active-set attacks [36], become more challenging, as the pointer proxies of other functions (which are assigned at load-time) must be leaked and then translated to the local function's proxies (which have no correlation even if the current function calls the intended target).

1.1 Contributions of This Paper

In this paper, we introduce ProxyCFI, a novel control flow integrity technology that thwarts recent advanced control flow attacks while incurring low performance overhead. Specifically, this paper makes the following contributions:

- We present ProxyCFI that provides an efficient and practical protection against advanced code reuse attacks, with a threat model notably more capable than that of traditional CFI techniques which must protect a shadow stack [32] or pointer encryption technologies which must protect encryption keys [17].
- We detail the implementation of ProxyCFI within the GNU GCC compiler toolchain.
- We demonstrate the efficiency of the approach running a wide range of CPU-centric and network-facing applications. In addition, we implement two compile-time optimizations, which ultimately reduce the slowdown of this technology to only 4% on average. In addition, our security analysis shows that the technology stops real-world advanced control flow attacks and demonstrates 100% coverage for the RIPE x86-64 control flow attack suite [40].

2 Protecting Control Flow with ProxyCFI

In this section, we detail our threat model and the broad ProxyCFI concept, then present how to build and verify programs (including shared libraries) with pointer proxies.

2.1 Threat Model

In this work, we assume a very powerful attacker who wants to redirect control flow to a code sequence that deviates from the programmer-specified CFG. In accomplishing their control flow attack, the attacker has read and write access to any data location, including globals, stack and heap variables, as well as data storage locations holding pointer proxies. The code segment of the program is assumed to be non-writable.

Given this powerful attacker, ProxyCFI works to prevent the attacker from hijacking control from the programmer-specified CFG. In addition, ProxyCFI also gives protection against non-gadget code reuse attacks (e.g., COOP where the attack does not leave the CFG of the program, but instead enlists the code in a CFG mimicry attack [33]).

2.2 Pointer Proxies

To stop control flow attacks, we replace all code pointers with pointer proxies. A pointer proxy is a random identifier (64-bits in our evaluated implementation), where pointer proxy P represents an edge from a particular code exit point to a code entry point. Wherever a code pointer would reside in the program (e.g. in a jump table or on the stack as a return address), it is replaced by its corresponding pointer proxy value P. Figure 1 illustrates a small code snippet in which the code pointers have been replaced by pointer proxies. As seen in the example, where code pointers would have been stored (e.g., on the stack for a return address), they are replaced with pointer proxies (denoted by a $).

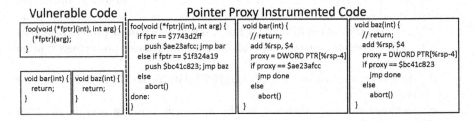

Fig. 1. Example Code Sequence using Pointer Proxies. Pointer proxies replace code pointers in a program with random identifiers associated with legal control flow edges. Multi-way direct branches translate pointer proxies into direct jumps.

At indirect jumps and returns, the pointer proxy is inspected, and then using a multi-way direct branch, the appropriate code entry point associated with the pointer proxy is jumped to. We call a multi-way direct branch, which matches a pointer proxy and then directly jumps to the associated code target, a *sled*. Direct jumps are not replaced with pointer proxies. Since our threat model assumes that code cannot be written, any direct jump is inherently write-protected, and thus no additional protections are required. Three multi-way branches can be seen in the example in Fig. 1. The indirect call to *bar()* and *baz()* in function *foo()* is implemented with a multi-way branch that jumps to *bar()* if the proxy *$7743d2ff* is encountered and jumps to *baz* when the pointer proxy is *$1f324a19*. Additionally, both of the returns from functions *bar()* and *baz()* are implemented with a multi-way branch.

Advanced control flow attacks such as control low bending [14] undermine CFI by using a code pointer copied from one function context to jump to addresses in some other function without violating CFI constraints. Pointer proxies are uniquely assigned to control flow edges (*i.e.*, a function of source and destination), thus, the pointer proxies of function X are meaningless to function Y. This powerful feature, which does not impact the usability of pointer proxies, thwarts large number of advanced control flow attacks. This aspect is shown in Fig. 1 in the returns of functions *bar()* and *baz()*. While both functions return to the same address (*i.e.*, label *done*), they each use a distinctly different pointer proxy. As such, if each of the functions were to disclose each other's pointer proxy and return upon it, it would not match any target in the return's multi-way branch, and would result in a violation. To stop the potential forgery of pointer proxies, all pointer proxy values are defined per-function, and they are re-randomized at program load time by the verifier as detailed in Sect. 2.4.

2.3   Building Code with Pointer Proxies

Building code to work with pointer proxies requires replacing every place in the program that uses a code pointer with a pointer proxy.

Hardening indirection with pointer proxies: All indirect branches (*e.g.*, switch jump tables, indirect calls, and returns), are replaced with a multi-way direct branch sleds. Of course, to know what targets must be tested for in a sled, we must fully anticipate all of the targets of each indirect branch. For locally sourced indirect jumps, such as a switch statement jump table, we can easily anticipate in the current compilation module all of the indirect jump targets. Indirect calls and returns require more analysis as the locations (and pointer proxy values) may come from another application module (*e.g.*, a shared library). Consequently, the compilation framework must support whole-program CFG analysis (including the call graph). In our prototype implementation in the GNU GCC toolchain, we utilize a two-pass compilation strategy, to first build the whole program CFG and then to compile programs with fully enumerated multi-way direct branch sleds. Details of this compilation strategy are covered in Sect. 3.

At indirect calls, the type of the target function is noted, and when the whole program CFG is constructed, an indirect function call is assumed to possibly happen to only same-typed function and has had its address taken. Similarly, the return address sleds target the instruction after all actual and potential calls to the returning function, where a potential call directly targets the function or indirectly targets the function through a compatible function pointer. This approach works quite well until a program declares a *void* * indirect function call, which could potentially call any function in the program. Fortunately, our optimizations detailed in Sect. 3.2 perform well to reduce the overall impact of these generic indirect function calls. To ensure that there is a valid sled entry for code pointers used across modules, ProxyCFI performs whole program analysis including shared libraries to ensure indirect calls from a shared library into the executable have a corresponding sled entry to handle callback functions. A similar analysis is performed for callbacks passed into shared libraries.

When Good Code Pointers Go Bad: ProxyCFI doesn't support dangerous code pointer operations such as code pointer arithmetic, which are characteristics of a buggy program [29]. For example, we could manufacture a code pointer to a private function in x86-64 GCC by simply adding the size of the preceding function (in bytes) to its code pointer. In our prototype GNU GCC C/C++ implementation of ProxyCFI, we issue a security warning for these operations at compile time and terminate the program if executed at runtime. While we were able to create these problems in test programs, none of our benchmark programs suffered from dangerous code pointer manipulations.

setjump() and *longjmp()* require special handling in the compiler because these functions implement a unique user-directed program control flow transition. Together, these functions implement a superset of function pointer behavior, such that a call to *setjmp()* can be the target of any other *longjmp()* in the program. Both function pointers and *longjmp()* share an indirect jump, but a *longjmp* should generate a multi-way direct branch including all of the pointer proxies assigned to the instruction immediately after each call to *setjmp()*. Thus, any tampering in the *setjmp()* control context cannot pull execution off the CFG.

2.4   Load-Time Program Verifier

Load-time Program Verifier ensures no legal control flow results in a violation and no deviation from the CFG is missed. It maintains this by verifying that programs utilize only pointer proxies for indirect branches via multi-way direct branches that are fully enumerated by the programmer-specified CFG. If an unexpected pointer proxy is encountered, the program is terminated.

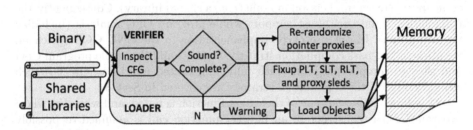

Fig. 2. ProxyCFI Program Loader. This figure illustrates the process of loading a ProxyCFI compliant program for execution.

Figure 2 shows how a binary or shared library is loaded and validated. The verifier ensures that only pointer proxies are used for control transitions. If a code object passes verification, the verifier generates load-time assigned pointer proxies, such that an attacker cannot anticipate any pointer proxy values even with an active read attack on the program.

The psuedocode for the ProxyCFI code verifier is shown in Algorithm 1. The verifier performs reachability analysis on the code object's CFG to validate that it is *(i)* free of indirect control transfers and *(ii)* all control transfers point to a valid instruction within the current code object or the entry point of another code object for calls. To this end, it performs a depth-first traversal of the CFG of the code object, inspecting all control transfer instructions. If indirect call/jump or return instructions are encountered in the code object, it immediately fails verification. For direct control transfer (*i.e.*, direct call, direct jump, loop instructions), it analyzes the target address for any possible violations.

For direct jump instructions, the verifier checks that the target address points to a valid instruction within the current code object. For direct function calls, the verifier validates that the target is a valid code object entry point. Finally, for multi-way branch sleds replacing an indirect call/jump and return, the verifier ensures that all targets are valid according to the CFG. Once the verifier completes reachability analysis of the CFG without failure, the code is safe to load and execute.

Algorithm 1. Load-time Program Verifier. The algorithm performs a reachability analysis of the CFG to identify any illegal jumps or uses of indirection.

```
 1: procedure VERIFY(obj)
 2:    for all f in obj do
 3:        ep ←f.entry_point
 4:        while ep ≠ ∅ do
 5:            e ← ep.pop()
 6:            if e.checked == True then
 7:                continue
 8:            else
 9:                br ←scan_for_next_branch(e)
10:                switch br do
11:                    case Indirect(br) or Invalid(br.target)
12:                        return Fail
13:                    case Direct_Branch
14:                        ep.push({br.target, e.next})
15:                    case sled
16:                        inspect ({sled.proxies})
17:                        ep.push({sled.targets})
18:    return Success
```

2.5   Deterring CFG Mimicry Attacks

A mimicry attack [38] on the CFG is one that implements attacker-directed control *without* leaving the programmer-specified CFG. With the introduction of powerful control flow integrity mechanisms, such as CFI [10] these non-gadget code reuse-based attacks have quickly grown in number, including counterfeit

OOP [33], control flow bending [14], and active-set backward-edge attacks [36]. ProxyCFI can provide protection against these attacks through per-function pointer proxy namespaces and load-time pointer proxy assignment.

Per-function pointer proxy namespaces: Traditional full-fledged code pointers represent a code location that is sharable with any other part of the program. It is this property that allows an adversary to copy a code pointer from one function and replay it in another, an approach that Counterfeit OOP [33] utilizes to implement method-level code reuse that does not leave the CFG. Pointer proxies deter CFG mimicry attacks by defining unique pointer proxy namespaces for each function. Thus, if a function copies the pointer proxy from another function, for example, by searching for pointer proxies up the stack, any attempt to use it will always result in a violation when the multi-way branch sled executes. In Fig. 1 although *bar*() and *baz*() both return to the same location, they each have their own proxy namespace which have different proxy-to-edge mappings (i.e., $ae23afcc and $bc41c823).

Load-time assignment of pointer proxies: Pointer proxies are re-randomized at load time to further deter mimicry attacks. This prevents offline analysis of code to generate a translation table from pointer proxies to source and target code addresses. Enforcing a non-readable code section and load-time assignment of proxies significantly complicates CFG mimicry attacks.

2.6   Shared Libraries with Pointer Proxies

Shared libraries are an attractive target for control flow attacks as they are used among multiple applications. Attacks that target libc, for example, can be reused on any application that links to it. The classic attack is *return-into-libc* [37], wherein the adversary overwrites the return address so that the program returns to an exploitable *libc* function such as *system()*. In addition, most shared libraries contain large enough codebases that leaves an attacker with wide selection of gadgets for all classes of code reuse attacks, such as ROP [18], JOP [12], LOP [25] and their variants [13, 22].

Connections into and out of shared libraries must be managed by unshared data or code that is generated dynamically. Returns and indirect calls are natural solutions to entering and exiting shared libraries because they draw on unshared data in the stack and the global offset table, respectively. As such, shared libraries require special handling with ProxyCFI which works to remove indirection. One solution to securing shared libraries is to forbid them – Intel chose this with SGX [23]. However, we want to retain their advantages: modularity, reduced page swapping, and simplified version management.

ProxyCFI compliant shared libraries use *unshared code* to manage control flow in and out because indirection must be replaced with multi-way branches. While calling a shared library function still goes through the *procedure linkage table (PLT)*, the indirect call within the PLT is dynamically replaced with a pointer proxy sled. At load time, extra space in the caller's address space is mmap'ed for the code that channels control flow on return. Shared library functions have their returns statically replaced with a relative jump down to the unshared multi-way branches.

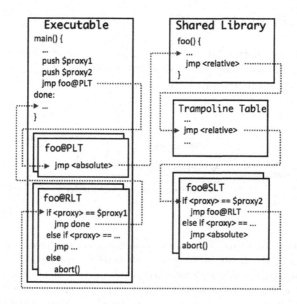

Fig. 3. Shared Library Control Flow. ProxyCFI allocates linkage tables using pointer proxies in the caller's address space, which permit safe entry and exit from a compliant shared library.

Our approach to deploying shared libraries with ProxyCFI is illustrated in Fig. 3. We split the process of returning from a shared library into two stages, which are associated with the *selection linkage table (SLT)* and the *return linkage table (RLT)*, respectively. Two pointer proxies are used to return from a shared library, one for the SLT and one for the RLT. For each PLT entry to a shared library function there is an RLT entry that contains a multi-way branch leading back to each call site in the code object. If shared library function *foo*() is called from multiple code objects, then each code object would have a separate RLT entry for *foo*() in its own address space. While the RLT specifies how to return within a code object, the SLT specifies which code object to return to. To accomplish this, SLT entries contain a multi-way branch of absolute jumps directed at RLT entries. Since the size of SLT entries varies based on the code objects that use the shared library, a trampoline table facilitates static generation of the relative jumps by forwarding control onto the appropriate SLT entry.

Using load-time assignment, it is possible to assign proxies for the tendrils into a shared library in a way that creates pointer proxies when the library is first loaded. Moreover, our approach allows the pointer proxies used to enter and exit the shared library function to be unique to each address space that utilizes the shared library. This ensures that an attacker cannot gather pointer proxy information from their own address space and use it to attack a program using the same shared library.

3   ProxyCFI in GNU GCC

In this section, we detail the implementation of ProxyCFI in the GNU GCC C/C++ toolchain. We present the overall compilation flow, and then dive into the details of the optimizations implemented.

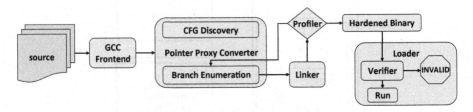

Fig. 4. Compilation Flow. The compilation occurs in two passes. In the first pass, the entire CFG of the program is discovered. In the second pass, all legal program entrypoints are assigned randomly selected pointer proxy identifiers, and all indirect jumps, indirect calls, and returns are replaced with fully enumerated multi-way direct branches. The linker resolves all jumps using compiler-generated global identifiers for all entry points. The profiler instruments the code to count the most frequent targets of multi-way branches, which is used for optimization. Finally, all code is passed through the pointer proxy verifier, assigned load-time random pointer proxies and loaded into execute-only pages before execution begins.

3.1   Compilation Flow

ProxyCFI instrumentation is done with a two-pass transformation on assembly generated mid-compilation by the existing GCC infrastructure. All sites of indirection are replaced with fully enumerated multi-way direct branches that validate CFG transitions with pointer proxies. Figure 4 describes the overall flow of ProxyCFI compilation.

- **Pass 1. CFG Discovery**: Assembly files are parsed for function labels, (direct or indirect) call sites, and return sites. Return edges are constructed by observing the target set for each direct and indirect call. Indirect call target sets include only functions that have had their addresses taken and have a matching type signature. Type information on function pointer calls are passed from the GCC frontend to the ProxyCFI compiler core.
- **Pass 2. Branch Enumeration**: Since the CFG contains all transitions between functions, multi-way branch targets are fully enumerated before the second pass begins. For return sites, pointer proxies are chosen in the

context of the called function and shared with the calling function's indirect call sled. Pointer proxies are generated in this way to deter CFG mimicry attacks (see Sect. 2.5 for details).

After generating a binary, runtime analytics, which are generated by the profiler, are passed back to the branch enumeration phase, at which point the multi-way branch sleds are rewritten with optimizations. Load time invocation of the verifier rewrites all pointer proxies before executing the program.

3.2 ProxyCFI Optimizations

Indirect jump and return sleds can become very long, especially for frequently called functions. To address these potential concerns, we implemented two optimizations: profile-guided sled sorting, and function cloning.

Profile-guided sled sorting. The main source of performance degradation with ProxyCFI is the overhead incurred by the repeated comparisons used to implement multi-way direct branch sleds. The number of checks required is directly proportional to the number of legitimate targets for the corresponding indirect control transfer instruction. Yet, we observed that these sleds were highly biased to only a few of the branch targets. Our sled sorting optimization takes advantage of the biased distribution of multi-way branch targets by sorting the entries in descending order of profiled execution count. As shown in Sect. 4, this optimization significantly reduces the average depth a program must traverse into a sled before finding the pointer proxy target.

Function Cloning. While profile-guided sorting of the sleds significantly reduces performance degradation associated with multi-way branch sleds, the improvements are limited for functions with more uniformly distributed sled profiles. To combat this, we adapted function cloning [15] – an optimization that creates specialized copies of functions – as a means to reduce overall sled lengths. For sleds with more uniform distributions, this optimization significantly reduces the performance overhead incurred by executing sleds. Figure 5 illustrates function cloning. A function with near uniform sled distribution is cloned (*e.g.*, function *f2* becomes identical functions *f2* and *f2_c1*). Then, half of the call sites to the cloned function are redirected to the cloned function.

This optimization also significantly reduces the attacker's agency in selecting CFG edges to exploit for CFG mimicry attacks [14,20,33].

4 Evaluation

In this section, we examine the performance and security of ProxyCFI. First, the performance impact of ProxyCFI is assessed by examining the slowdown incurred for many CPU-centric and network-facing benchmarks, with and without ProxyCFI optimizations. To gauge the security benefits, we performed penetration testing with the RIPE control flow attack suite [40] and recent advanced control flow attacks on real-world applications.

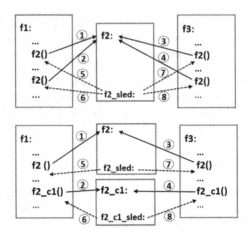

Fig. 5. Function Cloning. Function cloning cuts the number of legitimate edges by a factor of the number of clones. Legitimate edges 6 and 8 from f2_sled are no longer legitimate after cloning.

4.1   Evaluation Framework

ProxyCFI build framework. Our ProxyCFI compiler framework was built on GCC version 6.1.0. In our evaluations, we used Ubuntu 16.04 on x86-64. Using x86-64 is essential in our implementation because our shared libraries rely heavily on relative jumps to preserve code page sharing, which is significantly more efficient in 64-bit x86. We customized *GLIBC's* loader to handle ProxyCFI compliant shared libraries and mark code pages execute-only. Many modern processors have hardware support for execute-only memory. For example, recent Intel CPUs support unreadable code pages using the Memory Protection Keys (MPK) feature.[1] In our prototype implementation, we used this feature to make the code section execute-only (*i.e.*, disabled read/write access).

Benchmarks analyzed. We evaluated the performance and space overhead incurred by ProxyCFI using the SPEC CPU 2006 benchmarks. In addition, we evaluated the overhead on the network-facing application **redis-server**, running it with the standard *redis-benchmark* with 50 parallel clients and a 3-byte payload. To isolate the performance overhead incurred by ProxyCFI-hardened shared objects, we also ran microbenchmarks for varying shared library sled depths. To evaluate the security guarantees provided by ProxyCFI, we analyzed applications from all the major categories commonly targeted by control flow hijacking attacks including multimedia processing, Javascript engines, document rendering, network infrastructure and VM interpreters. Specifically, we analyzed the following commonly attacked applications (detailed in Sect. 4.3):

MuPDF is a light weight PDF XPS and EPUB parsing and rendering engine. MuPDF versions V1.3 and prior have a stack-based buffer overflow vulnerability

---

[1] Execute-only memory is also supported on ARMv8 and above.

(CVE-2014-2013) [8] that results in remote code execution via a maliciously crafted XPS document.

bladeenc is a cross-platform MP3 encoder which is also used as a daemon for encoding in distributed MP3 encoders/CDDB servers like *abcde*. *bladeenc* has several vulnerabilities that lead to CFG mimicry attacks that could be exploited remotely (CVE-2017-14648) [7].

dnsmasq is a DNS forwarder designed to provide DNS services to a small-scale networks, and it is included in most Linux distributions. Versions of *dnsmasq* prior to 2.78 have a stack-overflow vulnerability which enables a remote attacker to send a maliciously crafted DHCPv6 request to hijack control flow on the target system (CVE-2017-14493) [5].

Gravity is a dynamically typed concurrent scripting language written in C. The Gravity runtime contains a stack-based buffer overflow that leads to remote code execution (CVE-2017-1000437) [9].

4.2   Performance Analysis

We ran the SPEC CPU 2006 benchmarks performance analysis experiments on an Intel Xeon Gold 6126 Processor with 24 cores and 32GB RAM running Ubuntu 16.04 LTS Xenial Xerus

Figure 6 shows the performance overhead incurred by ProxyCFI instrumentation. For compute-intensive applications, the näive implementation's performance overheads are non-trivial, since these programs have high average sled depth. Average sled depth is a measure of how many pointer proxy tests are required in a sled, on average, before a direct branch is taken. Ideally, we would like this value to be close to 1 to lower the performance overhead for ProxyCFI. For applications with heavy use of function calls such as *perlbench*, *gobmk* and *sjeng*, the performance degradation for the unoptimized implementation is more pronounced, having a average return sled depth of 27 for *perlbench*.

With optimizations, the average sled depth drops dramatically, as do the performance overheads. For example, *perlbench* benefits significantly from profile-guided sled sorting optimization. *h264ref* also benefits significantly from optimizations, as it makes heavy use of generic function pointers with indirect functional call sleds having up to 855 entries, of which only two are frequently targeted. *gcc*, on the other hand, makes considerable use of both function calls (average sled depth of 32) and generic function pointer (with an average sled depth of 26 for indirect calls). The performance benefit of the function cloning optimization is more visible on *gcc*, as the probability distribution of taken branches falls off slower than the other applications. For network-facing applications, the performance overhead is insignificant due to their I/O-bound nature. The average performance overhead for *redis-server* is 0.25% and an average overhead of 0.93% for all of network-facing applications we evaluated.

Figure 7 shows the percentage increase in the binary size as a result of pointer proxy instrumentation, both with and without optimizations. On average the code size grows by 49% for our benchmarks with the worst case of 121% for *h264ref*, due to the large amount of instrumentation required for its generic function pointers. Finally, Fig. 8 shows the impact of ProxyCFI verification and load-time proxy randomization on program load times. As shown in the graph, it has approximately linear relationship with code size, the longest being 1200ms, consistent with previous works that perform load-time randomization [30,39].

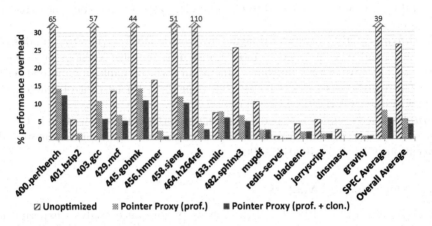

Fig. 6. Performance Overhead of ProxyCFI. Unoptimized shows the performance overhead without any optimization, while ProxyCFI (prof.) and ProxyCFI (prof. + clon.) show performance with profile-guided sled sorting (prof.) and function cloning (clon.) optimizations.

Shared library performance. We measured the cost of our shared library support infrastructure by microbenchmarking entries and exits to shared libraries and comparing it against unprotected shared library calls. The average percent slowdown for a shared library calls using optimized ProxyCFI compilation is 1.48% and 2.31% respectively for the best and worst-case average sled hit depths observed in our benchmark experiments.

4.3   Security Analysis

To assess the security strength of ProxyCFI, we first examine its ability to stop control flow attacks in the RIPE attack suite, then we examine to what extent ProxyCFI can stop real-world control flow attacks including CFG mimicry attacks.

Penetration testing with RIPE: RIPE is a control flow attack testbed that generates attacks by permuting five dimensions of attack: location (*e.g.*, stack, heap, ...), target (*e.g.*, return address, function pointers, ...), overflow technique (*e.g.*, direct/indirect), and function of abuse (*e.g.*, memcpy, ...) [40]. Native RIPE targets 32-bit x86 code, thus, with the help of a recently implemented low-fat pointer extension [19], we ported the RIPE test suite to x86-64. Our port supports the following five dimensions: location, target (including *setjmp()* and *longjmp()*), method, and overflow type. Permuting all RIPE dimensions totals up to 850 unique tests. **With ProxyCFI protections, 100% of the RIPE**

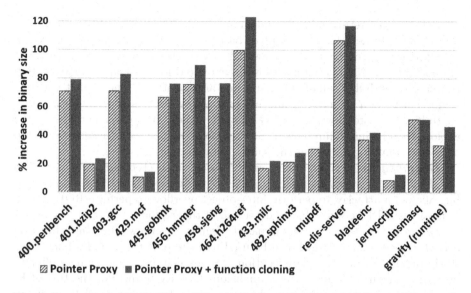

Fig. 7. Increase in Code Size. This graph shows the impact of ProxyCFI on code size. The blue bars (left) represent unoptimized ProxyCFI programs, while the green bar (right) represents optimized ProxyCFI programs. (Color figure online)

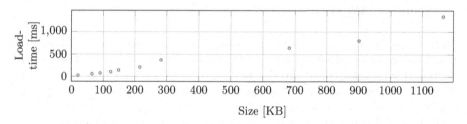

Fig. 8. Load-time overhead vs. code size This graph shows the impact of pointer proxy randomization on load-time.

attacks are stopped. In addition, ProxyCFI was able to detect the exact point at which attacks escape the CFG.

Real-world vulnerabilities. To evaluate the effectiveness of ProxyCFI against attacks on real-world applications, we included recent attacks reported on the National Vulnerability Database (NVD) in our evaluation.

With ProxyCFI, we were able to stop all of the following real-world attacks including CFG mimicry attacks.

In testing, we found that the declared violations enabled us to quickly identify the root cause of the vulnerability. We analyzed four attacks.

MuPDF has a stack-based buffer overflow vulnerability in the *xps_parse_color*() function which performs an unchecked *strcpy()* of a user supplied (via XPS input) array to a fixed size buffer [8]. The exploit uses this bug to overwrite the return address and jump to an ROP gadget. With ProxyCFI we were able to detect the stack pivot based on the corrupted pointer proxy.

Bladeenc's command line parser uses unchecked calls to *strcpy()* to copy parameters to a 256-byte buffer that are exploited for arbitrary code execution by using a carefully crafted command line arguments [7]. The exploit corrupts a function pointer to jump to another function which is also in its legal target set to hijack control flow via a CFG mimicry attack. We were able to detect the exploit when trying to jump using a forged pointer proxy (which was interpreted as invalid pointer proxy from the source address).

Dnsmasq has a vulnerability caused by an unchecked use of *memcpy()* in the *dhcp6_maybe_relay()* function to a 16-byte field of the variable *state*. This bug allows an attacker to perform inter-object overflow to perform ROP attack. Using ProxyCFI we were able to detect all of the exploits.

Gravity contains a stack-based buffer overflow in the function *operator_string _add()* which can be used to write past the end of a fixed-sized static buffer to achieve code execution. The exploit uses this vulnerability to overwrite a return address using a malicious Gravity script. For the ProxyCFI hardened version the attack was detected when the exploit tried to make an indirect jump based on forged pointer proxy.

5   Related Work

Memory safety. Memory corruption attacks have been often used to hijack control flow, either by injecting code or reusing existing code. Data execution prevention [1,2] is sidestepped entirely as reuse attacks need not inject code. Comprehensive memory safety techniques such as Softbound [31] can completely eradicate memory exploitation, but they suffer from high overhead or compatibility issues, deeming them as yet impractical for widespread adoption.

Control flow integrity: A new wave of practical defenses emerged with a focus on validating that execution adheres to a static, programmer specified CFG. Control flow integrity (CFI) [10] was the first of these CFG defenses. The defense inserts checks before indirect branches to make sure that all indirect control transfers are within the statically discovered CFG. Various coarse-gained variants have relaxed CFI constraints to achieve practical solutions through both software and hardware approaches [4,6,41,42].

CCFIR [41] uses a load-time randomized springboard section to redirect all indirect control flow transfers, which has been bypassed by a successive work [21]. Intel CET [6] provides rudimentary hardware protection for forward edges through its indirect branch tracking. Microsoft CFG enforces a weak form of CFI by restricting indirect function calls to function entry points [4]. While these techniques are valuable against straightforward code reuse techniques, CFG mimicry attacks effectively bypass this CFI techniques. Unlike these coarse grained CFI techniques, ProxyCFI provides fine grained protection, and also affords protection against CFG mimicry attacks. CCFI [28] is a fine-grained CFI technology that protects code pointers by storing hash based message authentication code

(MAC) alongside code pointers and checking the MAC before indirect branches. While CCFI can protect against CFG mimicry attacks, its high performance overhead (52% for SPEC'06) will undoubtedly limit its applicability in production environments. Like CCFI, ProxyCFI provides fine-grained control flow protection, while incurring significantly lower overheads (only 5.9% average slowdown for SPEC'06).

Other control flow integrity works have proposed to completely remove instructions employed for control flow hijacking attacks. Return-less kernels [26] avoid use of *ret* instruction by replacing them with a lookup into a static return table which provides protection solely against return-based attacks. Control-data isolation (CDI) [11] rewrites both forward and backward edges with exclusively direct branches. CDI would conceivably constrain execution to the programmer-specified CFG, if it were to verify that all binaries adhered to CDI compilation requirements. But since the approach still uses code pointers to identify program pointers, the approach is readily attackable with control flow attacks that do not leave the CFG, such as Counterfeit OOP [33]. Moreover, ProxyCFI addresses CFG mimicry attacks by replacing code pointers with pointer proxies that utilize per-function namespaces, which are assigned at program load-time to an execute-only memory.

Code-Pointer Integrity (CPI) [24] provides memory safety for code pointers by storing them in a safe region. CPI requires allocation of a safe data region inaccessible to an attacker. ProxyCFI does not require any special data region protections.

6   Conclusion

While significant effort has been spent to shut down control flow attacks, their existence and value persists today, even 40 years after the first buffer overflow attack. With ProxyCFI, we take the novel approach of replacing all of a program's code pointers with the much less powerful pointer proxy. A pointer proxy is a random identifier representing a specific program entry point from the context of a specific function. A control transfer with a pointer proxy utilizes a multi-way direct branch which fully anticipates all of the potential jump targets. As such, ProxyCFI provides much resistance to advanced control flow attacks because it is difficult to forge/swap pointer proxies to mimic a legitimate CFG transition. Our implementation of ProxyCFI is built into the GNU GCC C/C++ compiler toolchain, such that all code pointers are replaced with pointer proxies including those contained within shared libraries. Analysis of our pointer proxy implementation reveals that they introduce minimal slowdown when pointer-proxy specific optimizations are applied, only an average 4% slowdown across a wide range of benchmarks. Moreover, security analysis of ProxyCFI shows that it stops all of the control flow attacks we tested, including 100% of the attacks in the RIPE x86-64 attack suite and a wide range of real-world attacks including CFG mimicry attacks.

Looking ahead we see a number of avenues for growing the capabilities of ProxyCFI. In particular, we would like to implement support for reassigning pointer proxy values at runtime, and we would like to explore the use of pointer proxies for a limited set of data pointers.

Acknowledgement. This work was supported by DARPA under Contract HR0011-18-C-0019. Any opinions, findings and conclusions or recommendations expressed in this material are those of the authors and do not necessarily reflect the views of DARPA.

A   Redis-benchmark Results Breakdown

Table 2 shows the results of running *redis-server* with the standard *redis-benchmark* using 50 parallel clients and a 3-byte payload.

Table 2. Results of running *redis-benchmark* ProxyCFI compliant *redis-server* versus unhardened baseline

Command	Baseline (request/sec)	ProxyCFI (request/sec)
PING_INLINE	12320.9	12115.34
PING_BULK	12881.67	12926.58
SET	12469.83	12158.05
GET	12941.73	13010.67
INCR	10514.14	11189.44
LPUSH	12227.93	12997.47
RPUSH	11592.86	11828.72
LPOP	12659.83	12255.46
RPOP	12804.1	12604.8
SADD	12218.96	12055.46
HSET	12023.57	11872.7
SPOP	11855.36	11552.39
LPUSH (needed to benchmark LRANGE)	12968.49	12600.12
LRANGE_100 (first 100 elements)	6506.82	6325.19
LRANGE_300 (first 300 elements)	2788.99	2690.05
LRANGE_500 (first 450 elements)	2403.4	2211.26
LRANGE_600 (first 600 elements)	1730.2	1652.59
MSET (10 keys)	10409.08	9959.79

References

1. Data execution prevention (2003). Accessed 29 Feb 2018
2. Linux kernel 2.6.8 (2004). Accessed 29 Feb 2018
3. Windows ISV software security defenses (2010). Accessed 29 Feb 2018
4. Control flow guard (windows) - MSDN - Microsoft (2015). https://msdn.microsoft.com/en-us/library/dn919635.aspx. Accessed 13 Apr 2018
5. Cve-2017-14493 (2017). https://www.cvedetails.com/cve/CVE-2017-14493/. Accessed 12 Feb 2018
6. Intel control-flow enforcement technology (CET) (2017). https://software.intel.com/sites/default/files/managed/4d/2a/control-flow-enforcement-technology-preview.pdf. Accessed 13 Apr 2018
7. Bladeenc: Vulnerability statistics (2018). https://www.cvedetails.com/product/2851/Bladeenc-Bladeenc.html. Accessed 05 Jan 2018
8. Cve-2014-2013 (2018). https://www.cvedetails.com/cve/CVE-2014-2013/. Accessed 13 Apr 2018
9. Cve-2017-1000437 (2018). https://www.cvedetails.com/cve/CVE-2017-1000437/. Accessed 05 Jan 2018
10. Abadi, M., Budiu, M., Erlingsson, U., Ligatti, J.: Control-flow integrity. In: Proceedings of the 12th ACM Conference on Computer and Communications Security, pp. 340–353. ACM (2005)
11. Arthur, W., Mehne, B., Das, R, Austin, T.: Getting in control of your control flow with control-data isolation. In: Proceedings of the 13th Annual IEEE/ACM International Symposium on Code Generation and Optimization, pp. 79–90. IEEE Computer Society (2015)
12. Bletsch, T., Jiang, X., Freeh, V.W., Liang, Z.: Jump-oriented programming: a new class of code-reuse attack. In: Proceedings of the 6th ACM Symposium on Information, Computer and Communications Security, pp. 30–40. ACM (2011)
13. Buchanan, E., Roemer, R., Shacham, H., Savage, S.: When good instructions go bad: generalizing return-oriented programming to RISC. In: Proceedings of the 15th ACM Conference on Computer and Communications Security, pp. 27–38. ACM (2008)
14. Carlini, N., Barresi, A., Payer, M., Wagner, D., Gross, T.R.: Control-flow bending: on the effectiveness of control-flow integrity. In: 24th USENIX Security Symposium (USENIX Security 15), pp. 161–176. USENIX Association, Washington, DC (2015)
15. Cooper, K.D., Hall, M.W., Kennedy, K.: A methodology for procedure cloning. Comput. Lang. **19**(2), 105–117 (1993)
16. Cowan, C., et al.: Stackguard: automatic adaptive detection and prevention of buffer-overflow attacks. In: USENIX Security Symposium, San Antonio, TX, vol. 98, pp. 63–78 (1998)
17. Cowan, C., Beattie, S., Johansen, J., Wagle, P.: PointGuard TM: protecting pointers from buffer overflow vulnerabilities. In: Proceedings of the 12th Conference on USENIX Security Symposium, vol. 12, pp. 91–104 (2003)
18. Dai Zovi, D.: Practical return-oriented programming. In: SOURCE Boston (2010)
19. Duck, G.J., Yap, R.H.C, Cavallaro, L.: Stack bounds protection with low fat pointers (2017)
20. Evans, I., et al.: Control jujutsu: on the weaknesses of fine-grained control flow integrity. In: Proceedings of the 22nd ACM SIGSAC Conference on Computer and Communications Security, pp. 901–913. ACM (2015)
21. Gktas, E., Athanasopoulos, E., Bos, H., Portokalidis, G.: Out of control: overcoming control-flow integrity. In: 2014 IEEE Symposium on Security and Privacy, May, pp. 575–589 (2014)

22. Göktaş, E., Athanasopoulos, E., Polychronakis, M., Bos, H., Portokalidis, G.: Size does matter: why using gadget-chain length to prevent code-reuse attacks is hard. In: Proceedings of the 23rd USENIX Conference on Security Symposium, pp. 417–432. USENIX Association (2014)

23. Intel: Dynamic libraries (2015). Accessed 29 Feb 2018

24. Kuznetsov, V., Szekeres, L., Payer, M., Candea, G., Sekar, R., Song, D.: Code-pointer integrity. In: 11th USENIX Symposium on Operating Systems Design and Implementation (OSDI 2014), pp. 147–163. USENIX Association, Broomfield (2014)

25. Lan, B., Li, Y., Sun, H., Su, C., Liu, Y., Zeng, O.: Loop-oriented programming: a new code reuse attack to bypass modern defenses. In: 2015 IEEE Trustcom/BigDataSE/ISPA, vol. 1, pp. 190–197. IEEE (2015)

26. Li, J., Wang, Z., Jiang, X., Grace, M., Bahram, S.: Defeating return-oriented rootkits with return-less kernels. In: Proceedings of the 5th European Conference on Computer Systems, pp. 195–208. ACM (2010)

27. Liu, L., Han, J., Gao, D., Jing, J., Zha, D.: Launching return-oriented programming attacks against randomized relocatable executables. In: 2011 IEEE 10th International Conference on Trust, Security and Privacy in Computing and Communications (TrustCom), pp. 37–44. IEEE (2011)

28. Mashtizadeh, A.J., Bittau, A., Boneh, D., Mazières, D.: CCFI: cryptographically enforced control flow integrity. In: Proceedings of the 22nd ACM SIGSAC Conference on Computer and Communications Security, pp. 941–951. ACM (2015)

29. C+ MISRA: Guidelines for the use of the C/C++ language in critical systems. MIRA Limited, Warwickshire (2012)

30. Mohan, V., Larsen, P., Brunthaler, S., Hamlen, K.W., Franz, M.: Opaque control-flow integrity. In: NDSS, vol. 26, pp. 27–30 (2015)

31. Nagarakatte, S., Zhao, J., Martin, M.N.K., Zdancewic, S.: SoftBound: highly compatible and complete spatial memory safety for C. ACM SIGPLAN Not. **44**(6), 245–258 (2009)

32. Prasad, M., Chiueh, T.: A binary rewriting defense against stack based buffer overflow attacks. In: USENIX Annual Technical Conference, General Track, pp. 211–224 (2003)

33. Schuster, F., Tendyck, T., Liebchen, C., Davi, L., Sadeghi, A.-R., Holz, T.: Counterfeit object-oriented programming: On the difficulty of preventing code reuse attacks in C++ applications. In: 2015 IEEE Symposium on Security and Privacy (SP), pp. 745–762. IEEE (2015)

34. Shacham, H., Page, M., Pfaff, B., Goh, E.-J., Modadugu, N., Boneh, D.: On the effectiveness of address-space randomization. In: Proceedings of the 11th ACM Conference on Computer and Communications Security, pp. 298–307. ACM (2004)

35. Strackx, R., Younan, Y., Philippaerts, P., Piessens, F., Lachmund, S., Walter, T.: Breaking the memory secrecy assumption. In: Proceedings of the Second European Workshop on System Security, pp. 1–8. ACM (2009)

36. Theodorides, M., Wagner, D.: Breaking active-set backward-edge CFI. In: 2017 IEEE International Symposium on Hardware Oriented Security and Trust (HOST), pp. 85–89. IEEE (2017)

37. Tran, M., Etheridge, M., Bletsch, T., Jiang, X., Freeh, V., Ning, P.: On the expressiveness of return-into-libc attacks. In: Sommer, R., Balzarotti, D., Maier, Gregor (eds.) RAID 2011. LNCS, vol. 6961, pp. 121–141. Springer, Heidelberg (2011). https://doi.org/10.1007/978-3-642-23644-0_7

38. Wagner, D., Soto, P.: Mimicry attacks on host-based intrusion detection systems. In: Proceedings of the 9th ACM Conference on Computer and Communications Security, pp. 255–264. ACM (2002)

39. Wartell, R., Mohan, V., Hamlen, K.W., Lin, Z.: Binary stirring: self-randomizing instruction addresses of legacy x86 binary code. In: Proceedings of the 2012 ACM Conference on Computer and Communications Security, pp. 157–168. ACM (2012)

40. Wilander, J., Nikiforakis, N., Younan, Y., Kamkar, M., Joosen, W.: RIPE: runtime intrusion prevention evaluator. In: Proceedings of the 27th Annual Computer Security Applications Conference, ACSAC. ACM (2011)

41. Zhang, C., et al.: Practical control flow integrity and randomization for binary executables. In: 2013 IEEE Symposium on Security and Privacy (SP), pp. 559–573. IEEE (2013)

42. Zhang, M., Sekar, R.: Control flow integrity for cots binaries. In: USENIX Security Symposium, pp. 337–352 (2013)

CASFinder: Detecting Common Attack Surface

Mengyuan Zhang[1], Yue Xin[1], Lingyu Wang[1(✉)], Sushil Jajodia[2],
and Anoop Singhal[3]

[1] Concordia Institute for Information Systems Engineering, Concordia University,
Montreal, Canada
wang@ciise.concordia.ca

[2] Center for Secure Information Systems, George Mason University, Fairfax, USA
jajodia@gmu.edu

[3] Computer Security Division, National Institute of Standards and Technology,
Gaithersburg, USA
anoop.singhal@nist.gov

Abstract. Code reusing is a common practice in software development
due to its various benefits. Such a practice, however, may also cause
large scale security issues since one vulnerability may appear in many
different software due to cloned code fragments. The well known concept of relying on software diversity for security may also be compromised since seemingly different software may in fact share vulnerable
code fragments. Although there exist efforts on detecting cloned code
fragments, there lack solutions for formally characterizing their specific
impact on security. In this paper, we revisit the concept of software diversity from a security viewpoint. Specifically, we define the novel concept
of *common attack surface* to model the relative degree to which a pair
of software may be sharing potentially vulnerable code fragments. To
implement the concept, we develop an automated tool, *CASFinder*, in
order to efficiently identify common attack surface between any given
pair of software with minimum human intervention. Finally, we conduct
experiments by applying our tool to real world open source software
applications. Our results demonstrate many seemingly unrelated software applications indeed share significant common attack surface.

1 Introduction

Code reusing is a common practice in today's software industry due to the fact
that it may significantly accelerate the development process [7,10]. However,
such a practice also has the potential of leading to large scale security issues
because a vulnerability may be shared by many different software applications
due to the shared libraries or code fragments. A well known example is the *Heartbleed* vulnerability in *OpenSSL*, which caused widespread panic on the internet
since the vulnerable library was shared by many popular Web servers, including
Apache and Nginx [11]. In addition to shared libraries, the reusing of existing

S. N. Foley (Ed.): DBSec 2019, LNCS 11559, pp. 338–358, 2019.
https://doi.org/10.1007/978-3-030-22479-0_18

code fragments may also lead to similar vulnerabilities shared by different software applications. Unlike libraries, such reused codes are typically not traced by any official documentation, which makes it more difficult to understand their security impact. Finally, this phenomenon may also compromise the well known concept of relying on software diversity for security, since seemingly unrelated software applications made by different vendors may in fact share common weaknesses.

The issue of identifying and characterizing the security impact of shared code fragments has received little attention (a more detailed review of the related work will be given in Sect. 6). Most existing vulnerability detection tools focus on identifying vulnerabilities for a specific software application based on static and/or dynamic analysis, with no indication whether different software may be sharing similar vulnerabilities due to common libraries or reused codes [12]. On the other hand, existing efforts on software clone detection mostly focus on identifying reused code fragments based on either the textual similarity or functional similarity, with no indication of the security impact [41]. Clearly, there exists a gap between the two, i.e., *how can we leverage existing efforts on software clone detection to characterize the likelihood that given software applications may share similar vulnerabilities?*

In this paper, we address the above issue through defining the novel concept of *common attack surface* and developing an automated tool, *CASFinder*, to calculate the common attack surface of given software applications. Specifically, we first extend the well-known attack surface concept to model the relative degree to which a pair of software may be sharing potentially vulnerable code fragments. Such a formal model enables the quantification of software diversity from the security point of view, and its results may be used as inputs to higher level diversity methods (e.g., network diversity [47] and moving target defense [20]). Second, we develop *CASFinder* which is an automated tool that takes the source code of two software applications as the input and outputs their common attack surface result in an XML file or to a database. Third, we conduct experiments by applying our tool to a large number of real-world open source software applications belonging to seven different categories from *Github*. More than 80,000 combinations of software applications are analyzed, and our results demonstrate many seemingly unrelated software applications indeed share a significant level of common attack surface. In summary, the contribution of this paper is threefold.

- First, to the best of our knowledge, this is the first effort on formally modeling the security impact of reused code fragments. The common attack surface model may serve as a foundation and provide quantitative inputs to higher level security-through-diversity methods.
- Second, the *CASFinder* tool makes it feasible to evaluate the common attack surface between open source software applications, which may have many practical use cases, e.g., providing useful references for security practitioners to choose the right combinations of software applications in order to maximize the overall software diversity in their networks, and reusing the knowledge

about existing vulnerabilities in one software to potentially identify similar ones in other software.

- Third, our experimental results prove the possibility of similar vulnerabilities shared by seemingly unrelated software applications made by different vendors. We believe such a finding may help attract more interest to re-examining the concept of software diversity and its security implication.

The remainder of this paper is organized as follows. Section 2 provides a motivating example and background information. Section 3 defines the common attack surface model. Section 4 designs and implements the *CASFinder* tool. Section 5 evaluates the tool through experiments using real open source software. Section 6 reviews related work and Sect. 7 concludes the paper and provides future directions.

2   Preliminaries

In this section, we first present a motivating example in Sect. 2.1 and then provide background knowledge and highlight the challenge in Sect. 2.2.

2.1   Motivating Example

As an example, consider an enterprise network with Web servers running either the Apache HTTP server (*Apache*) or the Nginx HTTP server (*Nginx*), as well as a Cyrus IMAP server (*Cyrus*). Assume all three software applications are of the vulnerable versions that are affected by the *Heartbleed* vulnerability. This vulnerability has reportedly affected an estimated 24–55% of popular websites and gave attackers accesses to sensitive memory blocks on the affected servers, which potentially contain encryption keys, usernames, passwords, etc. [11]. The vulnerability is discovered inside the popular *OpenSSL* library, which is an extension of many Web and email server software applications for supporting the *https* connections.

Specifically, Fig. 1 demonstrates how this vulnerability functions in relation to the three software applications in our example. Those software simply hand the encryption tasks to the *OpenSSL* extension, and the vulnerability appears when the software make external calls to the *OpenSSL* extension. In establishing the SSL connections, the API invocation *SSL_CTX_new(method)* is a function for establishing SSL content, *SSL_new()* is for creating SSL sessions, and *SSL_connect()* for launching SSL handshakes. To exploit the *Heartbleed* vulnerability, attackers would craft a heartbeat request with a special length and send it to the servers. This request would cause different software applications to invoke the same library function *memcpy()* without any boundary check enabling attackers to extract sensitive memory blocks from the servers.

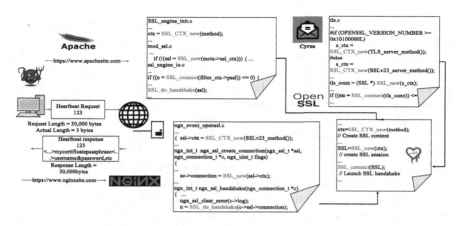

Fig. 1. An example of the *Heartbleed* vulnerability

The fact that this vulnerability exists inside the *OpenSSL* extension shared by all three software means an attacker can compromise those different software in a similar manner. This phenomenon is certainly not limited to this particular vulnerability. In this example, since both the *Apache* and *Nginx* projects are Web servers developed in C language, their similar functionality implies there is a high chance that the developers of both projects would not only import the same libraries, but also reuse the same or similar code fragments. In addition, as will be shown through our experimental results, code reusing also exists among software applications with very different functionalities. On the other hand, not all server software that use SSL connections are affected by this vulnerability, e.g., *Microsoft IIS* and *Jetty* are both immune to the vulnerability [11].

Clearly, there exists a need for identifying the software applications which may share such a common vulnerability, and for characterizing the level of such sharing since some software may share more than one such vulnerability. Such a desirable capability may have many practical use cases. For instance, it may allow similar software patches or fixes to be developed and applied to different software applications in order to mitigate a common vulnerability, which may significantly reduce the time and effort needed for developing such patches and fixes. This capability may also allow administrators to better judge the amount of software diversity in their networks, and to choose the right combinations of software applications (e.g., Apache and IIS w.r.t. this particular vulnerability) to increase the diversity. Finally, this capability would lead to a more refined approach to moving target defense (MTD) [13] since it could potentially allow us to quantify the amount of software diversity that is achieved by switching between different software resources under a MTD mechanism.

2.2   Background

We take two steps towards measuring the potential impact of cloned codes on security. The first step is to find similar code fragments in different software applications. The second step is to characterize the security impact of such code fragments. We first review some of the background concepts related to each step.

First, to detect similar code fragments between software, most clone detection methods are based on either the textual similarity or the functional similarity, and existing tools are mostly based on text, token, tree, graph, or metrics [41,42]. Among the existing tools, we have chosen *CCFinder* [23], a language-based source code clone detection tool, to find cloned code fragments within given software applications. As one of the leading token-based detection tools, *CCFinder* has received the Clone Award in 2002, and it supports multiple languages, including C, C++, Java, and COBOL. *CCFinder* first divides the given source code into tokens using a lexical analyzer. It then normalizes some of those tokens by replacing identifiers, constants and other basic tokens with generic tokens representing their language role. Finally, it uses a suffix-tree based sub-string matching algorithm to find common subsequences corresponding to clone pairs and classes [23]. A key advantage of such a token-based tool is that it can tolerate minor code changes, such as formatting, spacing and renaming, in the reused code.

However, the result from clone detection tools, including *CCFinder*, only reveals similar code fragments between source codes, without indicating any security impact. The primary challenge is therefore to model and quantify the potential impact of clone detection on security in terms of leading to potential vulnerabilities. To this end, a promising solution is to apply the attack surface concept [36], which is a well known software security metric that measures the degree of software security exposure. The measurement is taken as counts along three dimensions, the entry and exit points (i.e., methods calling I/O functions), channels (e.g., TCP and UDP), and untrusted data items (e.g., registry entries or configuration files), and the counting results are then aggregated through weighted summation. Attack surface measures the intrinsic properties of a software application, e.g., how many times does each method invoke I/O functions (which provides an estimate of security risks such as buffer overflow), regardless of external factors such as the discovery of the vulnerability or the existence of exploit code. Therefore, attack surface can potentially cover both known and unknown vulnerabilities.

Therefore, we will combine clone detection (i.e., *CCFinder*) with attack surface to quantify the likelihood that cloned code fragments may lead to potentially similar vulnerabilities shared between different software applications. For simplicity, we will focus on entry and exit points in this paper, and will consider channels and untrusted data items in our future work. We also note that, since it is not guaranteed that every entry or exit point will map to a vulnerability, the attack surface concept is only intended as an estimation of the relative abundance of vulnerabilities in software [36]. Consequently, our model and tool also inherit this limitation, and the results will only indicate the potential, instead of the actual existence, of common vulnerabilities.

Combining the result of clone detection with the attack surface concept is not a straightforward task. We discuss a key challenge in the following. In Fig. 2, function *handle_response()* and function *quicksand_mime()* are both entry points since they call I/O functions *fseek()* and *ftell* (from the standard C library). A naive application of the attack surface concept here would indicate each function count as one entry point and hence both have the same security implication. However, such a coarse-grained application ignores the exact number of I/O function calls (i.e., three calls in *handle_response()* and two in *quicksand_mime()*) whose difference may be significant in practice. In our model, we will take a more refined approach to address such issues.

```
1  fseek(fp, 0, SEEK_END);
2  size = ftell(fp);
3  fseek(fp, 0, SEEK_SET);
4  snprintf(fsize, 32, "Content−Length: %d\r\n\r\n", size);
```

```
1  long fsize = ftell(f);
2  fseek(f, 0, SEEK_SET);
3  free(decoded_mime);
```

Fig. 2. Examples of entry points: /Simple-Webserverche/server.c *handle_response()* (Top) and /quicksand_lite/libqs.c *quicksand_mime()* (Bottom)

3 The Model of Common Attack Surface

In this section, we model the security implication of cloned code fragments between software applications through two novel security metrics, namely, the *conditional common attack surface* (ccas) and the *probabilistic common attack surface* (pcas). Those two metrics are designed for different use cases as follows.

- The *conditional common attack surface* (ccas) is designed to be asymmetric for use cases in which one software is of particular interest and evaluated against all other software. For example, suppose a company has developed a new Web server application and wants to understand any similarity between their product and other existing Web servers such as *Apache* and *Nginx*. In such a case, the key is to rank those other software applications based on the relative percentage of shared attack surface, and the developer can apply the metric *ccas* for this purpose.
- Second, in a different scenario, suppose an administrator wants to understand the level of software diversity between all the software applications inside the same network. In such a case, both software in comparison are considered equally important, so the symmetric metric *pcas* would be more suitable, which will yield a unique measurement of shared attack surface between any pair of software. The following details the *ccas* and *pcas* metrics.

3.1   Conditional Common Attack Surface (CCAS) Metric

We first consider clone segments between two software applications identified using *CCFinder* [23] through an example.

Example 1. Figure 3 demonstrates clone segments between a Web server application *SimpleWebserver* and an ssh application *SSHBen*. In the figure, the *Clone id* is a unique number labelling a group of related clones inside both software applications. For instance, the code segments inside the solid line blocks indicate the clone segments with the same Clone id 28, and the dashed line blocks are for Clone id 78. Note that the same code may appear under different clone ids, e.g., line 146 and 147 in *Simple-Webserver* appear under both clone ids. Also note that, for Clone id 78, the matching between the two clone segments is *inexact* [23] since *strcat* does not exist in *SSHBen*.

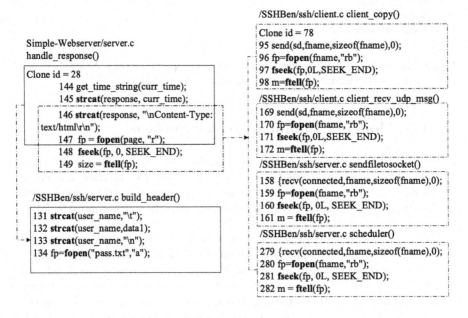

Fig. 3. An example of cloned segments

From the above example, it is clear that the clone segments belonging to the same Clone id are not identical between the two software applications. Therefore, the attack surface would be asymmetric as well. First, we define the *Common Attack Surface* as the collection of I/O function calls inside the clone segments as follows.

Definition 1 (Common Attack Surface). *Given two software applications A and B, the common attack surface of A w.r.t. B (or that of B w.r.t. A) under* *the Clone id i is defined as the multi-set (which preserves duplicates) of I/O function calls that exist inside the clone segments of A under the Clone id i, denoted as $cas_i(A|B)$ (or $cas_i(B|A)$).*

Example 2. To follow our example, we have

- cas_{28}(SimpleWebserver|SSHBen) $= \langle strcat, strcat, fopen \rangle$,
- cas_{28}(SSHBen|SimpleWebserver) $= \langle strcat, strcat, strcat, fopen \rangle$,
- cas_{78}(SimpleWebserver|SSHBen) $= \langle fopen, fseek, ftell \rangle$, and
- cas_{78}(SSHBen|SimpleWebserver) $= \langle fopen, fseek, ftell, fopen, fseek, ftell, fopen, fseek, ftell, fopen, fseek, ftell \rangle$.

Since the attack surface concept is based on the number of entry and exit points (i.e., methods invoking I/O functions), we follow the similar approach to calculate the size of common attack surface by counting the number of I/O function calls across different Clone ids, with those appearing under different Clone ids counted only once. We demonstrate this through an example.

Example 3. For Clone id 78, this gives three for *Simple-Webserver* and 12 for *SSHBen*. As to Clone id 28, we have three for *Simple-Webserver* and four for *SSHBen*. Note that *fopen* is considered under both Clone ids for *Simple-Webserver*, and hence we should count it only once. Based on those discussions, we can calculate the total number of I/O function calls for both Clone ids as five for *Simple-Webserver* and 16 for *SSHBen*.

Finally, we define the *Conditional Common Attack Surface* as the ratio between the size of the common attack surface of a software application (w.r.t. to another software) and the size of its entire attack surface (i.e., the total number of I/O function calls inside that software). This ratio indicates the degree to which the software shares with others similar I/O function calls (entry/exit points).

Definition 2 (Conditional Common Attack Surface). *Given two software applications A and B with totally n clone segments, and AS_A and AS_B as the total number of I/O function calls inside A and B, respectively, the conditional common attack surface of A w.r.t B (or that of B w.r.t. A), denoted as ccas(A|B) (or ccas(B|A)), is defined as:*

$$ccas(A \mid B) = \frac{| \bigcup_{i=1}^{n} cas(A \mid B) |}{AS_A}$$

$$ccas(B \mid A) = \frac{| \bigcup_{i=1}^{n} cas(B \mid A) |}{AS_B}$$

Example 4. The attack surface (i.e., the total number of I/O function calls) of *Simple-Webserver* and *SSHBen* are 16 and 182, respectively. We thus have $ccas(SSHBen \mid SimpleWebserver) = \frac{5}{16} = 0.3125$ and $ccas(Simple-Webserver \mid SSHBen) = \frac{16}{182} = 0.029$. The results show that *SSHBen* contains about 31% shared attack surface, whereas *SimpleWebserver* contains only

2.9%. By comparing a software application to many others, the developer of that application may gain useful insights from such results in terms of vulnerability discovery and security patch management.

3.2 Probabilistic Common Attack Surface Metric

The conditional common attack surface metric $ccas$ is designed for evaluating one software application against others. We now take a different approach of defining a symmetric probabilistic common attack surface metric for two software applications. Such a metric can be used to estimate the amount of effort that a potential attacker may reuse while attempting to compromise both software applications. The nature of such a use case implies the metric should be symmetric.

We apply Jaccard index for this purpose, which is commonly defined as $J(A, B) = \frac{A \cap B}{A \cup B}$ and used for analyzing the similarity and diversity between the two sets. To apply this metric in our case, we need to define both the intersection and union of the attack surface of two software applications. The common attack surface defined in previous section (Definition 1) can be considered as the intersection, but such a definition is not sufficient here since it is asymmetric in nature. Instead, we will define the intersection between the attack surface of two software applications using the standard multi-set intersection operation [43], which is described below.

Definition 3 (Intersection of Multi-Sets [43]). *Given two multi-sets $A = \langle A, f \rangle$ (where f is the multiplicity function such that for any $a \in A$, $f(a)$ gives the number of occurrences of a in the multiset) and $B = \langle A, g \rangle$, then their intersection, denoted as $A \cap B$, is the multi-set $\langle A, s \rangle$, where for all $a \in A$:*

$$s(a) = min(f(a), g(a)).$$

Example 5. Assume $U = \{a,a,a,b\}$ and $V = \{a,a,b,b\}$, if we apply the multi-set operation as defined above, we have $U \cap V = \{a,a,b\}$.

The union of the attack surface between two software applications can be defined as $AS_A \cup AS_B = AS_A + AS_B - cas(B \mid A) \cap cas(A \mid B)$. With both the union and intersection operations defined, we can now define the probabilistic common attack surface metric as follows.

Definition 4 (Probabilistic Common Attack Surface Metric). *Given two software applications A and B, with their attack surface AS_A and AS_B and the common attack surface $cas(B|A)$ and $cas(A|B)$, respectively, the probabilistic common attack surface of A and B is defined as:*

$$pcas(A.B) = \frac{\mid cas(B \mid A) \cap cas(A \mid B) \mid}{\mid AS_A \cup AS_B \mid}$$

Example 6. The size of attack surface in *Simple-Webserver* and *SSHBen* is 16 and 182, respectively. From our previous discussions, we have *cas* $(SSHBen \mid SimpleWebserver) \cap cas(SimpleWebserver \mid SSHBen) = \langle strcat,$ $strcat, fopen, fseek, ftell \rangle$ whose size is 5, and hence $pcas(SSHBen.Simple$ $Webserver) = \frac{5}{16+182-5} = 2.6\%$. Intuitively, this result indicates that, among all the I/O function calls, about 2.6% are shared between the two software applications. Such a result, when applied to all pairs of software applications inside a network, may allow administrators to estimate the degree of software diversity in the network from a security point of view.

4   Design and Implementation

To automate the evaluation of common attack surface between software applications, we design and implement a tool, *CASFinder*. Figure 4 depicts the architecture of *CASFinder*, which consists of three main components, the clone detection module, the source code labeling module, and the visualization module. The following describes those modules in more details.

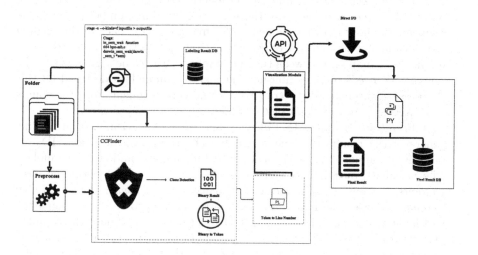

Fig. 4. The architecture

– *The Clone Detection Module.* As mentioned earlier, we choose *CCFinder* [23] as the basis of our clone detection module. The following details challenges and solutions for applying *CCFinder*. First, since our tool is developed and operated under Linux, we apply only the back end of *CCFinder*. One challenge is that, since the default Linux version of *CCFinder* is designed to work on *Ubuntu 9*, the newer versions of many libraries are no longer valid for *CCFinder*. Therefore, several libraries need to be installed separately, e.g., *libboost-dev* and *libicu-dev*, which will depend on the specific version of the Linux system and can be determined based on the warnings and errors produced by *CCFinder*. Second, various parameters can be fine tuned in *CCFinder* to customize its execution mode [22]. In particular, the most important parameters include b, the minimum length of the detected code clones, and t, the minimum number of types of tokens involved. We have chosen $b = 20$ and $t = 8$ based on experiences obtained through extensive experiments. In addition, parameter w is used to determine whether *CCFinder* will perform inner-file clone detection whose results contain clones between different parts of the same software application, which is not our focus, and therefore w is set to be *f-w-g+* to focus on inter-file clones. Finally, the default output of the *CCFinder* is stored in a binary file with *.ccfd* extension. Since we do not install any front end of *CCFinder*, we apply the command *./$PATH/ccfx -p name.ccfd* to translate the *.ccfd* file into a human-readable version. The resultant file contains only the token information, which cannot be directly mapped back to the source code files. Therefore, we have developed a script, *post-prettyprint.pl* [38], to convert the token information into corresponding line numbers in the source code.

- *The Source Code Labeling Module.* As mentioned above, the converted output of *CCFinder* provides only the file name and line number of the clone segments, without information needed for mapping them back to the original source code. For the purpose of generating traceable output with source code fragments, a mapping between the line number of the clone segments and the source code needs to be established. This second module is designed for this purpose by automatically retrieving a clone code segment from the source code according to the result of *CCFinder*.

- *The Visualization and CAS Calculation Module.* The visualization module generates the results of clone segments. The results include clone ID, file path, function name, clone segment, start line number, and end line number. The visualized output is organized as an *XML* tree with labels. The label *contents* contains the source clone segments from *CCFinder* outputs. Label *funcname* reveals the function names corresponding to the clone segments, and label *io* contains the common I/O functions. To calculate the common attack surface, we first need to identify the I/O functions. In our experiments, we have obtained the list of I/O functions from the GNU C library [40] (glibc), which is the GNU project's implementation of C standard library, as the database for examining the entry/exit points. In total, 256 I/O functions are stored in our database, e.g., function *memcpy* or *strcpy*, which could take user inputs as the source, and copy them directly to the memory block pointed to by the destination. Such functions have caused many serious security flaws including CVE-2014-0160 (i.e., the Heartbleed bug [8]). The final result of common attack surface is calculated based on the I/O functions shared among all software applications, and can be stored either in a file or into the database.

5 Experiments

This section presents experimental results on applying our tool *CASFinder* to real world open source software.

5.1 Dataset

To study the common attack surface among real world software applications, we need a large amount of open-source software to apply our tool. For this purpose, we have developed a script to automatically parse the download links at the open-source software hosts. Our research shows that *GitHub* [15] provides the customized API for users to search open-source software applications with customized requirements and to download them automatically. The results are presented in json code, which contains the download link of each application together with other information. In our experiments, we have set the parameter *language* to C programs, and use parameters *q*, *sort*, and *order* to specify the query conditions and to customize the sequence of results. We have developed the script to parse the json format output from the *GitHub* automatically and to store the information of the software download link, authors, publish time, size, and other descriptions into our local database. All the download links for each software application are stored separately. Since *Github* has a limitation with respect to the maximum requests in a certain amount of time, we design the process to sleep for certain time after each query. Our experimental environment is a virtual machine running Ubuntu 14.04, with the Intel core i3-4150 CPU and 8.0 GB of RAM. We have applied our tool to totally 293 different software applications belonging to seven categories. The software applications belong to several categories as follows: 32 in Databases, 62 in Web servers, 25 in ssh servers, 79 in FTP servers, 41 in TFTP servers, 6 in IMAP servers, and 48 in firewalls. Those amount to totally $\binom{293}{2} = 42,778$ pairs of software applications tested using our tool in the experiments.

5.2 Cross-Category Common Attack Surface

In this section, we apply the two proposed common attack surface metrics to totally 42,778 pairs of real world software. The first set of experiments reveal the existence of common attack surface between different categories of software applications. To convert the results to a comparable scale, we have normalized the absolute value of common attack surface reported by *CASFinder* by the size of the software. Figure 5 shows the existence of common attack surface across seven categories. The percentages on top of the bars inside each figure indicate the level of common attack surface between the category mentioned in the title of the figure and all the seven categories. We can observe that common attack surface exists in all of the category combinations. For example, the *DB* category has the highest level of common attack surface inside its own category (between different software inside that category), 27.9%, and it also shares more than 9% common attack surface with any other category.

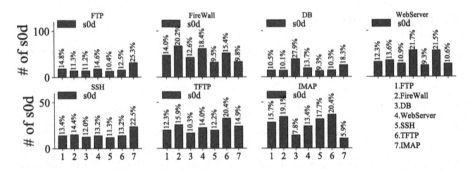

Fig. 5. Common attack surface across categories

In summary, the results across all categories are shown in the heat map in Table 1 where a darker color indicates a larger CAS value between the pair of categories. A visible diagonal with the darkest color in the heat map indicates the expected trend that different software in the same category yield the highest level of common attack surface, most likely due to their similar functionality, except for *SSH*. In fact, the category *SSH* has the lowest level of common attack surface within its category. The reason is that the *SSH* category only contains 25 software applications, which is not sufficiently large to produce any reliable trend. Due to similar reasons, we have omitted the results from the *IMAP* category in the heat-map.

Table 1. HeatMap for common attack surface in different categories

	FTP	FireWall	DB	WebServer	SSH	TFTP
FTP	18.2	13.8	13.7	17.9	12.8	15.3
FireWall	47.1	67.6	42.2	61.8	31.7	51.7
DB	14.4	13.9	38.4	18.8	12.8	14.1
WebServer	32.2	35.6	28.6	56.9	24.4	56.3
SSH	13.9	15.0	12.5	13.8	11.8	13.7
TFTP	19.8	25.5	16.5	22.4	19.6	32.6

After understanding the general existence of common attack surface among the seven categories of software applications, we aim to study more specific trends in our second sets of experiments. The left chart in Fig. 6 shows the accumulated number of pairs of software applications in the absolute value of common attack surface. The figure depicts only the results with a nonzero value, which include totally 9,852 pairs (which amounts to about 1/8 of the total number of pairs). We can observe that the accumulated number of pairs of software applications increases quickly before the value of common attack surface reaches about 12 and afterwards the accumulation flattens out. About 20% of software share common clone segments, and 56% of the clone segments contain at least one common attack surface. The right chart in Fig. 6 depicts the relationship between common attack surface and sizes of the software. We use the absolute values of common attack surface in this experiment. For the sizes, we use the normalized combined sizes $\log_{1000}(A^B)/1000$ when software A is compared with software B. We can observe that, with increasing sizes of the software, the value of common attack surface generally increases. This is as expected since the number of I/O functions would be roughly proportional to the size of the software.

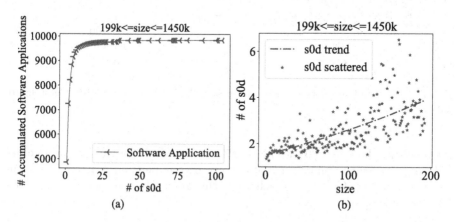

Fig. 6. CAS in accumulated software application pairs (a), CAS trend vs size (b) (Color figure online)

The left chart in Fig. 7 compares the average number of I/O functions and the average common attack surface over several years. The blue bars indicate the average number of I/O functions used in the software applications tested in our experiments based on the publishing year. The average number of I/O functions per software application does not have a simple trend and is used as a baseline for comparison. We can observe a clear downward trend in the average value of common attack surface over time, with software published around 2010 having a much higher value of common attack surface compared with more recent years, regardless of the number of average I/O functions. We believe this trend shows that code reusing plays a major role in common attack surface, since the trend can be easily explained by the backward nature of code reusing (i.e., programmers can only reuse older code). The right chart in Fig. 7 explores the trend of the probabilistic common attack surface metric versus the size. The value of the probabilistic common attack surface metric decreases since the increase of the number of I/O functions in software applications is faster than the increase of common attack surface.

In fact, those results match the results of existing vulnerability discovery models, which generally show that larger software applications typically have more vulnerabilities but a lower probability for having vulnerabilities per unit of software size. For example, Google Chrome (with the number of lines at 14,137,145 [2]) has 1,453 vulnerabilities over nine years [9], while Apache (with

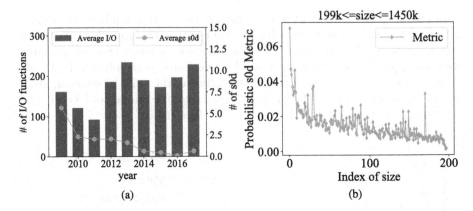

Fig. 7. CAS trend in years (a) and the probabilistic CAS metric (b) (Color figure online)

the number of lines at 1,800,402) has 815 over 19 years. However, the probability of having one vulnerability per unit of software size per year is $1.15 * 10^{-3}\%$ for Chrome and $2.4 * 10^{-3}\%$ for Apache (i.e., the larger Chrome has less vulnerabilities per unit of software size).

5.3   Common Attack Surface in the Same Category

We study the trend of common attack surface between software within the same category in this section. Figure 8 depicts the common attack surface for different sizes of software in the category *WebServer* and *FTP*, respectively, represented in both scattered and trending results. The orange scattered points and the dotted line indicate the result and the red dotted line is the same trend borrowed from Fig. 6 for comparison. We can observe that the trend of common attack surface in both categories increase with the size, which follows a similar trend as the cross category result. However, the trend of *WebServer* increases faster than the cross-category trend, which matches the results shown in Table 1. On the other hand, the trend in the *FTP* category grows slightly slower than the cross category trend, which can be explained by the fact that *FTP* shares a large amount of common attack surface with *WebServer* and *TFTP*.

The left chart in Fig. 9 depicts the trend of common attack surface over time in the same category. Each blue bar represents the average number of I/O functions in the years in the same category of the experiments. The red line shows the average number of common attack surface in those years. Compared to Fig. 7, the common attack surface in the same category has higher values, which also match the previous observations. The right chart in Fig. 9 reveals the trend of the probabilistic common attack surface metric versus the size in the same category, which shows a similar trend as the cross category result, although the trend within the same category starts from a higher value around 0.20 (in contrast, the cross-category metric starts from 0.06).

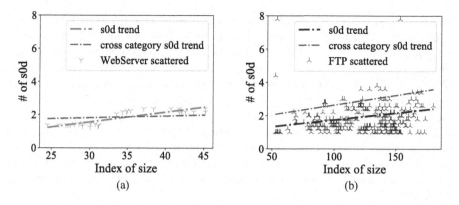

Fig. 8. Size trend in same category, WebServer (a) and FTP (b) (Color figure online)

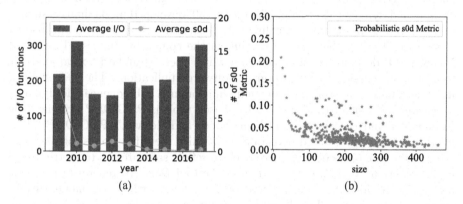

Fig. 9. Common attack surface over time and vs size (Color figure online)

6 Related Work

There exist extensive research on clone code detection although many of these tools are mainly for research purposes [42]. One of the popular tools in text-based clone detection is the *Dup* [3]; if two lines of code are identical after removing all whitespaces and comments, they are assigned as clone codes; the longest line matches are the output, but the minimum length of the reported code can be customized according to different needs. Another well-known approach [21] is applying the fingerprint in order to identify the redundancy on a substring of the source code. The fingerprinting calculation uses KARP-Rabins string matching approach [25, 26] to calculate the length of all *n* substrings. Ducasse developed [10] *duploc* which was designed to be a parsing free, language-independent tool which first reads the source file and sequences of the lines, then removes all comments and whitespace to create a set of condensed lines; afterward, a comparison is made based on the hash result, where scatter-plots indicate the visualization of a cloned result. Token-based clone detection is also one of the widely applied methods. One of the representative tools in token-based detection is *CCFinder* [23], which is applied in our work. Bakers Dup [3, 4] implements a similar approach as *CCFinder*. The detection process begins by tokenizing the source code, then using a suffix-tree algorithm to compare tokens. Unlike *CCFinder*, *Dup* does not apply transformation, but rather consistently renames the identifier. Raimar Falke [30] develops a tool called *iclones* [16], which uses suffix-trees to find clones in abstract syntax trees, which can operate in linear time and space. CP-Miner [32] as a well-designed token-based clone detector, uses frequent subsequence mining algorithms to detect tokenized segments. RTF [6] is a token-based clone detector that uses string algorithms for efficient detection; rather than using the more common suffix-tree, it utilizes more memory-efficient suffix array.

One of the leading tools using AST-based algorithm is the *CloneDR* developed by Baxter [7] which can detect exact and near-miss clone through applying hashing and dynamic algorithm. The *ccdiml* [39] developed by Bauhaus is similar to the *CloneDR* in the way of dealing with hash and code sequences, but instead of using AST, it applies IML algorithm in the comparing process. David and Nicholas [14] develop a tool named *Sim* which uses a standard lexical analyzer to generate a parsing-tree of two given software applications. The code similarity is determined by applying the maximum common subsequence and dynamic programming. One of the leading PDG-based tools is PDG-DUP presented by Komondoor and Horwit [27] and Komondoor and Horwitz's PDG-DUP [27] is another leading PDG-based detection tool, which identifies clones together and keeping the semantics of the source code to reflect software. As to metric-based clone detection, Mayrand et al. [37] uses the tool *Darix* to generate the metric and the clone identification is based on four values, which are name, layout, expression and control flow. Kontogiannis [28] uses Markov models to compute the dissimilarity of the code by applying the abstract pattern matching. Five widely used metrics are applied in a direct comparison in [29]. There are also some other approaches that using hybrid clone detections. In [30], the authors apply the suffix trees to find clones in AST; this approach can find clones in linear time and space.

The concept of attack surface is originally proposed for specific software, e.g., Windows, and requires domain-specific expertise to formulate and implement [17]. Later on, the concept is generalized using formal models and becomes applicable to all software [35]. Furthermore, it is refined and applied to large scale software, and its calculation can be assisted by automatically generated call graphs [33, 34]. Attack surface has attracted significant attentions over the years. It is used as a metric to evaluate Android's message-passing system [24], in kernel tailing [31], and also serves as a foundation in Moving Target Defense, which basically aims to change the attack surface over time so to make attackers' job harder [18, 19]. The study on automating the calculation of attack surface is another interesting domain, e.g., COPES uses static analysis from bytecode to calculate attack surface and to secure permission-based software [5]. Stack traces from user crash reports is used to approximate attack surface automatically [44].

The correlation between attack surface and vulnerabilities has also been investigated, such as using attack surface entry points and reachability to assess the risk of vulnerability [46]. A study about the relationship between attack surface and the vulnerability density is given in [45], although the result is only based on two releases of Apache HTTP Server. Despite such interest in attack surface, to the best of our knowledge, the common attack surface between different software has attracted little attention.

7 Conclusion

In this paper, we have defined the concept of common attack surface and implemented an automated tool for evaluating the common attack surface between given software applications. We have conducted experiments on real open source software and examined the common attack surface both within and between software categories. Our results have shown common attack surface to be pervasive among software. Our work still has some limitations which will lead to our future work. First, since we rely on *CCFinder* our tool also inherits its limitations, and one future direction is to explore other clone detection tools. Second, we have focused on entry/exit points of attack surface, and one future direction is to also consider channels and untrusted data items. Third, we have focused on the C language in this work, and extending it to other languages with different entry and exit libraries is an interesting future direction. Forth, we plan to extend the effort on correlating between common attack surface and known vulnerabilities. We have focused on reused codes only, and a future direction is to also consider their indirect impact on other parts of the software. Finally, one interesting future direction is to evaluate common attack surface between two binary files. Existing disassembling and de-compiling tools, such as IDA Pro [1], could reverse the binary code to source code for further common attack surface study.

Acknowledgment. Authors with Concordia University are partially supported by the Natural Sciences and Engineering Research Council of Canada under Discovery Grant N01035. Sushil Jajodia was supported in part by the National Institute of Standards and Technology grants 60NANB16D287 and 60NANB18D168, National Science Foundation under grant IIP-1266147, Army Research Office under grant W911NF-13-1-0421, and Office of Naval Research under grant N00014-15-1-2007.

References

1. Interactive disassembler. https://www.hex-rays.com/products/ida/
2. Open hub (2017). https://www.openhub.net/
3. Baker, B.S.: A program for identifying duplicated code. Comput. Sci. Stat. **24**, 49 (1993)
4. Baker, B.S.: On finding duplication and near-duplication in large software systems. In: Proceedings of 2nd Working Conference on Reverse Engineering, pp. 86–95. IEEE (1995)

5. Bartel, A., Klein, J., Le Traon, Y., Monperrus, M.: Automatically securing permission-based software by reducing the attack surface: an application to Android. In: Proceedings of the 27th IEEE/ACM International Conference on Automated Software Engineering, pp. 274–277. ACM (2012)

6. Basit, H.A., Jarzabek, S.: Efficient token based clone detection with flexible tokenization. In: Proceedings of the 6th Joint Meeting of the European Software Engineering Conference and the ACM SIGSOFT Symposium on the Foundations of Software Engineering, pp. 513–516. ACM (2007)

7. Baxter, I.D., Yahin, A., Moura, L., Sant'Anna, M., Bier, L.: Clone detection using abstract syntax trees. In: 1998 Proceedings of International Conference on Software Maintenance, pp. 368–377. IEEE (1998)

8. Carvalho, M., DeMott, J., Ford, R., Wheeler, D.A.: Heartbleed 101. IEEE Secur. Privacy **12**(4), 63–67 (2014)

9. CVE Community. Common vulnerabilities and exposures (1999). https://cve. mitre.org/

10. Ducasse, S., Rieger, M., Demeyer, S.: A language independent approach for detecting duplicated code. In: Proceedings of IEEE International Conference on Software Maintenance, ICSM 1999, pp. 109–118. IEEE (1999)

11. Durumeric, Z., et al.: The matter of heartbleed. In: Proceedings of the 2014 Conference on Internet Measurement Conference, pp. 475–488. ACM (2014)

12. Ghaffarian, S.M., Shahriari, H.R.: Software vulnerability analysis and discovery using machine-learning and data-mining techniques: a survey. ACM Comput. Surv. (CSUR) **50**(4), 56 (2017)

13. Ghosh, A.K., Pendarakis, D., Sanders, W.H.: Moving target defense co-chair's report-national cyber leap year summit 2009. Technical report, Federal Networking and Information Technology Research and Development (NITRD) Program (2009)

14. Gitchell, D., Tran, N.: Sim: a utility for detecting similarity in computer programs. In: ACM SIGCSE Bulletin, vol. 31, pp. 266–270. ACM (1999)

15. GitHub. Inc. A web-based hosting service for version control using Git. https:// github.com

16. Göde, N., Koschke, R.: Incremental clone detection. In: 13th European Conference on Software Maintenance and Reengineering, CSMR 2009, pp. 219–228. IEEE (2009)

17. Howard, M., Pincus, J., Wing, J.: Measuring relative attack surfaces. In: Workshop on Advanced Developments in Software and Systems Security (2003)

18. Jajodia, S., Ghosh, A.K., Subrahmanian, V.S., Swarup, V., Wang, C., Wang, X.S.: Moving Target Defense II: Application of Game Theory and Adversarial Modeling. Springer, Heidelberg (2012). https://doi.org/10.1007/978-1-4614-5416-8

19. Jajodia, S., Ghosh, A.K., Swarup, V., Wang, C., Wang, X.S.: Moving Target Defense: Creating Asymmetric Uncertainty for Cyber Threats, 1st edn. Springer, Heidelberg (2011). https://doi.org/10.1007/978-1-4614-0977-9

20. Jajodia, S., Ghosh, A.K., Swarup, V., Wang, C., Wang, X.S.: Moving Target Defense: Creating Asymmetric Uncertainty for Cyber Threats, vol. 54. Springer, Heidelberg (2011). https://doi.org/10.1007/978-1-4614-0977-9

21. Johnson, J.H.: Substring matching for clone detection and change tracking. In: ICSM, vol. 94, pp. 120–126 (1994)

22. Kamiya, T.: Tutorial of CLI tool ccfx (2008). http://www.ccfinder.net/doc/10.2/ en/tutorial-ccfx.html

23. Kamiya, T., Kusumoto, S., Inoue, K.: CCFinder: a multilinguistic token-based code clone detection system for large scale source code. IEEE Trans. Softw. Eng. **28**(7), 654–670 (2002)

24. Kantola, D., Chin, E., He, W., Wagner, D.: Reducing attack surfaces for intra-application communication in Android. In: Proceedings of the Second ACM Workshop on Security and Privacy in Smartphones and Mobile Devices, pp. 69–80. ACM (2012)
25. Karp, R.M.: Combinatorics, complexity, and randomness. Commun. ACM **29**(2), 97–109 (1986)
26. Karp, R.M., Rabin, M.O.: Efficient randomized pattern-matching algorithms. IBM J. Res. Dev. **31**(2), 249–260 (1987)
27. Komondoor, R., Horwitz, S.: Using slicing to identify duplication in source code. In: Cousot, P. (ed.) SAS 2001. LNCS, vol. 2126, pp. 40–56. Springer, Heidelberg (2001). https://doi.org/10.1007/3-540-47764-0_3
28. Kontogiannis, K., Galler, M., DeMori, R.: Detecting code similarity using patterns. In: Working Notes of 3rd Workshop on AI and Software Engineering, vol. 6 (1995)
29. Kontogiannis, K.A., DeMori, R., Merlo, E., Galler, M., Bernstein, M.: Pattern matching for clone and concept detection. Autom. Softw. Eng. **3**(1–2), 77–108 (1996)
30. Koschke, R., Falke, R., Frenzel, P.: Clone detection using abstract syntax suffix trees. In: 13th Working Conference on Reverse Engineering, WCRE 2006, pp. 253–262. IEEE (2006)
31. Kurmus, A., et al.: Attack surface metrics and automated compile-time OS kernel tailoring. In: NDSS (2013)
32. Li, Z., Shan, L., Myagmar, S., Zhou, Y.: CP-miner: finding copy-paste and related bugs in large-scale software code. IEEE Trans. Softw. Eng. **32**(3), 176–192 (2006)
33. Manadhata, P., Wing, J.: An attack surface metric. Technical report CMU-CS-05-155 (2005)
34. Manadhata, P., Wing, J.: An attack surface metric. IEEE Trans. Softw. Eng. **37**(3), 371–386 (2011)
35. Manadhata, P., Wing, J.: Measuring a system's attack surface. Technical report CMU-CS-04-102 (2004)
36. Manadhata, P.K., Wing, J.M.: An attack surface metric. IEEE Trans. Softw. Eng. **37**(3), 371–386 (2011)
37. Mayrand, J., Leblanc, C., Merlo, E.: Experiment on the automatic detection of function clones in a software system using metrics. In: ICSM, vol. 96, p. 244 (1996)
38. Petersenna. Ccfinder core. https://github.com/petersenna/ccfinderx-core
39. Raza, A., Vogel, G., Plödereder, E.: Bauhaus – a tool suite for program analysis and reverse engineering. In: Pinho, L.M., González Harbour, M. (eds.) Ada-Europe 2006. LNCS, vol. 4006, pp. 71–82. Springer, Heidelberg (2006). https://doi.org/10.1007/11767077_6
40. Rothwell, T.: The GNU C reference manual (2006). https://www.gnu.org/software/gnu-c-manual/
41. Roy, C.K., Cordy, J.R., Koschke, R.: Comparison and evaluation of code clone detection techniques and tools: a qualitative approach. Sci. Comput. Program. **74**(7), 470–495 (2009)
42. Roy, C.K., Cordy, J.R.: A survey on software clone detection research. Queen's Sch. Comput. TR **541**(115), 64–68 (2007)
43. Syropoulos, A.: Mathematics of multisets. In: Calude, C.S., PǍun, G., Rozenberg, G., Salomaa, A. (eds.) WMC 2000. LNCS, vol. 2235, pp. 347–358. Springer, Heidelberg (2001). https://doi.org/10.1007/3-540-45523-X_17
44. Theisen, C., Herzig, K., Morrison, P., Murphy, B., Williams, L.: Approximating attack surfaces with stack traces. In: Proceedings of the 37th International Conference on Software Engineering, vol. 2, pp. 199–208. IEEE Press (2015)

45. Younis, A.A., Malaiya, Y.K.: Relationship between attack surface and vulnerability density: a case study on apache HTTP server. In: Proceedings on the International Conference on Internet Computing (ICOMP), p. 1. The Steering Committee of the World Congress in Computer Science, Computer Engineering and Applied Computing (WorldComp) (2012)
46. Younis, A.A., Malaiya, Y.K., Ray, I.: Using attack surface entry points and reachability analysis to assess the risk of software vulnerability exploitability. In: 2014 IEEE 15th International Symposium on High-Assurance Systems Engineering (HASE), pp. 1–8. IEEE (2014)
47. Zhang, M., Wang, L., Jajodia, S., Singhal, A., Albanese, M.: Network diversity: a security metric for evaluating the resilience of networks against zero-day attacks. IEEE Trans. Inf. Forensics Secur. 11(5), 1071–1086 (2016)

Algorithm Diversity for Resilient Systems

Scott D. Stoller[✉] and Yanhong A. Liu

Department of Computer Science, Stony Brook University, New York, USA
{stoller,liu}@cs.stonybrook.edu

Abstract. Diversity can significantly increase the resilience of systems, by reducing the prevalence of shared vulnerabilities and making vulnerabilities harder to exploit. Work on software diversity for security typically creates variants of a program using low-level code transformations. This paper is the first to study *algorithm diversity* for resilience. We first describe how a method based on high-level invariants and systematic incrementalization can be used to create algorithm variants. Executing multiple variants in parallel and comparing their outputs provides greater resilience than executing one variant. To prevent different parallel schedules from causing variants' behaviors to diverge, we present a *synchronized execution* algorithm for DistAlgo, an extension of Python for high-level, precise, executable specifications of distributed algorithms. We propose static and dynamic metrics for measuring diversity. An experimental evaluation of algorithm diversity combined with implementation-level diversity for several sequential algorithms and distributed algorithms shows the benefits of algorithm diversity.

1 Introduction

Diversity can significantly increase the resilience of systems, by reducing the prevalence of shared vulnerabilities and making vulnerabilities harder to exploit. The idea of intentionally introducing software diversity as a defense mechanism has been around for decades, e.g., [5,6]. It is closely related to the well-known *moving target defense* (MTD) strategy: running different variants of a program at different times is MTD. Software diversity is an effective defense against attacks whose success depends on details of the victim software. Without knowing those details for the specific instance (variant) of the software being attacked, attackers can still attempt such attacks (e.g., by making random guesses at those details), but the probability of success is greatly reduced [16].

There is a large corpus of research on techniques for automatically introducing software diversity that increase resilience to various classes of attacks [16]. For example, Address Space Layout Randomization (ASLR), which randomizes the starting addresses of segments in a process's address space, is a classic form of software diversity that increases resilience to some types of memory corruption attacks.

The most common way to use software diversity to increase resilience is to run a randomly selected variant each time the program is executed. With this

© IFIP International Federation for Information Processing 2019
Published by Springer Nature Switzerland AG 2019
S. N. Foley (Ed.): DBSec 2019, LNCS 11559, pp. 359–378, 2019.
https://doi.org/10.1007/978-3-030-22479-0_19

approach, the use of diversity alters, with high probability, the effect of an attack, so the attack does not have the intended effect (e.g., gaining root privilege and installing a rootkit) [16]. The attack might still have a less malicious and less predictable but nevertheless undesirable effect (e.g., crash or incorrect output).

Another way is to run multiple variants of the application in parallel and compare their outputs. We call this *diversified replication*. Any difference in the outputs of the variants indicates misbehavior of one or more variants due to an attack; this triggers defensive action. This approach provides greater resilience, at the cost of more computational resources. It also provides greater resilience than traditional replication, in which replicas are identical and exhibit the same (incorrect) behavior when their vulnerabilities are exploited. Note that diversity may lead to different behavior (and therefore attack detection) in two ways: (1) it might cause a difference in the direct effect of the attack (e.g., which data structure is overwritten) or, (2) even if the direct effect of the attack is the same (e.g., the same data structure is overwritten), it might cause differences in subsequent behavior, due to differences in the algorithms or implementations used by the variants (e.g., one variant reads the affected data structure earlier in its computation and hence before the attack, and another reads the affected data structure later in its computation and hence after the attack).

This paper focuses on *algorithm diversity* for software resilience, in which different variants run different algorithms, i.e., perform different computations at a high level. In contrast, all of the work surveyed in [16] creates *implementation-level diversity*, changing details of the implementation without changing the algorithm. Algorithm diversity can introduce new and larger differences between variants than implementation-level diversity and hence can provide greater resilience, especially when used together with implementation-level diversity.

Algorithm variants may be obtained in a variety of ways, besides writing them manually. For standard problems (e.g., dictionary ADT), they can be obtained from algorithm libraries. A more general automated approach is to generate them by starting with a high-level algorithm (or specification) and applying different optimizations (algorithm improvements, automated using program analysis and transformation). In particular, we have used a method based on systematic incrementalization [18,19,22,27], which transforms programs to maintain high-level invariants incrementally, and related optimizations to generate multiple variants of many sequential algorithms and distributed algorithms [24–26].

Algorithm diversity and implementation-level diversity introduce different kinds of variation and together offer more diversity than either alone. We introduce *diversity metrics* that quantify the difference between—or equivalently, the similarity of—variants. We consider a static metric, *code diversity*, based on the instruction sequences in the compiled program, and two dynamic (behavioral) metrics: *trace diversity*, based on the sequence of instructions executed, and *input access diversity*, based on the sequence of accesses to input data. The latter dynamic metric is motivated by the fact that invalid inputs are the primary attack vector for external attackers. A direction for future work is to augment these broad diversity metrics with more specialized metrics that quantify resilience to specific classes of attacks.

Algorithm diversity can be applied to programs in any language. In this paper, we focus on Python and DistAlgo [24, 25], an extension of Python for high-level, precise, executable specifications of distributed algorithms. In contrast, existing work on automated software diversity primarily targets C programs or (disassembled) machine code.

Python is interpreted—more precisely, CPython, the predominant implementation of Python, compiles Python to bytecode and then runs the bytecode in an interpreter. Algorithm diversity applied to Python programs can be used together with implementation-level diversity applied to Python programs and the runtime system. This achieves greater total diversity and increases resilience to vulnerabilities in the runtime system, because vulnerabilities manifest only with specific inputs, and the runtime system's inputs include Python programs as well as network messages, UI events, etc. Diversity at the high-level language level provides additional protection from data-only attacks [4, 30], against which many runtime-system-level defenses are less effective. Algorithm diversity applied to Python programs can also provide resilience to functional faults in the runtime system, if the runtime system does not correctly implement the semantics of some built-in constructs or library functions in some (corner) cases.

Diversified replication requires *synchronized execution* (often called *N-version execution* [1]) of the variants; otherwise, their executions might diverge due to scheduling differences. Synchronized execution of distributed programs generally requires synchronization of message delivery order. DistAlgo's asynchronous message handling requires additional synchronization, to ensure that all variants handle corresponding messages at corresponding points in their executions. We developed a synchronized execution framework for DistAlgo that ensures this. Our framework can also suppport variants whose behaviors differ in prescribed ways.

Measuring dynamic diversity for Python and DistAlgo programs required development of new runtime monitoring tools, which are also more broadly useful. We designed and implemented a tool that intercepts accesses to fields of selected objects; we use it to log accesses to objects read as input, including objects received in messages. Handling built-in types such as integers and strings is tricky, because they are sometimes accessed directly by C code in the Python interpreter, but essential, because they are commonly used in program inputs.

We also designed and implemented a tracing tool that reconstructs the exact sequence of bytecode instructions executed by a Python program. It uses the standard Python tracing module to record the sequence of source lines executed, and then analyzes the compiled program to determine the sequence of bytecode instructions corresponding to each source line. Supporting DistAlgo requires some extra work, due to details of DistAlgo's implementation by translation to Python.

In summary, the contributions of this paper include:

- The first study of semi-automated algorithm diversity for software resilience, using a method based on systematic incrementalization to generate algorithm variants.
- A synchronized execution framework for DistAlgo and for high-level executable specifications of distributed algorithms.
- Static and dynamic metrics for measuring diversity.
- A runtime monitoring tool for Python and DistAlgo that logs accesses to fields of selected objects, including instances of built-in types.
- A tracing tool for Python and DistAlgo that reconstructs the exact sequence of executed bytecode instructions.
- Experimental evaluation of algorithm diversity combined with implementation-level diversity for several sequential algorithms and distributed algorithms, demonstrating that algorithm diversity can achieve more diversity than implementation-level diversity, and the two together can achieve even more.

2   Background on DistAlgo

Liu et al. [24,25] propose DistAlgo, a language for high-level, precise, executable specifications of distributed algorithms, and study its use for specification, implementation, optimization, and simplification of such algorithms. For expressing distributed algorithms at a high level, DistAlgo supports four main concepts by building on an object-oriented programming language, Python: (1) distributed processes that send messages, (2) control flow for handling received messages, (3) high-level queries for synchronization conditions, and (4) configuration for setting up and running. DistAlgo is specified precisely by a formal operational semantics [24].

Processes that Send Messages. A process type P is defined by a class definition for P that inherits from DistAlgo's built-in `process` class. The definition of P may contain, in addition to the usual definitions that may appear in Python classes, definition of a `setup` method for taking in and setting up the values used by the process, definition of a `run` method containing the main control flow of the process, and definitions of `receive` handlers for handling messages, as described below.

To create instances of P, DistAlgo provides a `new` P construct; it can optionally be preceded by the number of processes to create (the default is 1) and followed by "`at` h" where h identifies the host where the process(es) should be created (the default is the local host). After a new process has been created, and its `setup` method called to initialize it, invoking its `start` method causes execution of its `run` method.

Processes send messages using the statement `send` m `to` ps, where ps is a process or set of processes.

Control Flow for Handling Received Messages. Received messages can be handled asynchronously, using `receive` definitions, and synchronously, using `await` statements. A `receive` definition has the form `receive` m `from` p: *stmt*. It handles un-handled messages that match m `from` p, where m and p are patterns. If matching succeeds, unbound variables in m (and p) are bound to the corresponding component of the message (and the message's sender, respectively), and then *stmt* is executed.

To synchronize message handling with local computation, `receive` handlers are executed only at *yield points*. The program point before or after any statement can be declared as a yield point. In addition, there is an implicit yield point before each `await` statement, for handling messages while waiting. By default, any number of pending messages can be handled at a yield point.

An `await` statement has the form

$$\texttt{await } cond_1 : stmt_1 \texttt{ or } \ldots \texttt{ or } cond_k : stmt_k \texttt{ timeout } t : stmt$$

It waits until one of $cond_1$, ..., $cond_k$ is true or time t has elapsed, and then nondeterministically selects one of $stmt_1$, ..., $stmt_k$, $stmt$ whose condition is true and executes the selected statement. Each branch is optional.

High-Level Queries for Synchronization Conditions. DistAlgo provides constructs to express synchronization conditions in `await` statements as high-level queries over message histories (or other sets or sequences). A query can be an existential or universal quantification, a comprehension, or an aggregation. An existential quantification has the form `some` v_1 `in` s_1, ..., v_k `in` s_k | $cond$. It returns true iff $cond$ holds for some combination of values of variables that satisfies all v_i `in` s_i clauses. Universal quantification is similar, with keyword `each` instead of `some`.

A comprehension has the form {e: v_1 `in` s_1, ..., v_k `in` s_k, $cond$}. It returns the set of values of e for all combinations of values of variables that satisfy all v_i `in` s_i clauses and condition $cond$.

DistAlgo automatically maintains histories of messages sent and received by each process in variables `sent` and `received`; they are automatically eliminated if unused.

Configuration. Configuration for requirements such as use of logical clocks and use of reliable and FIFO channels can be specified using DistAlgo's `configure` statement. For example, `configure clock = Lamport` specifies that Lamport's logical clocks are used; it configures sending and receiving of a message to update the clock value, and defines a function `logical_time()` that returns the clock value.

3   Creating Variants Using Incrementalization

Algorithm variants differ from each other due to different high-level invariants they maintain and different ways of maintaining them. We describe the ideas

```
1 class P extends process:
2   def setup(s):
3     self.s := s        # set of all other processes
4     self.q := {}        # set of pending requests

5   def mutex(task):
6     self.t := logical_time()                              # request
7     send ('request', t, self) to s                        #
8     q.add(('request', t, self))                           #
9     await each ('request', t2, p2) in q
              | (t2,p2) != (t,self) implies (t,self) < (t2,p2)
10          and each p2 in s | some ('ack', t2, =p2) in received | t2 > t
11    task()
12    q.del(('request', t, self))                           # release
13    send ('release', logical_time(), self) to s           #

14  receive ('request', t2, p2):                            # receive request
15    q.add(('request', t2, p2))                            #
16    send ('ack', logical_time(), self) to p2              #

17  receive ('release', _, p2):                             # receive release
18    for ('request', t2, =p2) in q:                        #
19      q.del(('request', t2, p2))                          #
```

Fig. 1. Lamport's algorithm (lines 6–19) plus setup in DistAlgo.

of transforming expensive queries into high-level invariants and using systematic incrementalization to generate efficient algorithms that maintain the query results incrementally. Each resulting combination of ways of maintaining the invariants forms an algorithm variant.

Example. We use as an example Lamport's algorithm for distributed mutual exclusion, described in his seminal paper that proposed logical clocks [14]. The problem is for multiple processes to access a shared resource mutually exclusively, in what is called a critical section, i.e., there can be at most one process in a critical section at a time.

Each process can be expressed in DistAlgo as in Fig. 1 [24]. The process is set up with sets s and q (lines 3–4). To run a task mutually exclusively, the process sends a request and adds it to q (lines 6–8), waits for (i) own request (t,self) to be before each other request (t2,p2) in q and (ii) having received an ack with a time t2 later than t from each process p2 in s (lines 9–10) before doing the task in critical section (line 11), and then removes the request from q and sends a release (lines 12–13). When receiving a request or release, it sends back an ack and adds to or removes from q (lines 14–19).

The two conditions in await are key to the algorithm to ensure mutual exclusion, while the rest does basic sending and receiving of messages and bookkeeping of q.

Incrementalization. Incrementalization transforms queries and updates to maintain high-level invariants, including invariants for intermediate and auxiliary values, incrementally [22, 24, 26, 27]. It can yield diverse algorithms.

For the example in Fig. 1, the two conditions in `await` are queries, consisting of three quantifications including two that are nested; and assignments and bookkeeping for `s` and `q` and implicitly adding to `received` at `receive` handlers are updates.

The most direct algorithm can compute queries using iterations, in `for`-loops, whereas an incremental algorithm can maintain the results of queries at updates and look up the results as needed. An incrementalized algorithm maintains high-level invariants not only for the query results but also for intermediate and auxiliary values needed. Alternative invariants can often be used, yielding even greater diversity.

For example, the condition on line 10 in Fig. 1 can be transformed into

```
count {p2: p2 in s, ('ack', t2, p2) in received, t2 > t} = count s
```

and then—with variables `responded`, `number`, and `total` holding the set value and two `count` values, respectively, forming three invariants— transformed into:

```
number = total
```

Variable `total` is computed at set up of `s`, and `responded` and `number` at request, and the following `receive` handler is added:

```
receive ('ack', t2, p2):          # new message handler
    if t2 > t:                     # comparison in the condition
        if p2 in s:                # membership in the condition
            if p2 not in responded:  # test before adding
                responded.add(p2)    # add to responded
                number +:= 1         # increment number
```

The resulting algorithm differs significantly from direct iteration for the nested quantifications. The condition on line 10 could also be transformed into two nested `count` queries, and the condition on line 9 can be transformed into a `count` query also, or an aggregate query using `min`, yielding different algorithms for incremental maintenance. Details of these transformations are in [24].

In general, incrementalization can also transform nested loops that compute aggregate values such as sum and min [7, 23, 26]. For recursive functions as queries, the resulting incremental algorithm can still use recursion, forming an optimized recursive algorithm, or use iteration, forming an optimized iterative algorithm [20, 26]. Additionally, more refined data structures can be used to implement sets more efficiently [3, 10, 26], such as using one bit for each process in the set `responded` above. Incrementalization also enables new additional optimizations that are made possible as the results of systematic transformations [26].

4   Synchronized Execution for DistAlgo

A *diversified process* is a process with variants. A system may contain a mixture of diversified and un-diversified processes. A *gateway* process is created for each diversified process. It represents the variants of a diversified process to the rest of the system, making them appear as a single process. The gateway intercepts and forwards all inbound and outbound messages of all variants of the diversified process. We focus on synchronization of DistAlgo constructs; other I/O events, such as file accesses, can be synchronized using standard techniques.

Our synchronized execution framework consists of two parts: (1) an automated program transformation that (1a) ensures all messages are routed via the gateway, and (1b) inserts synchronization with the gateway at yield points, to ensure that all variants have yielded the same number of times before handling their copies of a given inbound message, despite differences in message latency and process execution speed; and (2) an algorithm run by the gateway that determines when to forward messages and when to report divergence (i.e., differences in behavior). When divergence is reported, the system may initiate application-specific defensive action.

We first present the core version of this approach, which assumes all variants of a process have the same communication pattern, i.e., send the same messages to the same destinations in the same order; we discuss later how to relax this assumption.

Handling Outbound Messages. To route outbound messages via the gateway, the transformation replaces all calls to DistAlgo's `send` method with calls to `send_sync`, and it inserts a definition of that method in every process class. `send_sync` sends the original message and its original destination to the gateway. Processes often send their own process id in messages. Since each variant of a diversified process has a unique process id, such messages will differ. To accommodate this as normal behavior, not divergence, `send_sync` replaces all occurrences of the variant's process id in the message with the gateway's process id. This also reflects the principle that the gateway represents the variants to the rest of the system. Pragmatically, it ensures that, if the recipient sends a reply to the process id contained in the message, the reply goes to the gateway, as desired.

The gateway stores un-forwarded outbound messages received from each variant in a separate FIFO queue. When all of the queues are non-empty, it compares the messages (including their destinations) at the heads of the queues. If they are identical, the gateway dequeues the message from all queues and forwards one copy to the destination, otherwise it reports divergence. To ensure liveness if some divergent variant fails to send a message, once one queue becomes non-empty, the gateway waits a limited amount of time for all queues to become non-empty; if this time limit is exceeded, the gateway reports divergence.

Synchronization at Yield Points and `await` *Statements.* The transformation inserts a call to `yield_sync(block, timeout)` at every yield point, and it inserts a definition of that method in every process class. The first argument *block* is a boolean that indicates whether the yield point is associated with an `await` statement. The second argument *timeout*, meaningful when the first argument is `True`, is a timeout duration if the `timeout` clause is present in that `await` statement and is `None` otherwise. The transformation also extends the `setup` method to initialize a variable `num_yields` to zero. `yield_sync` increments `num_yields`, sends a yield message containing *block*, *timeout*, and `num_yields` to the gateway and waits for a yield-reply message from the gateway before returning.

The transformation for an `await` statement with timeout ensures the total wait time is preserved, even though the waiting period may be split by interactions with the gateway. It transforms `await` $c_1: s_1$ `or` ... `or` $c_k : s_k$ `timeout` $t: s$ into

```
1    start_time = time.time()
2    while not (c₁ or ... or c_k):
3        elapsed = time.time() - start_time
4        remaining = t - elapsed
5        if remaining ≤ 0:
6            break
7        yield_sync(True, remaining)
8    if c₁: s₁
9    elif c₂: s₂
10   ...
11   elif c_k: s_k
12   else s
```

If the `await` statement has no timeout, then lines 1, 3–6, and 12 are omitted, and the second argument of `yield_sync` is `None`.

Handling Inbound Messages. When the gateway receives an inbound message m, it stores m in a queue of un-forwarded inbound messages, waits until it has received yield messages with the same `num_yields` from all variants, forwards to all variants and dequeues all un-forwarded inbound messages, and then sends a yield-reply message to all variants. The gateway communicates with the variants over FIFO channels, so all variants handle the forwarded messages before proceeding from the current yield point. In the copy of m to be forwarded to variant p, the gateway replaces all occurrences of its own process id with p's process id.

If the gateway has received a yield message from all variants, and has no inbound message to forward to them, its behavior depends on the values of *block* and *timeout* in the yield messages (if the values of *block* differ, or the values of *timeout* differ by more than a small amount, divergence is reported). If *block*=`False`, the gateway sends a yield-reply message to all variants, allowing them to proceed. If *block*=`True` and *timeout*=`None`, the gateway waits until it has received and forwarded an inbound message before sending a yield-reply message, since the conditions in the `await` statements will remain false until the variants' states are updated by handling of an inbound message. If *block*=`True` and *timeout* is a number, the gateway behaves as in the previous sentence, except it will also send a yield-reply message after time *timeout* has elapsed.

Process Creation. The program transformation reads a configuration file that specifies which process types are diversified and the types of their variants. For each diversified process type P, a gateway type GatewayP is generated (basically, this is done by instantiating template code with the type P and the types of its variants), and process creation statements with type P are transformed to create instances of GatewayP instead. The setup method of GatewayP creates an instance of each of the specified variant types, and passes the gateway's process id to the variants as an additional argument to their setup methods, which are transformed to accept this additional argument.

Relaxed Synchronization. The above approach effectively introduces a barrier synchronization for a diversified process's variants at each synchronization point. This ensures the most timely detection of divergence. An alternative approach, used in some other synchronized execution frameworks [13,33], is to allow one variant (the "leader") to get ahead, try to make the actions of the other processes (the "followers") consistent with the leader's actions (e.g., by delivering the same number of messages at the corresponding yield event), and reporting divergence when this is not possible. This may provide speedup but allows a divergent leader to perform divergent actions before the leader's divergence is detected; when this is unacceptable, such actions should not be allowed to have externally visible effects until the followers catch up and agree on the actions.

Allowing Differences in Message Pattern. It may be desirable to relax the requirement that corresponding messages sent by all variants of a process are identical, in order to allow greater diversity. For example, Lamport's distributed mutual exclusion algorithm [14] sends in ack messages the current value of the sender's logical clock, whereas the variant in [17, Fig. 3] sends in ack messages the logical time of the request being acknowledged. To support algorithm variants that have the same communication pattern but different message content, we modify the gateway to omit the equality check on outbound messages when the destination is a diversified process, in which case the gateway sends to the other gateway an array containing the message from each variant, which forwards each message in the array to its corresponding variant. The correspondence is determined by indexing variants in the order that their types are listed in the configuration file.

To support algorithm variants with different communication patterns, the configuration file can specify that certain types of messages are *un-synchronized*. When the gateway receives a message of an un-synchronized type from its i'th variant, it immediately forwards the message to the destination's gateway, which forwards the message to its i'th variant. For example, for synchronized execution of Lamport's distributed mutual exclusion algorithm and Ricart-Agrawala's distributed mutual exclusion algorithm [29], we specify that ack and release messages (used only in Lamport's algorithm) and response messages (used only in

Ricart-Agrawala's algorithm) are un-synchronized; the gateway still synchronizes messages of other types.

5   Diversity Metrics and Runtime Monitoring Tools

5.1   Code Diversity

Since diversity is the complement of similarity, we measure code diversity with a well-established document similarity metric, namely, n-gram similarity with winnowing [31], which is used in the popular software plagiarism detection tool Moss to measure similarity of program source code. We apply it to Python byte-code, specifically, the sequence of bytecode instructions in a compiled program. Bytecode similarity is more relevant than source-level similarity, because diversity at the Python level aims to increase resilience to flaws in the runtime system, and bytecode is the program representation used by the runtime system.

An n-gram is a sequence of n consecutive instructions, starting at any position. The algorithm computes the hash of every n-gram in the compiled program, and then (for scalability) selects a subset of those hashes and stores them in a set called the program's *fingerprint*. The number of selected hashes is controlled indirectly by an algorithm parameter w called the *window size*. A window of size w consists of the hashes of w consecutive n-grams in the program. The winnowing algorithm is guaranteed to select at least one hash from each window of size w, although it may select more.

A robust metric should have the property that a slightly modified program has high similarity to the original program. In Python bytecode, local variables and global variables are identified by index. Inserting one new global variable at the beginning of the program causes renumbering of all global variables; this could make the metric non-robust. To ensure robustness, we normalize variable indices within each n-gram: we re-index the first global variable accessed in the n-gram as 0, the second one as 1, etc., and similarly for local variables. For similar reasons, we replace absolute line numbers used as targets in jump instructions with a place holder.

We quantify code diversity (and similarity) of two programs as 1 minus the Jaccard similarity of their fingerprints. Recall that the Jaccard similarity of sets S and T is $|S \cap T|/|S \cup T|$. We use 1 minus Jaccard similarity so larger values indicate greater diversity.

An alternative to n-gram similarity is Levenshtein distance (a.k.a. edit distance, namely, the minimum number of single-element insertions, deletions, and substitutions needed to change one string to another) between the bytecode sequences in the compiled programs. Levenshtein distance is less suitable here, because it is sensitive to bytecode orderings in the compiled program that may be unimportant at runtime. For example, permuting the order in which function definitions appear in the compiled program has no effect on the program's runtime behavior but has a large effect on the Levenshtein distance. Similarly, swapping the branches in a conditional statement and negating the condition yields an equivalent program with high n-gram similarity to the original but (if the branches are large) a large Levenshtein distance from the original.

5.2   Trace Diversity

Trace diversity measures the similarity of the sequences of bytecode instructions executed by two programs. Our bytecode-level tracing tool uses the standard Python trace module to obtain a source-level trace, and then translates it to a bytecode trace. A "blacklist" of modules to be ignored during the conversion can be specified; in experiments, we blacklist some system modules, such as bootstrap and trace. For each source line mentioned in the trace, identified by filename and line number, the translator compiles that .py file to a .pyc file, loads the .pyc file using the marshal module to obtain a code object, repeatedly uses the dis (disassembler) module to obtain the bytecode for the entire program as a list of Instruction objects, and uses the source line number information in the Instruction objects to determine the sequence of instructions corresponding to each line of source code in that file. In the traces to be compared, we include only the opcode and argument attributes of each Instruction; other attributes (e.g., is_jump_target) are less important. We quantify similarity of two traces as the Levenshtein distance (edit distance) between them divided by their average length, for normalization.

5.3   Input Access Diversity

Input access diversity measures the similarity of sequences of accesses to input data by two programs, quantified as Levenshtein distance between the sequences divided by their average length, for normalization. The core of the implementation is a general tool to intercept accesses to attributes of selected objects, by overriding the __getattribute__ method of appropriate classes. In our use case, the overriding method logs the access and then calls the original __getattribute__ method. For user-defined classes, this is easily accomplished by inserting a definition of __getattribute__ in the class. This approach does not work for built-in types such as int, string, and tuple, which are common types of input data.

For each of these built-in classes, we define a new class, e.g. tracked_int for int, that inherits from the built-in class and overrides the __getattribute__ method. In the remainder of the description, we focus on int; other built-in types are handled similarly. The problem is that some accesses to attributes of tracked_int are not logged, because attributes of built-in types are sometimes accessed directly by C code in the CPython runtime system. For example, even if x is a tracked_int, the addition operator in x+y compiles to the bytecode instruction BINARY_ADD, which does not call __getattribute__ on either argument.

We overcome this problem by augmenting tracked_int to override all methods of int that access the integer value: __add__, __eq__, __le__, etc. If x is a tracked_int, an expression like x+y now compiles to bytecode that uses the

CALL_FUNCTION instruction to explicitly invoke x's __add__ method with argument y. The tracked_int.__add__ method logs the access to the first argument (self), calls __getattribute__ on the second argument (so the access to it will be logged, if it is a tracked_int), and then calls the built-in __add__ method. Since we need to override these operations anyway, we augment log entries to indicate which operation was performed on the accessed attribute.

If x is an int, not a tracked_int, then CALL_FUNCTION invokes the built-in __add__ method, which is implemented by C code that accesses the second argument without calling __getattribute__. Consequently, accesses to y are not logged, even if y is a tracked_int. To overcome this remaining problem, we modify the program to replace the remaining uses of int with a new class my_int, which inherits from int and overrides each two-argument method of int with a method that calls __getattribute__ on the second argument and then calls the original method. To accomplish this replacement, we bind the name int to our class my_int, using the assignment int = my_int. As a result, a constructor call such as int(1) returns an instance of my_int. The literal 1 still produces an int. Therefore, we transform the source program to replace literals with constructor calls, e.g., 1 with int(1).

The remaining aspects of input access tracking differ for Python and DistAlgo. These aspects are (1) determining which objects are tracked, and (2) creating meaningful identifiers for tracked objects. We could easily use the result of Python's built-in id function to identify objects, but it would be difficult to compare input access traces from different variants (or even different runs of the same variant), because the object identifiers in them would be unrelated. Instead, we create object identifiers that can be compared meaningfully with object identifiers in other logs, as described below. The identifier is stored in an attribute of each tracked object.

Python. For Python programs, the user specifies which objects should be tracked by modifying the program to make them instances of tracked classes. For convenience, our tracker class provides a method that recursively traverses an object or collection (dictionary, list, tuple, or set) and replaces all instances of trackable built-in types (i.e., types for which a corresponding tracked type exists) with instances of tracked types. In our benchmark programs, inserting one or two calls to this method suffices. Tracked objects are identified by a sequence number assigned in the order that the objects are created. When tracked objects are used for data read as input, these identifiers are meaningful across logs from different variants, because the variants are given the same inputs and hence read the inputs in the same order.

DistAlgo. For DistAlgo programs, all messages are automatically considered as inputs; additional inputs, if any, are handled as described above for Python programs. Instances of trackable built-in types in messages are automatically replaced with instances of tracked types. Our tracker class, which inherits from DistAlgo's process class, is automatically inserted as a parent class of every process class in the given program. It overrides process.send with a method that replaces all instances of trackable built-in types in the message with instances of tracked types.

To create meaningful identifiers for tracked objects received in messages, we observe that such an identifier should identify the message in which the object was received. Our identifier for such an object is a tuple (*host*, *procNum*, *msgNum*, *objNum*), where *host* is the host on which the sender is running, *procNum* identifies the sending process relative to the host, *msgNum* identifies the message relative to the sending process, and *objNum* identifies the object within the message.

To avoid dependence on standard process identifiers that cannot be meaningfully compared across executions, we identify processes by a sequence number assigned in the order in which the processes are created. The `tracker` class overrides `process.setup` with a method that assigns the process sequence number; `tracker.setup` stores the next available process sequence number in a local file. *msgNum* is a per-sender sequence number assigned in the order in which messages are sent. The object sequence number *objNum* is assigned to each object in the message in the order that the object is encountered in a depth-first traversal of the message.

Input access logs for DistAlgo programs also contain entries corresponding to `receive` events, so we can determine that a particular data item (possibly received in a previous message and stored in a data structure) was accessed while processing a particular message.

6   Evaluation

We evaluated our approach on several sequential and distributed algorithms, using Python 3.7.2 and DistAlgo 1.1.0b13. Our software is available at https://www.cs.stonybrook.edu/~stoller/software.

For each problem and each diversity metric, we measure the diversity achieved (1) by algorithm diversity alone by averaging the diversity metric for each pair of algorithms; (2) by implementation-level diversity (ILD) alone by averaging the diversity metric for each pair of an algorithm and its ILD variant (i.e., the variant obtained by applying ILD transformations to it); (3) by both forms of diversity together by averaging the diversity metric for each pair of an algorithm and the ILD variant of another algorithm. For code diversity, we used $n = 5$ (the value used in [31]), and we disabled winnowing (i.e., included all hashes in the fingerprint), because the bytecode for our examples is not too large. Library code is not included in our code diversity measurements.

Implementation-Level Diversity (ILD). We created ILD by applying these typical ILD transformations: (1) NOP insertion: after each line of code, insert a `pass` statement with probability 0.05; (2) instruction reordering: for each two adjacent independent lines of code, swap them with probability 0.5; (3) branch reordering: for each if-statement, swap the branches and negate the condition (if there is no `else` branch, pretend `else: pass` is present) with probability 0.5; (4) function

Table 1. Experimental results for sequential algorithms. In the "Level" column, "algo" and "impl" denote algorithm and implementation-level diversity, respectively. The last column contains averages.

Metric	Level	Reach	Hanoi	LCS	Pat. search	Sort	Tree search	Avg.
		3 variants	4 variants	3 variants	3 variants	4 variants	6 variants	
Code	algo	0.80	0.58	0.65	0.81	0.79	0.83	0.74
Diversity	impl	0.40	0.39	0.66	0.52	0.32	0.63	0.49
	both	0.80	0.65	0.82	0.83	0.80	0.89	0.80
Input Access	algo	1.04	0.54	0.58	0.28	0.77	0.35	0.59
Diversity	impl	0	0.18	0.82	0.21	0	0	0.20
	both	1.04	0.57	1.12	0.28	0.77	0.36	0.69
Trace	algo	1.45	0.42	1.22	0.69	0.81	0.80	0.90
Diversity	impl	0.05	0.30	0.60	0.23	0.11	0.14	0.23
	both	1.45	0.45	1.39	0.70	0.82	0.82	0.94

(including `receive` handler) reordering: for each two adjacent independent `def` statements, swap them with probability 0.5; (5) argument reordering: for each function (excluding `run`, `setup`, and receive handlers), swap the first two arguments, swap the third and fourth arguments, etc.; (6) field reordering: reorder the assignment statements that initialize the fields in each class, by swapping the first two, the third and fourth, etc. Applying more complicated implementation-level diversity techniques is future work; it will require significant effort, because existing implementations of those techniques do not handle Python.

6.1 Sequential Algorithms

Our experiments use these algorithms for these problems: (1) graph reachability: original (iterative) algorithm, incrementalized algorithm, and rule-based algorithm (generated from rules using the method in [21]); (2) Hanoi Tower: original recursive algorithm, optimized recursive algorithm, optimized iterative algorithm, and optimized iterative algorithm with swap; (3) longest common subsequence (LCS): original recursive algorithm, optimized recursive algorithm, and optimized iterative algorithm; (4) pattern searching: naive algorithm, Knuth Morris Pratt (KMP) algorithm, Rabin Karp algorithm; (5) sorting: heap sort, quicksort, insertion sort, and merge sort; (6) tree search: recursive and iterative algorithms for AVL trees, recursive algorithm for B-trees, iterative algorithm for red-black trees, and recursive and iterative algorithms for (unbalanced) binary search trees.

The results are in Table 1. We see from the last column that, for all three metrics, algorithm diversity creates more diversity than ILD, and that the two together create even more.

6.2 Distributed Algorithms

Our experiments use the following algorithms: (1) 2-phase commit (2PC); (2) Hirschberg-Sinclair's leader election (HSleader) [11]; (3) Lamport's distributed

Table 2. Experimental results for distributed algorithms, with 2 variants for each algorithm. In the "Level" column, "algo" and "impl" denote algorithm diversity and implementation-level diversity, respectively.

Metric	Level	2PC	HSleader	Lamutex	Paxos	RAmutex	Average
Code	algo	0.56	0.66	0.50	0.68	0.53	0.59
Diversity	impl	0.19	0.18	0.08	0.30	0.27	0.21
	both	0.59	0.68	0.53	0.68	0.54	0.60
Input Access	algo	1.10	0.47	0.21	0.28	0.61	0.53
Diversity	impl	0.08	0.04	0	0.03	0.17	0.06
	both	1.09	0.52	0.21	0.30	0.61	0.55
Trace	algo	0.20	0.35	0.13	0.54	0.21	0.29
Diversity	impl	0.06	0.03	0.02	0.13	0.04	0.06
	both	0.20	0.36	0.14	0.52	0.21	0.29

mutual exclusion (Lamutex) [14]; (4) Lamport's basic Paxos [15]; (5) Ricart-Agrawala's distributed mutual exclusion (RAmutex) [29]. We used configurations with 3 or 4 processes for each algorithm. There are two variants of each algorithm: one variant that uses high-level queries over message histories, and one that explicitly maintains the result of those queries (and related intermediate results and auxiliary values), updating them in assignment statements, especially in `receive` handlers.

When measuring the dynamic metrics, we avoid spurious differences between the variants due to the platform's scheduling variability by running all variants in parallel using synchronized execution (for programs other than 2PC, due to a bug that we are still resolving in the interaction between our program transformations for synchronized execution and input access tracing, when measuring input access diversity, we instead avoided such spurious differences by running the variants separately but each with the same pattern of injected message delays that are larger than the platform's scheduling variability and designed to avoid races in message delivery order).

The results are in Table 2. We see from the last column that, for all three metrics, algorithm diversity creates significantly more diversity than ILD. The trace diversity produced by ILD is considerably smaller than the input access diversity it creates. These results are not inconsistent, because both are measured as ratios, and input accesses constitute a small fraction of the program's full activity recorded in the bytecode trace. The results for trace diversity for algorithm diversity for distributed algorithms are notably smaller than for sequential algorithms, because the trace includes execution of DistAlgo runtime library for networking, which is not diversified.

7   Related Work

Existing techniques for automated software diversity, including all those surveyed in [16], create implementation-level diversity, changing details of the implementation without changing the algorithm. Typically this is done by applying relatively simple local transformations, like those used in our evaluation. There are also some complex global transformations, such as instruction set randomization. These transformations are fully automated and more easily applied to large programs, but they are limited in that they do not create algorithm diversity. For example, they do not change the pattern in which inputs are used by the program.

Most work on automated software diversity for resilience transforms C programs or (disassembled) machine code, for broader applicability to systems code. There is some work on automated diversity for programs in JIT-compiled high-level languages, which diversifies the machine code generated by the JIT compiler. For example, librando does this for Java and JavaScript [12], and INSeRT does this for JavaScript [32]. This low-level approach is suitable for creating implementation-level diversity. Our methodology diversifies the high-level program directly to create algorithm diversity.

In N-version programming [1], N versions of a system (or component) are created by separate and independent manual design and implementation efforts starting from the same requirements specification, and the versions are run in parallel with synchronized execution. The goal is resilience in the presence of design faults, since independent teams are less likely to make the same design mistakes. Our work, like other work on software diversity, aims to mitigate software vulnerabilities, not design errors. The two techniques could be used together to address both. N-version programming may introduce algorithm diversity, but not in a controlled way, and at the cost of significant manual effort. In contrast, our approach is to create variants using a program transformation and optimization method based on systematic incrementalization, which guides the process, helps control how much diversity is introduced, and helps ensure correctness compared to ad-hoc development of variants. Our program transformation system InvTS [9,19] provides semi-automated support for the method, significantly reducing manual effort.

Synchronized execution has been widely studied in the fault-tolerance community, where it is often called N-version execution. N-version execution frameworks typically work at the system-call level, so they can be applied to software running on a given operating system, regardless of the application programming language. Our synchronized execution framework is applicable only to applications written in DistAlgo, but it is more portable and lighter weight. It can be used on any OS supported by DistAlgo (Windows, macOS, Linux, and Android), while system-call based approaches are highly OS-specific, e.g., Varan [13] and Bunshin [33] are N-version execution frameworks for Ubuntu. It is lighter-weight because a single high-level synchronization event is typically implemented by multiple system calls.

7.1   Evaluation of Diversity Techniques

A few approaches are commonly used to evaluate implementation-level diversity techniques. One is to estimate the probability of a successful memory-related exploit (e.g., buffer overflow or format string attack) based on the information about the diversified program that the attacker would need to guess, more specifically, the type of information (e.g., the address of a specific object, or the difference between the addresses of two specific objects) and the number of possible values of that type of information due to the randomization in the diversity transformation. This approach is used in, e.g., [2,8].

Diversity techniques designed specifically to defend against ROP attacks are typically evaluated using a *coverage metric* that measures the fraction of ROP gadgets re-located by the transformation, and sometimes also an *entropy metric* that measures the number of possible new positions of the ROP gadgets, reflecting the probability of the attacker correctly guessing the new locations. This approach is used in, e.g., [12,28]).

These approaches based on specific vulnerabilities in low-level languages are unsuitable for evaluating diversity for interpreted languages, such as Java and Python.

Acknowledgements. This material is based on work supported in part by ONR Grant N00014-15-1-2208, NSF Grants CCF-1414078 and CNS-1421893, and DARPA Contract FA8650-15-C-7561. We thank Thang Bui, Rahul Gadi, Shikhar Sharma, Shalaka Sidmul, Shubham Singhal, and Swetha Tatavarthy for their contributions to the implementation and experiments.

References

1. Avizienis, A.: The N-version approach to fault-tolerant software. IEEE Trans. Softw. Eng. **11**(12), 1491–1501 (1985)
2. Bhatkar, S., DuVarney, D.C.: Efficient techniques for comprehensive protection from memory error exploits. In: 14th USENIX Security Symposium. USENIX Association (2005)
3. Cai, J., Facon, P., Henglein, F., Paige, R., Schonberg, E.: Type analysis and data structure selection. In: Möller, B. (ed.) Constructing Programs from Specifications, North-Holland, pp. 126–164 (1991)
4. Chen, S., Sezer, E.C., Gauriar, P., Iyer, R.K.: Non-control-data attacks are realistic threats. In: 14th USENIX Security Symposium. USENIX, August 2005
5. Cohen, F.B.: Operating system protection through program evolution. Comput. Secur. **12**(6), 565–584 (1993)
6. Forrest, S., Somayaji, A., Ackley, D.H.: Building diverse computer systems. In: 6th Workshop on Hot Topics in Operating Systems (HotOS), pp. 67–72 (1997)
7. Gautam, Rajopadhye, S.: Simplifying reductions. In: Conference Record of the 33rd ACM SIGPLAN-SIGACT Symposium on Principles of Programming Languages, pp. 30–41 (2006)
8. Giuffrida, C., Kuijsten, A., Tanenbaum, A.S.: Enhanced operating system security through efficient and fine-grained address space randomization. In: 21st USENIX Security Symposium, pp. 475–490. USENIX Association (2012)
9. Gorbovitski, M., Liu, Y.A., Stoller, S.D., Rothamel, T.: Composing transformations for instrumentation and optimization. In: Proceedings of the ACM SIGPLAN 2012 Workshop on Partial Evaluation and Program Manipulation, pp. 53–62 (2012)
10. Goyal, D.: A language theoretic approach to algorithms. Ph.D. thesis, Department of Computer Science, New York University (2000)

11. Hirschberg, D.S., Sinclair, J.B.: Decentralized extrema-finding in circular configurations of processors. Commun. ACM **23**(11), 627–628 (1980)
12. Homescu, A., Brunthaler, S., Larsen, P., Franz, M.: Librando: transparent code randomization for just-in-time compilers. In: ACM Conference on Computer and Communications Security, pp. 993–1004. ACM (2013)
13. Hosek, P., Cadar, C.: Varan the unbelievable: an efficient n-version execution framework. In: 20th International Conference on Architectural Support for Programming Languages and Operating Systems (ASPLOS 2015), pp. 339–353, March 2015
14. Lamport, L.: Time, clocks, and the ordering of events in a distributed system. Commun. ACM **21**(7), 558–565 (1978)
15. Lamport, L.: Paxos made simple. SIGACT News (Distrib. Comput. Column) **32**(4), 51–58 (2001)
16. Larsen, P., Homescu, A., Brunthaler, S., Franz, M.: SoK: automated software diversity. In: 2014 IEEE Symposium on Security and Privacy, SP 2014, Berkeley, CA, USA, 18–21 May 2014, pp. 276–291 (2014)
17. Liu, Y.A.: Logical clocks are not fair: what is fair? A case study of high-level language and optimization. In: Proceedings of the Workshop on Advanced Tools, Programming Languages, and Platforms for Implementing and Evaluating Algorithms for Distributed Systems, Egham, UK, July 2018
18. Liu, Y.A., Brandvein, J., Stoller, S.D., Lin, B.: Demand-driven incremental object queries. In: Proceedings of the 18th International Symposium on Principles and Practice of Declarative Programming, pp. 228–241. ACM Press (2016)
19. Liu, Y.A., Gorbovitski, M., Stoller, S.D.: A language and framework for invariant-driven transformations. In: Proceedings of the 8th International Conference on Generative Programming and Component Engineering, pp. 55–64. ACM Press (2009)
20. Liu, Y.A., Stoller, S.D.: From recursion to iteration: what are the optimizations? In: 2000 ACM SIGPLAN Symposium on Partial Evaluation and Semantics-Based Program Manipulation (PEPM), Boston, January 2000. Published in ACM SIGPLAN Notices, February 2000
21. Liu, Y.A., Stoller, S.D.: From datalog rules to efficient programs with time and space guarantees. ACM Trans. Program. Lang. Syst. **31**(6), 1–38 (2009)
22. Liu, Y.A., Stoller, S.D., Gorbovitski, M., Rothamel, T., Liu, Y.E.: Incrementalization across object abstraction. In: Proceedings of the 20th ACM Conference on Object-Oriented Programming, Systems, Languages, and Applications, pp. 473–486 (2005)
23. Liu, Y.A., Stoller, S.D., Li, N., Rothamel, T.: Optimizing aggregate array computations in loops. ACM Trans. Program. Lang. Syst. **27**(1), 91–125 (2005)
24. Liu, Y.A., Stoller, S.D., Lin, B.: From clarity to efficiency for distributed algorithms. ACM Trans. Program. Lang. Syst. **39**(3), 12:1–12:41 (2017)
25. Liu, Y.A., Stoller, S.D., Lin, B., Gorbovitski, M.: From clarity to efficiency for distributed algorithms. In: Proceedings of the 27th ACM SIGPLAN Conference on Object-Oriented Programming, Systems, Languages and Applications, pp. 395–410 (2012)
26. Liu, Y.A.: Systematic Program Design: From Clarity To Efficiency. Cambridge University Press, Cambridge (2013)
27. Paige, R., Koenig, S.: Finite differencing of computable expressions. ACM Trans. Program. Lang. Syst. **4**(3), 402–454 (1982)
28. Pappas, V., Polychronakis, M., Keromytis, A.D.: Smashing the gadgets: hindering return-oriented programming using in-place code randomization. In: 33rd IEEE Symposium on Security and Privacy, pp. 601–615. IEEE Computer Society (2012)

29. Ricart, G., Agrawala, A.K.: An optimal algorithm for mutual exclusion in computer networks. Commun. ACM **24**(1), 9–17 (1981)
30. Rogowski, R., Morton, M., Li, F., Monrose, F., Snow, K.Z., Polychronakis, M.: Revisiting browser security in the modern era: new data-only attacks and defenses. In: 2017 IEEE European Symposium on Security and Privacy, EuroS&P 2017, Paris, France, 26–28 April 2017, pp. 366–381. IEEE (2017)
31. Schleimer, S., Wilkerson, D.S., Aiken, A.: Winnowing: local algorithms for document fingerprinting. In: 2003 ACM SIGMOD International Conference on Management of Data, pp. 76–85. ACM (2003)
32. Wei, T., Wang, T., Duan, L., Lu, J.: INSeRT: protect dynamic code generation against spraying. In: International Conference on Information Science and Technology, pp. 323–328. IEEE, March 2011
33. Xu, M., Lu, K., Kim, T., Lee, W.: BUNSHIN: compositing security mechanisms through diversification. In: USENIX Annual Technical Conference, pp. 271–283. USENIX Association (2017)

Malware

Online Malware Detection in Cloud Auto-scaling Systems Using Shallow Convolutional Neural Networks

Mahmoud Abdelsalam[1,2(✉)], Ram Krishnan[1,3], and Ravi Sandhu[1,2]

[1] Institute for Cyber Security and Center for Security and Privacy Enhanced
Cloud Computing, University of Texas at San Antonio,
San Antonio, TX, USA
{mahmoud.abdelsalam,ram.krishnan,ravi.sandhu}@utsa.edu
[2] Department of Computer Science, University of Texas at San Antonio,
San Antonio, TX, USA
[3] Department of Electrical and Computer Engineering,
University of Texas at San Antonio, San Antonio, TX, USA

Abstract. This paper introduces a novel online malware detection approach in cloud by leveraging one of its unique characteristics—auto-scaling. Auto-scaling in cloud allows for maintaining an optimal number of running VMs based on load, by dynamically adding or terminating VMs. Our detection system is online because it detects malicious behavior while the system is running. Malware detection is performed by utilizing process-level performance metrics to model a Convolutional Neural Network (CNN). We initially employ a 2d CNN approach which trains on individual samples of each of the VMs in an auto-scaling scenario. That is, there is no correlation between samples from different VMs during the training phase. We enhance the detection accuracy by considering the correlations between multiple VMs through a sample pairing approach. Experiments are performed by injecting malware inside one of the VMs in an auto-scaling scenario. We show that our standard 2d CNN approach reaches an accuracy of ≃90%. However, our sample pairing approach significantly improves the accuracy to ≃97%.

Keywords: Security · Auto-scaling · Online malware detection ·
Cloud IaaS · Deep learning · Convolutional Neural Networks

1 Introduction

Cloud computing characteristics [15] enable novel attacks and malware [5, 9–12, 22]. In particular, cloud has become a major target for malware developers since a large number of Virtual Machines (VMs) are similarly configured. Automatic provisioning and auto configuration tools have led to the widespread use of auto-scaling, where VMs scale-in/out on demand. Applications utilizing auto-scaling architectures[1] is one of the most prevalent in cloud. As a result, a malware

---

[1] Amazon architecture references. https://aws.amazon.com/architecture/.

© IFIP International Federation for Information Processing 2019
Published by Springer Nature Switzerland AG 2019
S. N. Foley (Ed.): DBSec 2019, LNCS 11559, pp. 381–397, 2019.
https://doi.org/10.1007/978-3-030-22479-0_20

that infects one VM can be easily reused to infect other VMs that are similarly configured or imaged. To that end, cloud has become a very interesting target to most attackers.

In malware analysis, files are scanned before execution on the actual system either through static or dynamic analysis. Once an executable/application is deemed to be benign, it executes on the system without further monitoring. Such methods often fall short in the case of cloud malware injection [11], a threat where an attacker injects a malware to manipulate the victim's VMs, because the initial scan is usually bypassed or malware is injected into an already scanned benign application. Consequentially, the need for online malware detection, where you continuously monitor the whole system for malware, has become a necessity.

Few works [1,6,17,20,21] exist in the domain of *online* malware detection in cloud. Typically, machine learning is used in online malware detection. First, a set of system features are selected and used to build a model. Then, this model is used for malware detection. Some works use system calls while others use performance metrics. Although such works target cloud systems in some sense, there is no real difference between standard online malware detection methods and cloud-specific methods except in the features selected for machine learning, where cloud-specific methods restrict the selection of features to those that can only be fetched through the hypervisor. One can argue that such works focus on malware detection in VMs running on a hypervisor.

However, what makes cloud computing powerful is the novel characteristics that they support [15] such as on-demand self-service, resource pooling and rapid elasticity via auto-scaling. In this paper, we explore malware detection approaches that can leverage specific cloud characteristics. In particular, we focus on auto-scaling. The high-level idea is that in an auto-scaling scenario, where multiple VMs are spawned based on demand, each of those VMs is typically a replica. This means the "behavior" of those VMs need to closely correspond with each other. If a malware were to be injected online into one of those VMs, the infected VM's behavior will likely deviate at some point in time. Our work seeks to detect such deviations when they occur. A sophisticated attacker can attempt to simultaneously inject malware into multiple VMs, which could induce similar behavior across those VMs, and thereby escape our detection mechanisms. This is an interesting challenge and we plan this for future work. This paper focuses on malware detection when exactly one of the VMs in an auto-scaling environment is compromised.

In terms of the approach, first, we introduce and discuss a cloud-specific online malware detection approach. It applies 2d CNN, a deep learning approach, for online malware detection by utilizing system process-level performance metrics. A 2d input matrix/image is represented as the *unique processes* × *selected features*. We assume that similarly configured VMs should have similar behavior, so we train a single model for VMs that belongs to the same group such as the group of application servers in a 3-tier auto-scaling web architecture of web servers, application servers and database servers. Next, we introduce a new approach that leverages auto-scaling. Here, we consider correlations between multiple VMs by pairing samples from pairs of those VMs. Samples collected at the same time from multiple VMs are paired and fed into CNN as a single sample.

CNN is chosen because of its simplicity and training speed as opposed to other deep learning approach (e.g. Recurrent Neural Networks). Also, for the sake of practicality, we show that even a shallow CNN (LeNet-5) trained only for a few epochs can be effective for online malware detection. In summary the contributions of this paper are two-fold:

- We introduce a 2d CNN based online malware detection approach for *multiple* VMs.
- We improve 2d CNN by introducing a new approach by pairing samples from different VMs to accommodate for correlations between those VMs.

To the best of our knowledge, our work is the first to leverage cloud-specific characteristics for online malware detection. The remainder of this paper is organized as follows. Section 2 discusses related work on cloud-specific online malware detection. Section 3 explains the key intuition about the idea presented. Section 4 outlines the methodology including the architecture of the CNN models. Section 5 describes the experiments conducted and results. Finally, Sect. 6 summarizes and concludes this paper.

2   Related Work

Many research works address the problem of online malware detection using different set of features and machine learning algorithms. Some works [6,8,14] focus on using systems calls while others [3,18,19] focus on using API calls. Others [16,23] focus on using memory features or performance counters [7].

Only a few research works address the problem of cloud malware detection since many of the standalone malware detection approaches work for detecting malware in single VMs in the cloud as well. Most, if not all, of the cloud-specific malware detection techniques falls under the online malware detection category (including anomaly detection approaches). Furthermore, they all focus on extracting features from the hypervisor since it adds another security layer.

Dawson et al. [6] focus on rootkits and intercept system calls through the hypervisor to be used as features. Their system call analysis is based on a non linear phase-space algorithm to detect anomalous behavior. Evaluation is based on the dissimilarity among phase-space graphs over time.

Wang [20] introduced Entropy based Anomaly Testing (EbAT), an online analysis system of multiple system-level metrics (e.g. CPU utilization and memory utilization) for anomaly detection. The proposed system used a light-weight analysis approach and showed a good potential in detection accuracy and monitoring scalability. However, the evaluation used did not show pragmatic and realistic cloud scenarios.

Azmandian et al. [4] propose an anomaly detection approach where all features are extracted directly from the hypervisor. Various performance metrics are collected per process (e.g., disk i/o, network i/o) and unsupervised machine learning techniques like K-NN and Local Outlier Factor (LOF) are used.

Classification of VMs is used for anomaly detection. Pannu et al. [17] propose an adaptive anomaly detection system for cloud. It focused mainly on various faults within the cloud infrastructure. Although this work is not directly addressing malware, such technique is valid for malware detection since malware can cause faults in VMs, thus worth mentioning. It used a realistic testbed experimentation comprising 362-node cloud in a university campus. The results showed a good potential with over 87% of anomaly detection sensitivity. One of the drawbacks of this work lies within using two-class SVM. Therefore, it suffers from a data imbalance problem (where there is an imbalance of data from various classes during the training period), which led to several false classification of new anomalies.

The work by Watson et al. [21] is similar to [17] but directly addressed detecting malicious behavior in the cloud. It tried to overcome the drawbacks in [17] by using one class Support Vector Machine (SVM) for detection of malware in cloud infrastructure. The approach gathers features at the system and network levels. The system level features are gathered per process which includes memory usage, memory usage peak, number of threads and number of handles. The network level features are gathered using the CAIDA's CoralReef[2] tool. The study shows high accuracy results; however, it uses known-to-be highly active malware that easily skew the system's resource utilization (e.g., by forking many processes).

In our earlier work [1], we showed that malware can be effectively detected using black-box VM-level performance and resource utilization metrics (such as CPU and memory utilization). Although, the work showed promising results for highly active malware (e.g., ransomware), it is not as effective for low-profile malware that would not impact black-box level resource utilization significantly. Subsequently, we introduced a CNN based online malware detection method for low-profile malware [2]. This work utilized resource utilization metrics for various processes within a VM. The method was able to detect low-profile malware with accuracy of $\simeq 90\%$. Although, this work yielded good results, it targeted a single VM much like other related works. Unlike our prior work and other related works, this paper targets malware detection when multiple VMs are running, while leveraging specific cloud characteristics such as auto-scaling.

3   Key Intuition

In classification-based process-level online malware detection methods, a machine learning model is trained on benign and malicious samples of processes where the goal is to classify a new input sample. The data collection phase, usually, works by running a VM for some time (benign phase) and then injecting a malware (malicious phase) while logging the required data. This is referred to as a single run. The data set includes multiple runs with same/different malware

---

[2] CoralReef Suite: https://www.caida.org/tools/measurement/coralreef/.

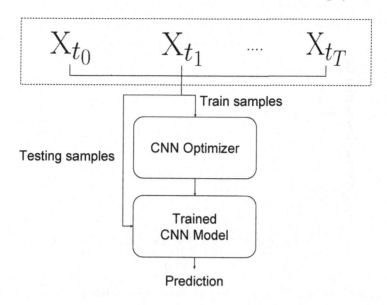

Fig. 1. Single VMs Single Samples (SVSS)

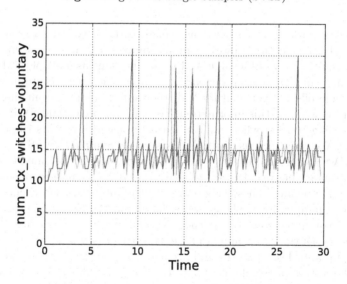

Fig. 2. Number of used voluntary context switches over 30 min for two different runs of the same unique process.

which is later divided into training and test data sets. In other words, given sample X at time t (X_t), the task is to compare X_t to previously seen samples of the training data set. For a single run, we deal with individual samples of a single VM. Thus, we refer to this approach as Single VMs Single Samples (SVSS) which is shown in Fig. 1.

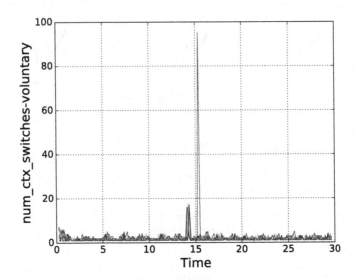

Fig. 3. Number of used voluntary context switches over 30 min for one run of 10 VMs in an auto-scaling scenario.

SVSS can work in an auto-scaling scenario where we have a trained model for each auto-scaling tier; however, input samples will lose some information. Note that multiple runs of a single VM is not the same as multiple VMs running at the same time. The reason depends mostly on the architecture in place. If a VM has some effects over another VM, then input samples from single VM in multiple runs will lose this information. To that end, we extend SVSS and build an auto-scaling testbed where we can learn from multiple VMs running at the same time. We refer to it as Multiple VMs Single Sample (MVSS).

The MVSS approach, however, has a disadvantage in the context of process-level performance metrics. Processes have a very dynamic nature, meaning spikes are always happening. These spikes are mostly due to sudden events or traffic surges. For example, Fig. 2 shows two different runs of the same process for the number of voluntary context switches. No malware is running inside either of the two VMs. During the training phase, two patterns will be learned, a smooth recurring up and down pattern and a pattern where there can be some spikes. During the testing phase, if either pattern is seen, it will be regarded as benign.

On the other hand, Fig. 3 shows one run of the same unique process in 10 VMs (belongs to the same group of VMs in an auto-scaling scenario). VMs are running at the same time in an auto-scaling scenario. The red colored process belongs to a VM where a malware was injected. There are two major spikes in the figure. The first spike happened in the same unique process of all the VMs. If one of the processes did not have that spike and it was classified as benign, it might be a misclassification since such spike should happen to all VMs at the same time. The second spike (caused by the malware injected) is only observed in the infected VM which should be classified as malicious. In simple words, observing a noticeable enough spike by a particular process should be classified as malicious (i.e., second spike). However, sudden behavior changes can happen (i.e., first spike) and flagging an observed spike always as malicious can cause many misclassifications. As such, considering multiple VMs, spikes that are observed in a particular process in all VMs at the same time shouldn't be classified as malicious since it can be caused by any sudden change in behavior (e.g. sudden increase in the number requests to web server). MVSS and SVSS will lose such correlations between VMs since they learn from individual samples regardless the scenario.

Consequentially, we introduce a new approach where the correlation of multiple VMs is utilized by pairing samples (at the same time). In other words, given sample X of VM vm_i at time t ($X_{vm_{i_t}}$), the idea is to compare $X_{vm_{i_t}}$ to previously seen paired samples of multiple VMs. We refer to this approach as Multiple VMs Paired Samples (MVPS).

Table 1. Process-level performance metrics

Metric category	Description
CPU information	CPU usage percent, CPU times in user space & kernel space, CPU times of children processes in user space & system space
Context switches	Number of context switches voluntary & involuntary
IO counters	Number of read requests, write requests, read bytes, written bytes, read chars, written chars
Memory information	Amount of memory swapped out to disk, Proportional set size, Resident set size, Unique set size, Virtual memory size, Number of dirty pages, Amount of physical memory, text resident set, Memory used by shared libraries
Network information	Number of received bytes, Number of sent bytes
Others	Process status, Number of used threads, Number of opened file descriptors

4   Methodology

Detailed explanation of CNN is left out of this paper. However, it will suffice to say that CNN is a deep learning approach used extensively in image recognition. Hence, it takes 2d images as input. In our work, the first dimension represents the processes in the system and the second dimension represents the features collected for each process. Consider a sample X at a particular time t, that records n features (performance metrics) per process for m processes in VM vm, such that:

$$
\mathbf{X}_{vm_t} = \begin{bmatrix}
 & f_1 & f_2 & \cdots & f_n \\
p_1 & \vdots & \vdots & \cdots & \vdots \\
 & \vdots & \vdots & \vdots & \ddots & \vdots \\
p_m & \vdots & \vdots & \cdots & \vdots
\end{bmatrix}
$$

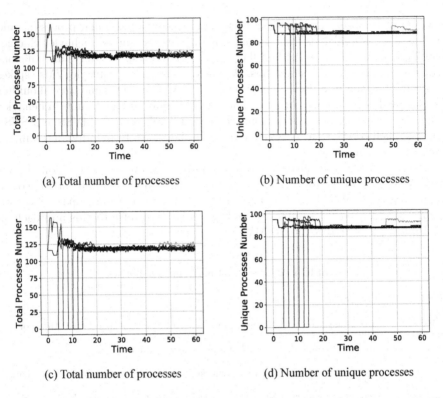

(a) Total number of processes

(b) Number of unique processes

(c) Total number of processes

(d) Number of unique processes

Fig. 4. Total number of standard processes versus the number of unique processes in VMs in an auto-scaling scenario. (Color figure online)

Table 1 shows the process-level performance features which can be fetched through the hypervisor. CNN requires a specific process to remain in the same row (in the input matrix) for all inputs in a single run. This means that process ID (PID) can not simply be used directly. Processes get killed and get created frequently so a PID identifying one process might identify a different process later on. For that reason, we define a *unique process* which is identified by three elements: process name (name), command line used to execute the process (cmd) and hash of binary executable (if applicable). In addition, unique processes help in smoothing the number of processes in a highly active server since most malware creates new non-unique processes. Figure 4 (a)–(b) and (c)–(d) show two different experiments (each with a different malware) where the number of total standard processes are compared to the number of unique processes. Red portions are the start of malware execution. As shown in the figure, the total number of processes in such a highly active VM does not help much in revealing the malware behavior as opposed to the number of unique processes. Throughout this paper the terms process and unique process are used interchangeably where both refer to unique process.

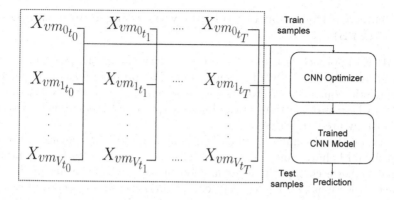

Fig. 5. Multiple VMs Single Sample (MVSS)

4.1 Malware Detection in Multiple VMs Using Single Samples (MVSS)

This is a relatively straight-forward task. We target multiple VMs in an auto-scaling scenario. Figure 5 shows the approach used in MVSS. In MVSS we have samples $X_{vm_{i_{t_k}}}$ from multiple VMs running at the same time, where X is a sample of VM vm_i at time t_k. Samples from many runs are collected and are fed to the CNN optimizer where the learning process takes place. Then the trained CNN model is used for predictions.

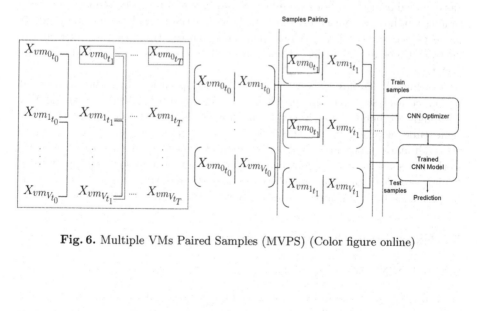

Fig. 6. Multiple VMs Paired Samples (MVPS) (Color figure online)

4.2 Malware Detection in Multiple VMs Using Paired Samples (MVPS)

The MVPS approach is inspired by the duplicate questions problem in online Q&A forums like Stack Overflow and Quora. The problem focuses on determining semantic equivalence between questions pairs. It is a binary classification problem where two questions Q_1 and Q_2 are given and the task is to determine whether they are duplicates.

Based on the aforementioned assumption that VMs that belong to the same group should behave similarly, we use the same analogy to tackle our problem. To that end, we change the formalization of our problem by using the above duplicate questions problem concept except, in our case, we are given two samples $X_{vm_{i_{t_k}}}$ and $X_{vm_{j_{t_k}}}$ from different VMs, where $X_{vm_{i_{t_k}}}$ is a 2d matrix (image in CNN terminology) that belongs to vm_i at time t_k and $X_{vm_{j_{t_k}}}$ is a 2d matrix that belongs to vm_j at the same time t_k. Figure 6 shows the pairing samples approach. Our goal is to find whether $X_{vm_{i_{t_k}}}$ and $X_{vm_{j_{t_k}}}$ are duplicates (similar). This is done by pairing the two samples as an input to CNN. Two samples are considered similar if they are benign, whereas two samples are considered not similar if either one of them is malicious (red bordered samples are malicious).

By pairing samples, we are actually taking into account the correlations between samples of different VMs. This is due to the fact that CNN works by finding spatial correlation within images. MVPS works in an auto-scaling scenario where there are at least two VMs of the same group. Note that it is important that we only pair samples of the same time as pairing samples of different times might have completely different values if the behavior of the VMs has changed over time. For example, a web server handling one request per sec at time t_1 will have a completely different behavior than a web server handling 100 requests per sec at time t_2. Consequentially, pairing two samples taken at two different times might mislead the classifier if the behavior of the VMs changed over time.

Pairing all samples is a very time consuming operation. In addition, that will introduce a class imbalance problem since we are only infecting a single VM. Although, we believe that infecting multiple VMs is hard to occur at the exact same time in practice, not the least because a malware needs time to infect other similarly configured VMs. Like mentioned earlier we set this for future work. Consequentially, as shown in Fig. 6, we pair a malicious sample with all benign samples from other machines at a particular time. On the other hand, we pair each benign sample sequentially with the sample of the following VM.

5   Experiment Setup and Results

5.1   CNN Model Architecture

A deep CNN model would require considerably larger processing power. In reality, this might not be affordable. For the sake of practicality, we chose to work

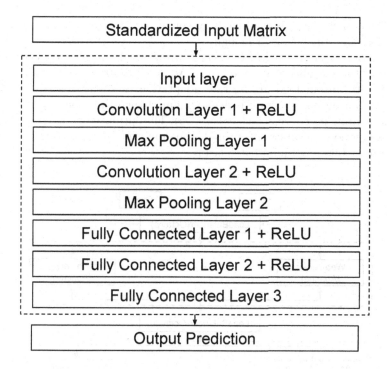

Fig. 7. CNN Model (LeNet-5)

with a shallow CNN. We show that even a shallow CNN can achieve near optimal results in our pairing approach. Figure 7 shows the CNN model used in this work. We chose LeNet-5 [13] CNN model. Although, it is currently by no means one of the state-of-the-art CNN models, its shallowness makes it one of the best candidates in practice. Note that in the context of online malware detection, the model might need to be trained multiple times based on the deployed workloads in place. For example, a 3-tier web architecture and a Hadoop architecture might need different trained models.

The CNN model receives a standardized 2d matrix. Lenet-5 CNN consists of 7 layers (excluding the input layer). The input layer is a 2d matrix of 120×45 (120×90 for MVPS), representing a sample of maximum 120 processes and 45 features. For empty processes (i.e., processes that do not run at the start time but might start in the future), rows are padded with zeros. The first layer is a convolutional layer with 32 kernels of size 5×5 with zero padding ending. This results in a 32 feature maps of size 120×45. The second layer is a max pool layer of size 2×2 which downsizes each dimension by a magnitude of 2, resulting in a 32 feature maps of size 60×23 (60×45 for MVPS). The third layer, another convolutional layer with 64 kernels of size 60×23, is followed by a max pool layer which results in 64 down sized feature maps of size 30×12 (30×23 for MVPS). Fifth and sixth layers are fully connected layers of size 1024 and 512, respectively. Last layer is another fully connected layer of size 2, representing prediction class (malicious or benign).

ReLU activation is used after every convolutional and fully connected layer (excluding the last fully connected layer). Adam Optimizer, a stochastic gradient descent with automatic learning rate adaptation, is used to train the model. Adam optimizer learning rate is a maximum change threshold to control how fast the learning process can be (set to $1e-5$). The optimizer works by minimizing the loss function (mean cross entropy). Random grid search is used to tune the CNN parameters (e.g., mini batch size).

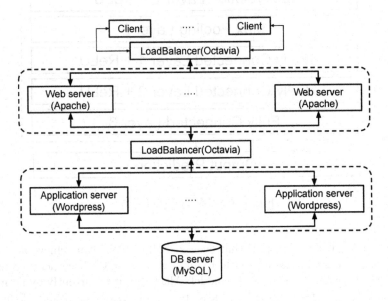

Fig. 8. 3-tier Web Application

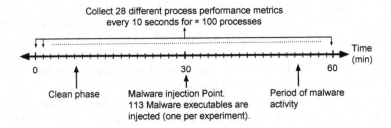

Fig. 9. Data collection overview

5.2   Experimental Setup

Our experiments were conducted on an Openstack testbed. Figure 8 shows the 3-tier web architecture built on top of our testbed, with auto-scaling enabled on the web and application server layers. The scalability policy is based on the average CPU utilization of the total VMs of each tier (scalability group). It scales-out if the average CPU utilization is above 70% and scales-in if the average CPU utilization is less than 30%. The number of servers spawned in each tier were between 2 and 10 based on the traffic load. Traffic was generated based on ON/OFF Pareto distribution with parameters set according to the NS2[3] tool defaults.

The data collection process is shown in Fig. 9. Each of our experiments was 1 h long. The first 30 min is the clean phase. The second 30 min is malicious phase where a malware is injected. A set of 113 malware were used for each of the different experiments. Malware binaries were randomly obtained from VirusTotal[4]. All firewalls were disabled and an internet connection was provided to avoid any hindrance to the malware's malicious intentions. Samples were collected at 10 s intervals, so during a single experiment 360 samples were collected for one VM.

5.3   Evaluation

We use four evaluation metrics.

$$Precision = \frac{TP}{TP + FP}$$

$$Recall = \frac{TP}{TP + FN}$$

$$Accuracy = \frac{TP + TN}{TP + TN + FP + FN}$$

$$Fscore = 2 \times \frac{Precision \times Recall}{Precision + Recall}$$

Precision is the number of correct malware predictions. Recall is the number of correct malware predictions over the number of true malicious samples. Accuracy is the measure of correct classification. F score is the harmonic mean of precision and recall. True Positive (TP) refers to malicious activity that occurred and was correctly predicted. False Positive (FP) refers to malicious activity that did not occur but was wrongly predicted. True Negative (TN) refers to malicious activity that did not occur and was correctly predicted. False Negative (FN) refers to malicious activity that occurred but was wrongly predicted.

---

[3] NS2 manual. http://www.isi.edu/nsnam/ns/doc/node509.html.
[4] VirusTotal website. https://www.virustotal.com.

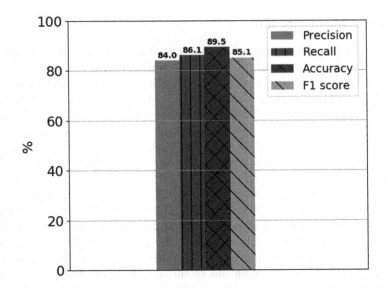

Fig. 10. Optimized MVSS CNN classifier results

5.4   MVSS and MVPS Results

Like most standard machine learning classification problems, data was split into three sets: training (60%), validation (20%) and testing (20%) sets. We split on the 113 experiments to 67, 23 and 23 respectively. This ensures that validation and testing phases are exposed to unseen malware. After training the model on the training set, validation set is used to tune the model parameters as well as choosing the highest accuracy model. The model is evaluated on the validation set after each epoch and the highest accuracy model is chosen. Then the testing set is used to test the chosen model (optimized classifier).

Figure 10 shows the results of MVSS optimized classifier. The optimized classifier yields accuracy of ≃90% while precision, recall and fscore are ≃85% on the test data set. This approach achieved good results compared to the similar simple 2d CNN approach in [2]. There are two reasons for this improvement. First, increasing the number of data (113 malware experiments as opposed to 25). Second, using data from multiple VMs as opposed to a single VM; however, we still had to filter part of the data to balance our data sets (i.e., balance the ratio of benign to malicious samples).

Figure 11 shows the results of MVPS optimized classifier. There is a significant increase in the four evaluation metrics when compared to the MVSS classifier. The optimized chosen MVPS classifier had a highest accuracy of ≃98.2% during the validation phase. It yielded a ≃96.9% accuracy on the test data set. Fscore, recall and precision all jumped to ≃91% on the test data set. The main reason for this high improvement is that the MVPS approach finds correlations between the multiple VMs running at the same time which is very beneficial in an auto-scaling scenario.

In both cases, mini-batch size of size 64 and learning rate of $1e - 5$ yielded the best results. The CNN model was trained only for 20 epochs. Note that we do not use a dropout layer (to avoid over-fitting) since it is not useful when using a shallow CNN trained for only a few epochs.

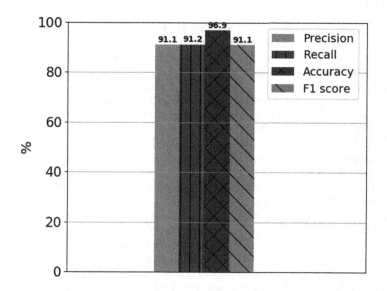

Fig. 11. Optimized MVPS CNN classifier results

6   Conclusion and Future Work

In this paper, we introduced an online malware detection approach to leverage the behavior correlation between multiple VMs in an auto-scaling scenario. The approaches introduced used 2d CNN for malware detection. First, we introduced the MVSS method which targets multiple VMs using single individual samples. MVSS achieved good results with an accuracy of $\simeq 90\%$. Then, we introduced MVPS which targets multiple VMs using paired samples. MVPS takes the previous approach a step forward by pairing samples from multiple VMs which helps in finding correlations between the VMs. MVPS showed a considerable improvement over MVSS with an accuracy of $\simeq 96.9\%$. In the future, we plan to use different use case scenarios such as Hadoop and Containers as well as perform an analysis using different CNN models architecture. We also plan to perform an analysis to evaluate the effectiveness of ordering the processes and features in the input matrix. Finally, we plan to develop techniques to handle the situation when multiple VMs are infected simultaneously by an attacker. One direction, instead of using pairs of samples, is to use tuples of samples (3-tuple, 4-tuple or more).

Acknowledgment. This work is partially supported by NSF CREST Grant HRD-1736209, DoD ARL Grant W911NF-15-1-0518, and NSF CAREER Grant CNS-1553696.

References

1. Abdelsalam, M., Krishnan, R., Sandhu, R.: Clustering-based IaaS cloud monitoring. In: 10th IEEE CLOUD. IEEE (2017)
2. Abdelsalam, M., Krishnan, R., Sandhu, R.: Malware detection in cloud infrastructures using convolutional neural networks. In: 11th IEEE CLOUD. IEEE (2018)
3. Alazab, M., Venkatraman, S., Watters, P., Alazab, M.: Zero-day malware detection based on supervised learning algorithms of API call signatures. In: Proceedings of the Ninth Australasian Data Mining Conference, vol. 121, pp. 171–182. Australian Computer Society, Inc. (2011)
4. Azmandian, F., Moffie, M., Alshawabkeh, M., Dy, J., Aslam, J., Kaeli, D.: Virtual machine monitor-based lightweight intrusion detection. ACM SIGOPS Oper. Syst. Rev. **45**, 38–53 (2011)
5. Dahbur, K., Mohammad, B., Tarakji, A.B.: A survey of risks, threats and vulnerabilities in cloud computing. In: ISWSA (2011)
6. Dawson, J.A., McDonald, J.T., Hively, L., Andel, T.R., Yampolskiy, M., Hubbard, C.: Phase space detection of virtual machine cyber events through hypervisor-level system call analysis. In: 2018 1st International Conference on Data Intelligence and Security (ICDIS), pp. 159–167. IEEE (2018)
7. Demme, J., et al.: On the feasibility of online malware detection with performance counters. In: ACM SIGARCH Computer Architecture News, vol. 41. ACM (2013)
8. Dini, G., Martinelli, F., Saracino, A., Sgandurra, D.: MADAM: a multi-level anomaly detector for Android malware. In: Kotenko, I., Skormin, V. (eds.) MMM-ACNS 2012. LNCS, vol. 7531, pp. 240–253. Springer, Heidelberg (2012). https://doi.org/10.1007/978-3-642-33704-8_21
9. Gholami, A., Laure, E.: Security and privacy of sensitive data in cloud computing: a survey of recent developments. arXiv preprint arXiv:1601.01498 (2016)
10. Grobauer, B., Walloschek, T., Stocker, E.: Understanding cloud computing vulnerabilities. IEEE Secur. Privacy **9**, 50–57 (2011)
11. Gruschka, N., Jensen, M.: Attack surfaces: a taxonomy for attacks on cloud services. In: IEEE CLOUD, pp. 276–279 (2010)
12. Jensen, M., Schwenk, J., Gruschka, N., Iacono, L.L.: On technical security issues in cloud computing. In: IEEE CLOUD (2009)
13. LeCun, Y., Bottou, L., Bengio, Y., Haffner, P.: Gradient-based learning applied to document recognition. Proc. IEEE **86**(11), 2278–2324 (1998)
14. Luckett, P., McDonald, J.T., Dawson, J.: Neural network analysis of system call timing for rootkit detection. In: 2016 Cybersecurity Symposium (CYBERSEC), pp. 1–6. IEEE (2016)
15. Mell, P., Grance, T., et al.: The NIST definition of cloud computing (2011)
16. Ozsoy, M., Donovick, C., Gorelik, I., Abu-Ghazaleh, N., Ponomarev, D.: Malware-aware processors: a framework for efficient online malware detection. In: 2015 IEEE 21st International Symposium on High Performance Computer Architecture (HPCA), pp. 651–661. IEEE (2015)
17. Pannu, H.S., Liu, J., Fu, S.: Aad: adaptive anomaly detection system for cloud computing infrastructures. In: 2012 IEEE 31st Symposium on Reliable Distributed Systems (SRDS), pp. 396–397. IEEE (2012)

18. Pirscoveanu, R.S., Hansen, S.S., Larsen, T.M., Stevanovic, M., Pedersen, J.M., Czech, A.: Analysis of malware behavior: type classification using machine learning. In: 2015 International Conference on Cyber Situational Awareness, Data Analytics and Assessment (CyberSA), pp. 1–7. IEEE (2015)
19. Tobiyama, S., Yamaguchi, Y., Shimada, H., Ikuse, T., Yagi, T.: Malware detection with deep neural network using process behavior. In: COMPSAC, vol. 2. IEEE (2016)
20. Wang, C.: EbAT: online methods for detecting utility cloud anomalies. In: Proceedings of the 6th Middleware Doctoral Symposium. ACM (2009)
21. Watson, M.R., et al.: Malware detection in cloud computing infrastructures. IEEE TDSC **13**, 192–205 (2016)
22. Xiao, Z., Xiao, Y.: Security and privacy in cloud computing. IEEE Commun. Surv. Tutorials **15**, 843–859 (2013)
23. Xu, Z., Ray, S., Subramanyan, P., Malik, S.: Malware detection using machine learning based analysis of virtual memory access patterns. In: 2017 Design, Automation & Test in Europe Conference & Exhibition (DATE). IEEE (2017)

Redirecting Malware's Target Selection with Decoy Processes

Sara Sutton[✉], Garret Michilli[✉], and Julian Rrushi[✉]

Department of Computer Science and Engineering, Oakland University,
Rochester, MI 48309, USA
{smsutton2,gdmichilli,rrushi}@oakland.edu

Abstract. Honeypots attained the highest accuracy in detecting malware among all proposed anti-malware approaches. Their strength lies in the fact that they have no activity of their own, therefore any system or network activity on a honeypot is unequivocally detected as malicious. We found that the very strength of honeypots can be turned into their main weakness, namely the absence of activity can be leveraged to easily detect a honeypot. To that end, we describe a practical approach that uses live performance counters to detect a honeypot, as well as decoy I/O on machines in production. To counter this weakness, we designed and implemented the existence of decoy processes through operating system (OS) techniques that make safe interventions in the OS kernel. We also explored deep learning to characterize and build the performance fingerprint of a real process, which is then used to support its decoy counterpart against active probes by malware. We validated the effectiveness of decoy processes as integrated with a decoy Object Linking and Embedding for Process Control (OPC) server, and thus discuss our findings in the paper.

Keywords: Malware interception · Decoy processes ·
Operating system kernel · Deep learning

1 Introduction

Malware keep wreaking havoc in both general-purpose computing and industrial control systems, despite various types of defense tools deployed against them. Amongst those tools, honeypots showed exceptional promise. Malware detection on honeypots is straightforward and unequivocal, since they have no activity of their own. Any system or network operation is indicative of intrusion. High interaction honeypots, in particular, provide the utmost protection. They run operating system (OS) services that are identical to those on machines in production. They also intentionally allow malware to run on the decoy machine. These factors contribute to a deep insight into malware's exploits and rootkit operations, which the defender can turn into signature or rule-based detectors to protect machines in production from the same or somewhat similar malware.

© IFIP International Federation for Information Processing 2019
Published by Springer Nature Switzerland AG 2019
S. N. Foley (Ed.): DBSec 2019, LNCS 11559, pp. 398–417, 2019.
https://doi.org/10.1007/978-3-030-22479-0_21

Nevertheless, advanced malware select their targets wisely. They probe their targets for inconsistencies that reveal decoys. In order to design better decoys, we experimented with a practical approach that uses live performance counters to detect a honeypot, as well as decoy I/O on machines in production. Decoy I/O consists of phantom I/O devices and supporting mechanisms that are deployed on machines in production [16]. Performance counters are data that characterize the performance of a process, kernel driver, or the entire OS. Their intended use is to help determine performance bottlenecks and fine-tune machine performance. Performance counters are provided by the OS and hardware devices [3].

Contribution. This work defensively affects malware's target selection by means of decoy processes. It causes changes in malware's findings in order to enable a decoy to qualify as a valid target of attack. The existence of a decoy process is projected onto a machine via instrumentation of data structures related to performance counters in the OS kernel. The performance consistency of a decoy process is attained via deep learning. We design and train a convolutional neural network that can learn the performance profile of a real process, which we use to support its decoy counterpart against active probes by malware. The OS of reference in this work is Microsoft Windows.

Novelty. To the best of our knowledge, this work is the first to leverage OS-level performance data to project a decoy process and protect it from adversarial probes. We explored data structure instrumentation in our previous work to emulate the existence of a decoy process [17]. Nevertheless, those data structures were strictly related to processes and threads and hence only created partial decoy process existence without run-time performance dynamics.

Saldanha and Mohanta from Juniper Networks proposed a deception methodology based on decoy processes called HoneyProcs, with a patent pending [1]. HoneyProcs aims at detecting malware that inject code into other processes. HoneyProcs works by creating a real process, which tries to mimic a legitimate process. Once the decoy process reaches a steady state, it stops making progress with its execution, which leaves its state immutable. HoneyProcs uses such fixed state as a baseline against any changes, including those caused by code injection. HoneyProcs is vulnerable to the very same decoy detection technique that we used in our honeypot experiment, which is discussed in detail later on in this paper.

Real-time performance counters show that the resource utilization of a decoy process freezes to constant or 0 values. For example, a simple analysis of the working set of a decoy process reveals that its size, namely the number of its memory pages that are currently present in physical main memory, remains constant or decreases due to the global memory frame replacement algorithm. This is abnormal, given that the working set is a moving window representing memory localities. Similarly, the page fault rate of a decoy process swiftly drops to 0, while a continuous 100% page hit rate simply is not possible due to demand paging in virtual memory.

Organization. The remaining of this paper is organized as follows. Section 2 describes an experiment that reveals a detectability weakness of honeypots stemming from their complete lack of activity. Section 3 visits the OS mechanisms behind the display of decoy processes on a machine. Section 4 describes the deep learning approach that protects a decoy process from malware probes. Section 5 reports on implementation, testing, and validation of this work. In Sect. 6 we discuss research related to various aspects of this work. Section 7 summarizes our findings and concludes the paper. The appendices desribe the threat model, define what is out of scope, and discuss additional related works.

2   Honeypot Experiment

Stress Testing Decoy Covertness. The purpose of this experiment was to assess the ability of honeypots and decoy I/O to protect their decoy function in practice. The experiment was performed from a red team's perspective. It was done separately on a Windows honeypot, and then on a Windows machine in production equipped with decoy I/O. The testbed was comprised of two desktops and a laptop machine, all three of which were connected to a local area network that was logically and physically isolated from any other networks.

0-Value Exploit. We simply ran metasploit [2] to launch a publicly known exploit against the honeypot, leveraging a publicly known vulnerability. The exploit returned a command prompt, i.e. a shell, which was usable to fetch and run additional code. The test was detected as soon as the first packet reached the honeypot machine. Nevertheless, none of the exploit, nor the vulnerability, was of any value to the defender by virtue of all of this material being public and hence already well known. What was left for the defender was to wait for the testers' next steps, namely operations like those referenced in the threat model.

Engaging Performance Probes. At this point, we actively collected performance data regarding host processor utilization, memory use, and secondary storage activity. We wrote a PowerShell script with the purpose of gathering those performance data live and in real-time. A large number of samples are collected every second until a data repository is filled. The script enabled us to view a table of the names and process identifiers of all processes currently running on the system, as well as view and store all the details and attributes of a specific process of our choice.

Searching for Patterns of Absent/Low Resource Utilization. We found that per-process performance analysis is much more accurate in spotting inactivity than machine-wide performance analysis. The honeypot was characterized by host processor time that was somewhat comparable to a machine in production in low use. Processor time refers to the percentage of elapsed time that the processor spends executing an active thread. A similar observation holds for the amounts of time the processor spent executing user space code and kernel space

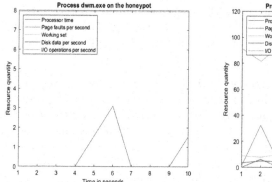
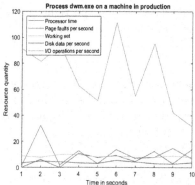

Fig. 1. Visual comparison between a process' performance on the honeypot and its performance on a machine in production.

code. In this experiment, the code that generated all this machine-wide activity consisted of our own script in user space, and honeypot monitoring tools in kernel space.

The execution of all this code overall also generated interrupt arrival rates and page fault rates that were hardly distinguishable from their counterparts on a machine in production in low use. A complicating factor is that, at times, multiple independent threat actors may land on a honeypot. Often cases malware even compete with each-other. The lack of attribution in machine-wide performance parameters hinders honeypot inactivity detection. These findings informed our decision to direct deep learning towards the performance profile of specific processes rather than the machine as a whole.

When directing our script towards specific processes, in most cases we obtained performance counters that indicated a total lack of any resource utilization. In a few cases, performance counters revealed existent but low resource utilization, which we deemed to be related to our own moves on the honeypot. Processes on a honeypot simply do not make progress with their execution, consequently their processor time is 0. New pages in memory are not referenced, consequently no page faults occur. Human-machine interaction is absent and thus interrupts do not occur. Secondary storage is not accessed, consequently the data rate and the number of I/O operations per second are both null.

Our findings are illustrated in Fig. 1. The data plot on the left shows some of the performance parameters of the Desktop Windows Manager (DWM) on the honeypot, where patterns of absent or low resource utilization are clearly evident. The visualized performance parameters, except the working set, are constantly 0. There are a few processor time spikes, however those are very minimal. It is interesting to see how the working set, which is represented by the flat horizontal line at the very top, never changed from 87420928. Of course, with no page faults occurring, the working set could not change. In the data plot, we have applied a log_{10} reduction of the working set to make it fit within the same plot as the other performance parameters.

The data plot on the right shows the same performance parameters of the DWM process, but this time on a machine in production. The working set and the disk data per second have both been reduced log_{10}. They are high and variable.

Experiment Repeated on Decoy I/O. This time, the probes were directed against a decoy process amongst real processes on a machine in production. Decoy I/O consisted of a decoy network interface controller, which projected a decoy network providing connectivity to a decoy Object Linking and Embedding for Process Control (OPC) server, as in [17]. The decoy process was an OPC client, which, just like HoneyProcs, maintained a consistent appearance. More specifically, it appeared to load the same libraries, had the same size on disk, and created the same number of threads, as its real counterpart. Nevertheless, when probed over performance counters, the decoy OPC process was immediately detected as in the honeypot experiment.

Multiple Performance Samples are Needed for Accuracy. It is normal for a machine in production, and hence a valid malware target, to have periods of inactivity or low use, which may be quite common. In this experiment, performance probes were collected over an extensive time window to make sure that production activity was observed if existent.

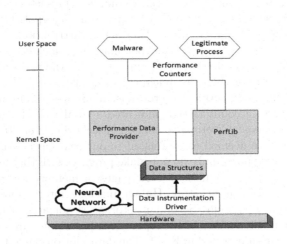

Fig. 2. Decoy process visibility via performance data instrumentation.

3   Decoy Processes

Make Visible, but Do Not Create. In this work, the idea is to only project or display the existence of a process for malware to see rather than concretely create that existence. In other words, we aim at making a process that is as visible as its real counterpart, without having to spawn it. The rationale is simple. Spawning a real process, albeit for use as a decoy, consumes real resources on the machine, which adds to the overhead of running detection tools specific to decoy processes. The data, code, and libraries of a decoy process would need to be stored in real frames in main memory. As we discussed earlier in this paper, freezing execution to create an immutable state is ineffective, therefore there would have to be real activity, which would consume CPU cycles and secondary storage.

Own real activity comes with its own complexities, since it needs to be distinguished from malicious activity, thus bringing the malware detection challenge almost in a form similar to conventional intrusion detection. The browsing presence of a decoy process is achieved by inserting a synthetic entry in the process table in the OS kernel. Furthermore, most of the known techniques to hide a malicious process are usable to create exactly the opposite effect, namely show the existence of a process that does not exist. The task manager tool, the tasklist command, and the ps command, all display an entry for the nonexistent process in their output.

The visibility of a decoy process is attained as illustrated in Fig. 2. Performance counters originate in drivers in the OS kernel. These drivers operate as performance counter library (PERFLIB) providers, which furnish performance data in response to queries. Performance data are accumulated in data structures, such as linked lists. We have written data structure instrumentation code, which deposits synthetic performance data for a decoy process in the repository of performance counters. These performance data, real and synthetic, are then provided to a consumer in user space, including possible malware. This way we project the existence of a decoy process by means of synthetic resource utilization dynamics. Clearly the synthetic performance data need to be consistent, which we address via a neural network and discuss in detail later on in this paper.

Timing the Replies to Performance Queries. Performance data are counted in the OS kernel during specific time windows as related events occur. For example, a counter of page faults is incremented each time a trap to the OS kernel is made as a result of a reference to a page that is not present in physical main memory. The counter's value is not stored immediately in the repository of performance counters. Instead, it is buffered until the counting period is complete. Consumers of performance data in user space will not receive fresh counter data until after the counting period. It is of paramount importance that the data instrumentation driver depicted in Fig. 2 does not deposit the synthetic performance data too fast or too slow in the repository of performance counters.

In this work, the synthetic performance data are decided and produced by a neural network. This process takes relatively little time, since the neural network is already fully trained at the time it is utilized as a source of such data. The neural network needs the performance counters that pertain to all real processes on the machine in order to function. The data instrumentation driver collects these performance counters by accessing directly the repository where they are stored. Once the neural network delivers the performance counters for a decoy process to the data instrumentation driver, the latter buffers them until the end of the counting period, at which point it stores them in the repository.

Safety. Making a nonexistent and hence a decoy process visible via synthetic performance data is safe on honeypots. There are no humans who interact with honeypots while the latter are in operation, consequently the risk of a user interacting with a decoy process is null. The risk on a machine in production equipped with decoy I/O is considerable. We rely on a safety measure from related previous work [17], which is a filter driver integrated into the driver stack of the monitor device. The driver filters out decoy entries from frames of bytes bound for the monitor, before those data have traveled far enough to be displayed on the monitor. Since we know the name and the performance data of the decoy process a priori, we can have them filtered out from the user's visual.

4   Performance Support for a Decoy Process

We now discuss how our approach learns the performance fingerprint of a real process, to be able to perform performance recognition tasks in support of the process' decoy counterpart. We express details of our approach through the lenses of deep learning. The reader is referred to [7] for a detailed discussion of deep learning. In this paper, we base our reasoning on an OPC client process on a machine in production. The rationale for selecting an OPC client as a subject of deep learning and decoy process is connected to its integration with a decoy I/O capability, which we developed in our previous work [17]. Nevertheless, we deem that this work is applicable to all processes.

The rationale for solving the decoy process performance challenge on a machine in production is that the latter presents an environment that is much more complex than the environment of a honeypot. On a honeypot, most or all processes can be configured to be decoy processes, whose performance parameters we can choose ourselves. This makes it easier to calculate their projected resource utilization. On a machine in production, the performance of real processes is beyond our control, therefore our approach needs to be robust enough to work with any possible values they may have. And all this while malware are probing for performance inconsistencies.

4.1   Heatmaps

Heatmap Design. We model the machine's resource utilization as a heatmap, where performance parameters are represented as a color with a given strength. An example heatmap used in this work is depicted in Fig. 3. The higher a performance parameter, the stronger its color in the heatmap. An excerpt from the set of performance parameters that we used in this work is given in Table 1. These parameters are taken from the whole resource utilization spectrum, in the hope that they can enable our approach to learn the performance fingerprint of a process. With performance parameters and real processes aligned along the ordinate and abscissa, respectively, each heatmap cell visually indicates the value of a performance parameter for a specific real process.

Table 1. Some of the performance counters visually assembled in heatmaps.

CPU	Memory	Secondary storage
User Time	Page Faults/sec	IO Read Operations/sec
Privileged Time	Working Set	IO Write Operations/sec
-	Working Set Peak	IO Other Operations/sec
-	Pool Paged Bytes	IO Read Bytes/sec
-	Pool Nonpaged Bytes	IO Write Bytes/sec

The idea is to train the neural network by feeding it a large number of images generated by heatmaps. Each image of heatmap is labeled in such a way that its class label, i.e. an object type associated with the heatmap, is an array of color strengths, namely one color strength for each performance parameter of the decoy process. If training succeeds, the neural network can be used for heatmap recognition. The neural network reads a heatmap, which most likely was not seen during the training phase, and produces in output a class label. The class label, in turn, informs our approach as to what specific values to give the performance counters of the decoy process.

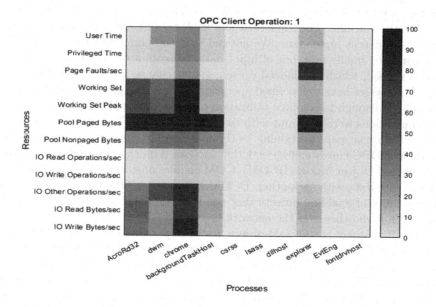

Fig. 3. A performance heatmap for neural network consumption.

Adapting to the Performance of Real Processes. The decoy process, in our case the decoy counterpart of an OPCExplorer process, exhibits performance parameters that depend directly on the resource utilization of real processes on the machine. When probed by malware, we take a screenshot of the performance counters of real processes, metaphorically speaking, and turn them into a heatmap for recognition by the neural network. All processes are taken into account when estimating the performance parameters of the decoy process. We only show a few in Fig. 3 to make the heatmap fit within the page borders. In reality, heatmaps are much larger.

The resource utilizations of any processes created by malware are also included in the heatmaps. Those processes are referred to as foreign_process$_n$ in the heatmaps, regardless of how they are named by the threat actors. Standard internal names for such processes prevent the neural network from getting confused. Multiple processes may be created off the same executable file. For example, the user may be running several chrome tabs, each of which runs as a separate process. Of course, we include the performance of all such processes in the heatmaps. This is what makes our approach cognizant of the current resource utilization load on the machine.

Performance Correlation with Input Data. The heatmap of Fig. 3 contains an explicit process activity indication at the very top. This is for the neural network to include in its internal heatmap processing. The activity indicator ties the performance fingerprint of the process to be mimicked with the input data of that process. When fed with different input data, a real OPCExplorer process may have totally different resource utilizations for the same resource utilization load on the machine. The same heatmap leads to different performance parameters for different input data. We noticed that the performance of a process is insensitive to small variations of input data. Instead, a better input categorization is needed, which can indeed cause a visible change in resource utilization.

All processes have well defined operations in their design, which we find to be meaningful enough to resource utilization to cause changes. In this work, we use such operations to relate input with resource utilization. The operations that we used within heatmaps pertaining to the OPCExplorer process are summarized in Table 2. OPC consists of objects that are based on the Microsoft Distributed Component Object Model (DCOM). COM enables objects on the same machine to exchange data with each-other. DCOM is basically COM, but with the added functionality of enabling objects that reside on different machines to exchange data with each-other over the network.

An OPC object is a DCOM object. As all objects, an OPC object has methods and attributes. The attributes are also known as tags, or data points, which represent parameters of a physical system. Examples include voltage, phase, and current. An OPC server hosts OPC objects, which an OPC client can access in reading or writing over the network. The reader is referred to [10] for a detailed specification of OPC. Some of the operations in Table 2 refer to groups. These are sets of tags, possibly from different OPC objects, which the system operator has reasons to gather together when performing a given OPC task.

We assign numerical values to OPC operations, which we refer to as opcodes. These are identifiers that we use to differentiate OPC operations from each-other. Opcodes are then included in heatmaps for the neural network to process along with the other data. For example, the heatmap of Fig. 3 shows an opcode of 1, which corresponds to viewing OPC server properties on the OPC client.

During testing experience we noticed the I/O performance parameters including I/O write and read operations from secondary storage are always 0 value as those processes didn't consume any I/O resources during our testing experiment. These counters are included in the label construction, however they are always of value zero and therefore do not effect classification. The collected label measurements are used for training our neural network.

Table 2. Categories of operations on an OPC client as used in heatmaps.

OPC Server Ops	Group-level Ops	Tag Related Ops
View OPC server properties	View group properties	List tags of an OPC object
Add an alarm	Change group properties	Add a tag to an OPC object
-	Create a new group	Include a tag in a group
-	Delete an existing group	Remove a tag from a group
-	-	Read a tag
-	-	Write a tag

4.2   Deep Learning of Performance Fingerprints

Training Set and Labeling. As we run the OPCExplorer process to perform the operations of Table 2 one at a time, we collect the performance counters of all other processes on the machine. Those performance data enable us to build heatmaps. We also collect the performance counters of the OPCExplorer process, which collectively enable us to unequivocally establish a class label for each heatmap. All these labeled heatmaps are used to train the neural network.

Test Set. We repeat the previous steps, but this time do not include the labeled heatmaps in the actual training of the neural network. We set aside these labeled heatmaps for later use, once the neural network is fully trained.

Algorithmic Approach. A convolutional neural network has multiple layers of neurons which include at least one input layer and one output layer and some number of hidden layers including Rectified liner unit, pooling, fully connected and softmax. The hidden layers are used to adjust and scale the activation of given features from the heatmap images. Thus, the number of layers are critical.

The inner workings of the convolutional neural network are given in Algorithm 1. One of the most critical steps is the configuration of the layers of this neural network. We add a standard input layer to load and initialize the heatmaps from the training set for further processing. Several rectified linear unit (ReLU) layers are also added to the neural network. ReLU layers increase the pace and effectiveness of the performance fingerprint learning. They zero out negative values and maintain positive values in convolved heatmaps undergoing processing. We also add several pooling layers, which reduce the number of heatmap image parameters that the neural network needs to work with.

The neural network includes several batch normalization layers, which adjust and scale the activations of given features from the heatmap images. The fully connected layer produces a vector with size equal to the number of class labels. Each element of this vector is the probability for a class label of the heatmap image that was just processed by the neural network. Some of these probabilities may be negative. Furthermore, the sum of all these probabilities may not be 1.0. The softmax layer corrects such anomalies, and thus normalizes the vector in question into a probability distribution. The classification layer assigns the correct class label to a heatmap image that was just processed, on the basis of that probability distribution.

Once the training is complete, we run the neural network to classify heatmaps from the test set. We compare the known class labels for those heatmaps with the class labels produced by the neural network, in order to calculate the heatmap recognition accuracy. If the attained level of accuracy is low, we add more layers to the neural network and retrain it from scratch. We keep revising the neural network design until we attain a satisfactory accuracy.

Algorithm 1. Algorithm to train and test a convolutional neural network for heatmap classification.

---

1 Function Learn-Performance-Fingerprint (G, V);
    Input   : Training set of heatmaps G, testing set of heatmaps V.
    Output: Convolutional neural network Π, heatmap recognition accuracy δ.
2 $\delta \leftarrow 0$
3 **for** $\forall$ *heatmap* $\nu \in G$ **do**
4   Read ν into array α in memory;
5   Add Label(ν) to α;
6 **end**
7 **while** $\delta < 90$ **do**
8   Empty Π if any layers present;
9   Define the input layer of Π;
10   Add $count_1$ ReLU layers to Π;
11   Add $count_2$ pooling layers to Π;
12   Add $count_3$ batch normalization layers to Π;
13   Add a fully connected layer to Π;
14   Add a softmax layer to Π;
15   Add a classification layer to Π;
16   Select Π's training options;
17   trainNetwork(Π);
18   **for** $\forall$ *heatmap* $\epsilon \in V$ **do**
19         $\delta \leftarrow \Pi(\epsilon)$
20   **end**
21   Increase $count_1$, $count_2$, and $count_3$;
22 **end**

Usable Oracle. At this point, a fully trained neural network with high accuracy can be queried by the data structure instrumentation code. A query contains a heatmap that is representative of the resource utilization of all processes on the machine, of course excluding the decoy OPCExplorer process. The response by the neural network contains a class label, which the data structure instrumentation code can easily convert into performance data for the decoy OPCExplorer process. Those data are reported to malware in the form of performance counters in response to their probes.

5   Experimental Testing and Validation

Implementation. We wrote Matlab code to implement the deep learning approach. We also wrote other Matlab code to generate heatmaps. We extended the PowerShell script that we used in the honeypot experiment to collect live performance data from all processes running on the machine. These data are stored in files, which are then read by Matlab code to produce heatmaps. The sample interval and number of samples collected are specified by the operator. Increasing the number of samples collected per interval creates a heatmap with greater density of data points. Labeling the heatmaps was a tedious task, which we completed manually one heatmap at a time. To that end, we exercised the OPC client operations referenced in Table 2 manually by interacting with the OPC client software similarly to a system operator. As we ran those operations, we measured the performance counters of the real OPCExplorer process, which we then used for labeling heatmaps.

The need for manual and hence time consuming work limited the number of heatmaps that we could label and use to train the neural network, which in turn affects negatively the accuracy of the neural network itself. As an aside note, in terms of future work, an artificial intelligence approach that uses a virtual keyboard and mouse to drive the functionality of the OPC client software would be most useful to improve the feasibility of this work.

Testing Against Live Malware. A large set of OPC malware samples involved in the Dragonfly malware campaign have been publicly available for academic research for quite some time. Those malware samples come in many versions. Nevertheless, none of these samples appeared to analyze system or network activity on the compromised machine prior to attacking an OPC server. They perform a network search for servers, identify those specific servers that host OPC objects, and then simply pursue the tags in those objects over the network. BlackEnergy style of malware attacks also seem to ignore system or network activity prior to initiating keystroke interception, or prior to making VPN connections over the network.

Extended/Revised Honeypot Experiment. Since the use of performance counters to detect decoys is new, and thus there are no malware that use it, we do not seem to have the means of testing this work against real-world malicious code, as we have in our previous works. We had to return to the honeypot experiment, which, at the beginning, had succeeded to detect the honeypot and decoy I/O. We repeated the various experimental trials on a machine in production equipped with decoy I/O. This time, the red team approach was equipped with the details of the entire contribution made in this paper. In other words, the red team was assumed to have awareness of the fact that system activity on the machine might be due to decoy processes, and that the performance data of those decoy processes are regulated by a deep learning algorithm based on heatmap recognition.

Visually, decoy processes resemble their real counterparts. Most importantly, they also have performance dynamics, which we assessed by putting the red team in the best attack conditions possible. The thought processes are illustrated in Fig. 4. More specifically, in our red team role, we had a replica of the machine to be protected, with the only difference being that the OPCExplorer process was real. We measured empirically the performance parameters of all processes, including those of the real OPCExplorer process. As we were performing those measurements on the replica, in some cases we intentionally left all processes in low or moderate use, except the real OPCExplorer process.

Since most of the performance parameters of the other processes were low or near constant, they perturbed the performance of the real OPCExplorer process by a lesser amount than on a usual machine in production. We called these performance measurements group 1 (G1). In other cases, we drove the other processes such as to perform average or higher load tasks, and called the corresponding performance measurements group 2 (G2) and 3 (G3), respectively. In G2 and G3 conditions, the other processes affected the performance of the real OPCExplorer process by a larger amount than in G1 conditions. Overall, these maneuvers enabled us in our red team role to collect measurements that statically tied the performance of all processes on the replica with the performance of the real OPCExplorer process.

Our convolutional neural network is trained to cope with any arbitrary amount of effect that the other processes may have on the performance of the real OPCExplorer process, and thus reproduce consistent performance parameters on its decoy counterpart under all circumstances. The reason we are emphasizing specific G1, G2, and G3 conditions, is that, from a testing or probing perspective, our experience with this research suggests that the likelihood of seeing performance parameters on a machine in production that are similar to those observed and recorded on the replica varies across the G1, G2, and G3 spectra. After all, the feasibility of these testing/probing techniques depends on the portability of resource utilization dynamics from the replica onto the compromised machine, i.e. the machine in production.

G1 conditions are the most favorable to a threat actor, since their occurrence is statistically more common, especially on client machines in production. At times, users commonly interact with a few application programs at a time. Some users place higher demand on their machine, in which case G2 and G3 conditions take place. Nevertheless, we found that, even when G2 and G3 conditions occur, they are hardly stationary enough to resemble a specific predefined resource utilization pattern characterized on the replica. A threat actor may attempt to interact with processes in order to force their resource utilization to get close to a precalculated resource utilization signature. However, we deem the following adversarial actions to be out of reach:

- Non-invasively reduce the resource utilization of a process that is taking input from the legitimate user. Thus, a threat actor may be able to adjust the resource utilization of a process by increasing it. If the adjustment requires a decrease, the threat actor is impotent.
- Interact with a process that has a graphical user interface (GUI). The reason is simple, namely the legitimate user will certainly notice. Making process AcroRd32 load a portable document format (PDF) file and scrolling over the pages, or making Chrome browse a website, will display the respective GUI components on the screen.

Instead of requiring our red team approach to adjust the resource utilization of processes, if possible, and/or wait for a lucky resource utilization combination to occur, we facilitated the red team assessment by creating usable G1, G2, and G3 dynamics on the compromised machine. This is what we meant with putting the red team in the best attack conditions possible earlier in this section. The appearance of usable G1, G2, and G3 dynamics on the compromised machine may be a rare event, but we assume it to be possible in order to favor the highest strength of the red team assessment.

In some of the tests, we left most of the processes on the compromised machine in low or moderate use. In other tests, we used same or similar stimuli as on the replica to create G2 and G3 conditions that were close enough to the performance signatures taken on the replica.

This is the culmination of the target validation on the compromised machine. Since in our red team role we had prior knowledge of the performance of the real OPCExplorer process within a precalculated resource utilization signature, we could simply compare the expected performance of the OPCExplorer process with the performance collected on the compromised machine. If the two diverged by a non-negligible amount, the conclusion would be that we had landed on a decoy. Some of the findings of these trials are depicted in Fig. 5.

The Performance Fingerprint of a Process is not Fixed. The data plots in Figs. 5 and 6 show that, often cases, we get slightly different performance data for a real process such as OPCExplorer, although the performance data of the other processes do not change or change minimally. For example, under identical or similar underlying performance dynamics, we measured a processor user space time for real process OPCExplorer equal to 3.61. A few seconds later, without any change of conditions, we measured 3.05. Consequently, a decoy performance inconsistency has to be a considerable departure from patterns of resource utilization, since small departures are normal.

Overall, our work is able to keep the performance data of a decoy process within the normal variability of the performance fingerprint of its real counter-

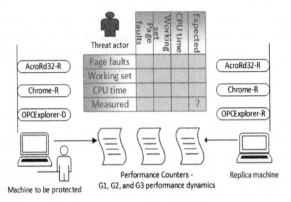

Fig. 4. Assessing the accuracy of our convolutional neural network in protecting a decoy OPCExplorer process from malware probes. (Color figure online)

part. We had cases of incorrect class labels produced by the neural network, however those were relatively infrequent. We deem that those misses were due to the small number of heatmaps in the training set. With a larger training set, this work may attain a higher accuracy. We also had a few challenges during the actual measurements of performance data. The OPC client software that we worked with displayed hints and other help via pictures and other graphics on its graphical user interface. We noticed that the reading and displaying of those graphics one at a time, and for specific periods of time, did affect the performance parameters that we were measuring.

Load Disturbance Attempts. In our red team role, we created processes that requested large amounts of memory, consisted mostly of CPU bursts, or generated heavy I/O traffic. We also created processes that changed the amount of resources abruptly and quickly, from very high to very low, and then back to very high. The neural network tolerated these disturbances, with no noticeable class label changes.

6   Related Work

Several works have explored prediction models to estimate resource utilization at runtime. Matsunaga et al. surveyed supervised machine learning to train data points and predict execution time. However, the authors only attain detailed estimated in relation to fixed input data [12]. In contrast, in our work we consider any input data. Miu et al. examined features extracted from input data to find specific instances that maximize the accuracy of predicted execution time of a process. They used a combination of input features to learn regression models using C4.5 decision tree builders. Their method depends on learning from historical data [13]. Li et al. predicted scheduling in a Round Robin manner in the distributed stream data processing. For scheduling, a greedy algorithm is

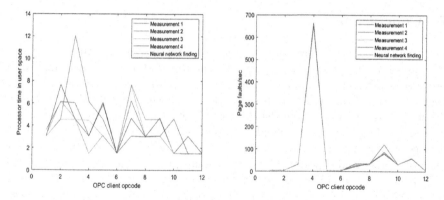

Fig. 5. Sample I - Empirical measurements of performance data versus deep learning class labels.

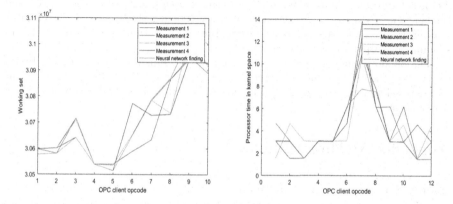

Fig. 6. Sample II - Empirical measurements of performance data versus deep learning class labels.

used to assign threads to machine under the guidance of prediction results [11]. Amiri et al. reviewed prediction models, including machine learning methods, to estimate performance and workload in the cloud [4].

Pietri et al. proposed a method to predict execution time for a parallel work-flow based on its structure in the cloud [14]. Their approach divides tasks to various levels based on their data dependency. Other related works focus on using machine learning to build a framework for mobile devices that can find features related to computational resource consumption from the input data that are given to a program [9]. These works use program slicing and sparse regression to extract pertinent information from program execution. Our work is different in that it is load dependent, and hence can predict the resource utilization parameters of a decoy process as other real processes continuously change their own performance parameters.

7   Conclusions

Live real-time performance counters enable a deep insight into the performance of a process. Our honeypot experiment showed that performance analysis of processes can catch the many inconsistencies of high interaction honeypots and decoy real processes, and can also be a threat to decoy I/O if left unaddressed. We described interventions in the OS kernel that project the existence of a decoy process, without having to spend resources on creating an actual process. We devised a convolutional neural network that can learn the performance fingerprint of a process in support of its decoy counterpart. In conclusion, we validated and quantified the ability of such decoy processes to sustain a realistic resemblance with a valid target of attack, thus possibly causing changes to malware's target selection.

Acknowledgment. This material is based on research sponsored by the USAFA and Oakland University under agreement number FA7000-18-2-0022. The U.S. Government is authorized to reproduce and distribute reprints for Governmental purposes notwithstanding any copyright notation thereon.

The opinions, findings, views, conclusions or recommendations contained herein are those of the authors and should not be interpreted as necessarily representing the official policies or endorsements, either expressed or implied, of the USAFA or the U.S. Government.

Appendix A: Threat Model

Probes Originate from the Inside. A malware sample has compromised a machine, and is now assessing whether or not it is a decoy. We have observed that this target validation assessment is commonly a precursor to attack operations such as the following:

- Launching local exploits to escalate the current privilege.
- Installing rootkits to preserve access.
- Installing I/O interceptors to capture keystrokes, webcam traffic, file system and network traffic.
- Accessing data and sending them to a threat actor over the network.
- Launching an exploit on the compromised machine against another target over the network.

These operations are typically implemented as separate malware modules, which follow the initial exploit. A multi-stage dropper downloads them onto the compromised machine over the network from another machine under threat actor's control. A single-stage dropper comes with these modules incorporated in it. The dropper itself is downloaded over the network similarly to the malware modules.

The Initial Exploit May Yield Partial or 0 Value. On several occasions, the initial exploit may go undetected, consequently the malware operations referenced previously are the defender's opportunity to detect the malware based on its contact with decoys. A common case of this occurrence is when the initial exploit leverages a 0-day vulnerability on a machine in production equipped with decoy I/O. When targeting a honeypot, the same exploit is certainly detected upfront. Nevertheless, as we wrote earlier in this paper, it is possible to avoid making network contact with a honeypot on the basis of its lack of network activity. Some initial exploits yield no value to the defender, as in our honeypot experiment.

Withstanding Probes is of Significance. Decoy processes and their performance consistency, along with other types of consistency, are decisive on whether malware fall into a trap, or step away from a decoy target, erase themselves and hence disappear even before the defender sees any cues at all. An ineffective decoy results in none of the malware modules or even the dropper ever being brought onto the machine.

Appendix B: Out of Scope

The deep learning in this work needs to be hidden and protected from malware, otherwise threat actors may manipulate its computations and evade it. One solution is to run the deep learning on a virtual machine (VM), which is managed by a hypervisor and is isolated from the host machine. The overhead of a VM solution needs to be carefully assessed. Another solution is to run the deep learning on a hardware sideboard physically isolated from the host machine. This other solution comes with an added cost, which could be kept as low as under $50 with the right hardware design.

A honeypot's lack of network activity can be leveraged remotely to avoid attacking it. A threat actor operating on a compromised machine in production may select the next targets to be only those machines that the compromised machine is observed to communicate with. Since, by definition, no machine in production communicates with a honeypot, the threat actor will never hit a honeypot.

Because of room limitations, and to be able to describe the main contribution thoroughly, we do not include these efforts in this paper.

Appendix C: Additional Related Works

Several related works use stealth techniques to hide computer resources. Hooking, which prevents a request from accessing resource usage, and Direct Kernel Object manipulation (DKOM), which manipulates specific data in the OS kernel. Butler et al. described a non-hooking method to implement a device to hide and unlinked processes in EPROCESS blocks on Microsoft Windows [5]. On the other hand, Tsai et al. identified DKOM that can target all resources of an object directory, and thus alter and hide kernel objects that are commonly used by the OS in memory [18].

Jones et al. presented a technique to detect a virtual machine monitor (VMM)-based process that is maliciously hidden. This technique uses cross view validation, and then patches the executable code in order to affect its execution. The authors can detect any hidden processes that are running within a guest virtual machine. Their technique leverages CPU inflation, which is the CPU time consumed by each process within VMM and the guest operating system [8]. Unlike all these related works that we just discussed, our research stands out through the emphasis placed on creating a decoy process in EPROCESS to appeal to a threat actor, while hiding the decoy process from legitimate users.

As far as resource utilization prediction goes, we also use machine learning to predict performance parameters for a decoy process. However, our work is different than related works. As we mentioned earlier in this paper, our approach is load dependent. The other related works do not make load dependent estimations. Secondly, our work leverages input categorization based on process operations. Along with heatmap design and deep learning, these factors provide for a high level of accuracy, which is adequate to withstand malware probes.

Malware have a history of validating their targets prior to carrying out their operations. Some of these malware detect debuggers and/or virtual machines. An active debuger may be indicative of an execution environment operated by defenders in support of dynamic code analysis. Furthermore, a virtual execution environment is commonly used to host honeypots [6]. Similarly, some malware detect CPU emulators, which are also used for dynamic code analysis and honeypots [15]. As we wrote earlier in this paper, no other works appear to leverage OS-level performance data to detect decoys as of this writing.

References

1. Honeyprocs: Going beyond honeyfiles for deception on endpoints. https://forums.juniper.net/t5/Threat-Research/HoneyProcs-Going-Beyond-Honeyfiles-for-Deception-on-Endpoints/ba-p/385830. Accessed 23 Feb 2019
2. Metasploit framework. https://www.metasploit.com/. Accessed 23 Feb 2019
3. Performance counters. https://docs.microsoft.com/. Accessed 23 Feb 2019
4. Amiri, M., Mohammad-Khanli, L.: Survey on prediction models of applications for resources provisioning in cloud. J. Netw. Comput. Appl. **82**(C), 93–113 (2017)
5. Butler, J., Undercoffer, J.L., Pinkston, J.: Hidden processes: the implication for intrusion detection. In: IEEE Systems, Man and Cybernetics SocietyInformation Assurance Workshop, 2003, West Point, NY, USA, pp. 116–121, June 2003
6. Chen, X., Andersen, J., Mao, Z.M., Bailey, M., Nazario, J.: Towards an understanding of anti-virtualization and anti-debugging behavior in modern malware. In: Proceedings of the IEEE/IFIP International Conference on Dependable Systems and Networks, pp. 177–186 (2008)
7. Goodfellow, I., Bengio, Y., Courville, A.: Deep Learning. MIT Press, Cambridge (2016). http://www.deeplearningbook.org

8. Jones, S.T., Arpaci-Dusseau, A.C., Arpaci-Dusseau, R.H.: VMM-based hidden process detection and identification using Lycosid. In: Proceedings of the Fourth ACM SIGPLAN/SIGOPS International Conference on Virtual Execution Environments, New York, NY, USA, pp. 91–100 (2008)
9. Kwon, Y., et al.: Mantis: efficient predictions of execution time, energy usage, memory usage and network usage on smart mobile devices. IEEE Trans. Mob. Comput. **14**(10), 2059–2072 (2015)
10. Lange, J., Iwanitz, F., Burke, T.: OPC - From Data Access to Unified Architecture. VDE VERLAG GMBH, 4th edn. (2010)
11. Li, T., Tang, J., Xu, J.: Performance modeling and predictive scheduling for distributed stream data processing. IEEE Trans. Big Data **2**(4), 353–364 (2016)
12. Matsunaga, A., Fortes, J.A.B.: On the use of machine learning to predict the time and resources consumed by applications. In: Proceedings of the 10th IEEE/ACM International Conference on Cluster, Cloud and Grid Computing, Washington, DC, USA, pp. 495–504 (2010)
13. Miu, T., Missier, P.: Predicting the execution time of workflow activities based on their input features. In: Proceedings of the 2012 SC Companion: High Performance Computing, Networking Storage and Analysis, Washington, DC, USA, pp. 64–72 (2012)
14. Pietri, I., Juve, G., Deelman, E., Sakellariou, R.: A performance model to estimate execution time of scientific workflows on the cloud. In: 9th Workshop on Workflows in Support of Large-Scale Science, pp. 11–19, November 2014
15. Raffetseder, T., Kruegel, C., Kirda, E.: Detecting system emulators. In: Garay, J.A., Lenstra, A.K., Mambo, M., Peralta, R. (eds.) ISC 2007. LNCS, vol. 4779, pp. 1–18. Springer, Heidelberg (2007). https://doi.org/10.1007/978-3-540-75496-1_1
16. Rrushi, J.: Phantom I/O projector: entrapping malware on machines in production. In: 12th International Conference on Malicious and Unwanted Software (MALWARE), Fajardo, Puerto Rico, USA, pp. 57–66, October 2017
17. Rrushi, J.: DNIC architectural developments for 0-knowledge detection of OPC malware. IEEE Trans. Dependable Secure Comput. 1 (2018)
18. Tsaur, W.-J., Chen, Y.-C., Tsai, B.-Y.: A new windows driver-hidden rootkit based on direct kernel object manipulation. In: Hua, A., Chang, S.-L. (eds.) ICA3PP 2009. LNCS, vol. 5574, pp. 202–213. Springer, Heidelberg (2009). https://doi.org/10.1007/978-3-642-03095-6_21

Author Index